K.-W. Bieger

Kreiszylinderschalen unter radialen Einzellasten

Circular Cylindrical Shells Subjected to Concentrated Radial Loads

Klaus-Wolfgang Bieger

Kreiszylinderschalen unter radialen Einzellasten

Berechnungsverfahren und Katalog von Einflußflächen

Circular Cylindrical Shells Subjected to Concentrated Radial Loads

Computation Methods and Charts of Influence Surfaces

English Translation by
Heinz G. Juhl

Springer-Verlag Berlin Heidelberg New York

Dr.-Ing. KLAUS-WOLFGANG BIEGER
o. Professor an der Technischen Universität Hannover

Englische Übersetzung von — English Translation by
Dr.-Ing. HEINZ. G. JUHL
Pa. Dep. of Transportation, Harrisburg, U.S.A.

Mit 18 Abbildungen und 108 Tafeln — With 18 figures and 108 charts

ISBN 3-540-07848-7 Springer-Verlag Berlin Heidelberg New York
ISBN 0-387-07848-7 Springer-Verlag New York Heidelberg Berlin

Library of Congress Cataloging in Publication Data
Bieger, Klaus-Wolfgang, 1927– Kreiszylinderschalen unter radialen Einzellasten. English and German.
Bibliography: p. Includes index. 1. Shells (Engineering) 2. Cylinders. I. Title.
II. Title: Circular cylindrical shells subjected to concentrated radial loads.
TA660.S5B5 624'.1776 76-39787

Gesamtherstellung: Universitätsdruckerei H. Stürtz AG, Würzburg.

Vorwort

Zylinderschalen werden häufig durch konzentrierte Belastungen beansprucht. Der Querschnitt wird bei geschlossenen Schalen fast ausschließlich und bei Tonnendächern meistens kreisbogenförmig ausgebildet. Bei Überdachungen in Form von Tonnenreihen, Shedschalen u.ä. ergeben sich Einzellasten durch angehängte leichte Kranbahnen, schwere Heizungs-, Lüftungs- oder Beleuchtungseinrichtungen. Für größere Kräfte müssen Rippen oder Aussteifungsringe vorgesehen werden [1, 2]. Geschlossene Kreiszylinderschalen als Behälter, Silos, Rohre u.ä. können konzentrierte Beanspruchungen durch Einbauten, Auflagerung oder spezielle Lastansätze erhalten (z.B. Rohreinpressungen bei Brunnen oder Katastrophen-Lasten bei Reaktorsicherheitsbehältern [3]). Selbst Naturzug-Kühltürme werden für Einzellastangriff näherungsweise als Kreiszylinderschale aufgefaßt [4]. Ähnliche Probleme treten im Maschinen- und Schiffbau auf.

Die Berechnung der Kreiszylinderschalen unter Einzellasten ist sehr zeitraubend, da eine geschlossene Lösung – analog der klassischen Singularität der Platte – nicht bekannt ist. Zwar läßt sich die Plattensingularität bei Lösungen mit Doppelreihen [5] und einfachen Reihen [6] abspalten, der numerische Aufwand ist jedoch vor allem bei genaueren Theorien noch recht erheblich. Es lag daher nahe, Einfluß- oder Zustandsflächen infolge Einzellasten anzufertigen.

Wählt man für deren Ermittlung das geschlossene Rohr unter einer im Gleichgewicht stehenden Einzellastgruppe, dann werden wegen der fehlenden Randbedingungen die Einflußflächen bis auf das Vorzeichen identisch mit den entsprechenden Zustandsflächen und können als eine Darstellung des partikulären Integrals der Differentialgleichung aufgefaßt werden.

Diese in Form von Höhenschichtplänen im Anhang dargestellten Einfluß- bzw. Zustandsflächen für die Verformungen und Schnittgrößen entstanden noch während meiner Tätigkeit bei Herrn Professor Dr.-Ing. W. Koepcke an der Technischen Universität Berlin mit dankenswerter Unterstützung der Deutschen Forschungsgemeinschaft.

Für die englische Übersetzung sei Herrn Dr.-Ing. H. G. Juhl ebenso gedankt, wie dem Springer-Verlag für die sorgfältige Herstellung des Werkes.

Hannover, im Frühjahr 1976

Klaus-Wolfgang Bieger

Preface

Cylindrical shells are subjected often to concentrated loads. A circular shape of the cross section is chosen almost exclusively for closed shells, and for most oft the cases of roofs of shell rows. For roofs designed as shell rows, north light shells etc. concentrated loads occur due to reactions from suspended light cranes and supports of heavy heating, air conditioning or lighting installations. For larger forces, ribs or stiffener rings have to be provided [1, 2]. Closed circular cylindrical shells, used for containers, silos, pipes etc., may be subjected to concentrated loadings from built-in installations, supports or special loads (e.g., jacking of well pipes or catastrophic loads of safety reactor container vessels [3]). Even natural-draft cooling towers are assumed, for concentrated load application, as a circular cylindrical shell as approximation [4]. Similar problems occur in ship building and machinery construction.

The analysis of circular cylindrical shells, subjected to concentrated loads, is very time-consuming because a closed solution – analogous to the classic singularity method for plates – is not known. Although the plate singularity can be split off by solutions with double [5] or simple series [6], the effort for the numerical solution is still large, in particular, for the more exact theories. Thus, it was natural to produce influence or condition surfaces for concentrated loads.

If one chooses for their determination the closed pipe, subjected to a group of concentrated loads in equilibrium, the influence surfaces, because of the missing boundary conditions, become identical, except for the sign, with the corresponding condition surfaces which can be considered as a representation of the particular integral of the differential equation.

These influence or condition surfaces, respectively, for the deformations and section stress resultants, presented in contour-line form in the Appendix, originate from my work with Professor Dr.-Ing. W. Koepcke at the Technische Universität Berlin, and with the support by the Deutsche Forschungsgemeinschaft.

Equal gratitude is expressed to Dr.-Ing. H. G. Juhl for the English translation, and to the Springer-Verlag for the careful production of the book.

Hannover, in the spring of 1976

Klaus-Wolfgang Bieger

Comment of the translator:

To avoid confusion of the multiplication symbol " · " between numbers with the decimal location symbol " · ", applied in the English literature, the use of the comma as the decimal symbol was not changed.

Inhalt

Contents

1. Einleitung

Die erste Arbeit über Einzellasten auf Kreiszylinderschalen ist m.W. von Aas-Jakobsen [1] veröffentlicht worden, der jedoch grundsätzlich Ringrippen oder Versteifungsringe vorsieht. Koepcke [7] löst das Problem durch Überlagerung zweier Lastfälle am unendlich langen unausgesteiften Rohr. Die Lösung der drei gekoppelten partiellen Differentialgleichungen für die drei Verschiebungskomponenten nach der „genauen" Theorie der Kreiszylinderschale von Flügge [8] ist sehr mühevoll, so daß im folgenden meist von der sogenannten Donnellschen Näherung Gebrauch gemacht wird [9, 10, 5, 11, und 12]. Bei Streichung einiger Glieder in den Schnittkraft-Verformungs-Beziehungen lassen sich nämlich die drei Differentialgleichungen auf eine einzige achter Ordnung für die Radialverschiebung w zurückführen. Für das unendlich lange Rohr ist diese Näherung jedoch zu ungenau (vgl. [13–15, 6]). Auch für die Lösung des Einzellastproblems wurde eine Verbesserung der Donnellschen Näherung angestrebt [14, 16–20].

Bei Einführung einer Lösungsfunktion [6, 21] lassen sich jedoch auch ohne Streichung irgendwelcher Glieder die Flüggeschen Differentialgleichungen auf eine einzige partielle Differentialgleichung achter Ordnung für diese Lösungsfunktion zurückführen. Es müssen lediglich die drei Lastkomponenten X, Y und Z (vgl. Bild 1) getrennt untersucht werden, da die Zusammenhänge zwischen den Verschiebungen und der Lösungsfunktion für jede Lastkomponente verschieden sind. Wählt man einen gleichen funktionalen Aufbau für jede Lastkomponente, so unterscheiden sich die Lösungsfunktionen wegen der bis auf das Lastglied identischen Differentialgleichungen nur durch eine Konstante. Somit können dann die Verschiebungsanteile einfach summiert werden. Im vorliegenden Fall interessiert jedoch allein die radiale Belastung, so daß nur die Zusammenhänge für diese Lösungskomponente angegeben werden. Der numerische Aufwand ist bei dieser Lösung nur unwesentlich größer als nach der Donnellschen Näherung, da alle Verschiebungen und Schnittgrößen durch einfache Differentiationen aus der Lösungsfunktion gewonnen werden.

Wählt man einen Fourier-Reihenansatz in einer Richtung – z.B. der Ringrichtung – und einen Fourier-Integralansatz in der anderen – z.B. der Längsrichtung –, dann läßt sich auch für eine Einzellast das Fourier-Integral geschlossen lösen. Man erhält schließlich eine einfache Fourier-Reihe für die Lösungsfunktion und somit für alle Verformungen und Schnittgrößen.

Da auch bei der Kreiszylinderschale der Satz von der Gegenseitigkeit der Verschiebungen und Lasten (Maxwell-Betti) gilt, werden bei dem von Randbedingungen freien, unendlich langen, geschlossenen Rohr die Einflußflächen der statischen Größen für eine im Gleichgewicht stehende Einzellastgruppe identisch mit den entsprechenden Zustandsflächen. Lediglich die Vorzeichen kehren sich um, wenn die Summe der Ableitungen der Lösungsfunktion nach den beiden Koordinatenrichtungen ungerade ist. Dies ist der Fall bei den Verschiebungen in Längs- (u) und Umfangsrichtung (v), bei der Verdrehung in Umfangsrichtung (χ_φ) und bei den Randquerkräften ($\bar{q}_x$ und $\bar{q}_\varphi$).

1. Introduction

The first paper on circular cylindrical shells, subjected to concentrated loads, has been published, to my knowledge, by Aas-Jakobsen [1], who, however, provides, as a principle, ring ribs or stiffener rings. Koepcke [7] solves the problem by superposition of two load cases of the infinite, non-stiffened pipe. The solution of the three coupled partial differential equations, for the three displacement components, with the "exact" theory of the circular cylindrical shell by Flügge [8], is very tedious; therefore, the so-called Donell's approximation has mostly been used in the following [9, 10, 5, 11, 12]. By cancelling some of the terms in the stress-strain relations, the three differential equations can be reduced to a single equation of the eighth order in the radial displacement w. However, for the infinite pipe, this approximation is not accurate enough (see [13–15, 6]). Also, for the solution of the concentrated load problem, an improvement of Donnell's approximation was sought [14, 16–20].

However, by introduction of a solution function [6, 21], Flügge's differential equations can be reduced also to a single partial differential equation of eighth order in this solution function, without the cancellation of any term. It is only necessary to investigate separately for the three load components X, Y and Z (see Fig. 1), because the relations between the displacements and the solution function for each load component are different. When selecting the same function structure for each load component, the solution functions are different only by a constant, because the differential equations are identical except for the load terms. Thus, the displacement contributions can simply be summarized. Only the radial load is of interest, for the here considered case, thus, only the relations for this component of the solution will be given. The computational effort for this solution method is only slightly larger than with Donnell's approximation, because all of the displacements and section stress resultants are obtained from the solution function by simple differentiations.

If one chooses a Fourier series assumption in one direction – for instance the hoop direction –, and a Fourier integral assumption in the other – for instance the longitudinal direction –, a closed solution of the Fourier integral can be obtained, even for a concentrated load. One obtains a simple Fourier series for the solution function, and therefore for all deformations and section stress resultants.

Because the theorem of equivalence of displacements and loads (Maxwell-Betti) applies also to circular cylindrical shells, the influence surfaces of the statical quantities, due to a concentrated load group in equilibrium, for the infinitely long closed pipe, free of boundary conditions, become indentical with the corresponding condition surfaces. Only the signs are reversed if the total number of derivations of the solution function into the two coordinate directions is odd. This is the case for the displacements, (u) and (v), into the longitudinal and the hoop direction, respectively, for the rotation into the hoop direction (χ_φ) and for the transverse boundary forces ($\bar{q}_x$ and $\bar{q}_\varphi$).

Aus diesem Grunde wurde für die Ermittlung der hier vorliegenden Flächen das von Randbedingungen freie, geschlossene Rohr gewählt. Bei der genaueren Flüggeschen Schalentheorie sind dann die Ergebnisse nur von einem Steifigkeitsparameter (k) abhängig. Sollten bestimmte Randbedingungen, z.B. an den Binderscheiben, bereits erfüllt sein, so müßte jeweils eine entsprechende Anzahl zusätzlicher Flächen in Abhängigkeit von einem weiteren Spannweitenparameter und von den gewählten Randbedingungen vertafelt werden. Lediglich unter Zuhilfenahme der Donnellschen Näherung gelingt es, nur mit einem kombinierten Parameter bei der an den Binderscheiben gelenkig gelagerten Einfeldschale auszukommen [22]. Holand hat so in [11] Einflußflächen für die Biegemomente der in Ringrichtung unendlich ausgedehnten Einfeldschale vertafelt. Es bereitet jedoch mit Hilfe des Lastumordnungsverfahrens keine Schwierigkeiten, auch aus dem hier vertafelten unendlich langen Rohr einfeldrige, gelenkig gelagerte Schalen zu gewinnen (vgl. Abschnitt 5.1.2).

Bei den Zylinderschalen treten im Gegensatz zu Platten selten wandernde Einzellasten auf, es interessieren also hauptsächlich Zustandsflächen. Lediglich zur Bestimmung der maximalen Schnittgrößen unter den im allgemeinen auf einen bestimmten Bereich verteilten konzentrierten Belastungen benötigt man die Einflußflächen, denn bei den Schalen werden wie bei den Platten die Einflußordinaten der Momente und Querkräfte im Aufpunkt unendlich groß. Auch bei den Normalkräften ergeben sich unter der Einzellast, hier jedoch in der Höhe begrenzte, enge Schläuche. Die Volumen oberhalb der letzten Ordinaten der im Anhang dargestellten Einflußflächen der Biegemomente und Querkräfte können nahezu genau ermittelt werden (vgl. Kap. 6). Wegen der relativ kleinen Ordinaten vor allem bei den Biegemomenten m_x ist der Inhalt des nicht dargestellten Schlauches zum Teil in der Größenordnung des Volumens unter der letzten angegebenen Höhenschichtlinie. Ähnlich ungünstig sind die Verhältnisse bei den Randquerkräften. Es ist daher für eine genaue Ermittlung der Schnittgrößen beliebiger konzentrierter Belastungen meist erforderlich, die Volumen der nicht darstellbaren Bereiche zu berechnen.

For this reason, the closed pipe, free of boundary conditions, was chosen for the computation of the here presented surfaces. Hence, with Flügge's, more exact theory of shells, the results are depending only on a stiffness parameter (k). If certain boundary conditions, for instance at transverse frame supports, are satisfied already, a corresponding number of additional surfaces, depending on an additional span length parameter and on the chosen boundary conditions, would have to be tabulated. For the single-span shell, hinge supported at the transverse frames, only one combination parameter will suffice when using Donnell's approximation [22]. Holand tabulated this way, in [11], the influence surfaces for the bending moments of the single-span shell, extending infinitely in the hoop direction. However, by using the load rearrangement method, single-span hinge supported shells can be derived, without difficulties, from the infinitely long pipe, tabulated herein (see Section 5.1.2).

Contrary to plate structures, moving concentrated loads on cylindrical shells occur rarely, thus, mainly the condition surfaces are of interest. Only for the computation of the maximum section stress resultants, due to concentrated loadings which are distributed over a certain area, influence surfaces are needed, because the influence ordinates of moments and transverse forces of shells, as for plates, become infinitely large at the reference point. Narrow trunks are obtained also for the normal forces under the concentrated load, however, here limited in height. The volumes above the last ordinate of the influence surfaces for bending moments and transverse forces, shown in the Appendix, can be obtained almost exact (see Chapter 6). Because of the relatively small ordinates, in particular for the bending moments m_x, the volume of the trunk, not shown, is in the order of magnitude of the volume under the last contour line shown. The conditions are similarily unfavorable for the transverse boundary forces. Thus, for an exact computation of the section stress resultants due to arbitrary concentrated loadings, it is necessary, in most cases, to compute the volumes of the ranges which cannot be shown.

2. Grundlagen zur Berechnung von Kreiszylinderschalen

Da es immer erforderlich sein wird, die Randbedingungen an den Quer- und Längsrändern nachträglich zu erfüllen, sollen hier die wichtigsten Grundlagen mitgeteilt werden.

Die *Voraussetzungen* für die Ermittlung der Zustands- bzw. Einflußflächen sind die nach der Flüggeschen Schalentheorie üblichen:

a) Der Baustoff sei homogen und isotrop und folge dem Hookeschen Gesetz. Dies ist vor allem beim Stahlbeton nicht gegeben. Für den Gebrauchslastzustand ist jedoch die mit dieser Annahme hergeleitete Schnittgrößenermittlung hinreichend genau.

b) Die Deformationen seien klein im Verhältnis zur konstanten Schalendicke und diese wiederum klein gegenüber den anderen Abmessungen. Stabilitätsfälle und sogenannte dickwandige Schalen werden also ausgeschlossen.

c) Die Normale zur Mittelfläche der Schale soll auch nach der Verformung erhalten bleiben. Diese Bedingung ist höchstens in unmittelbarer Nähe von Einzellasten und am Schalenrand nicht eingehalten und wird sich nur in einem Bereich von der Größenordnung der Schalendicke auswirken.

d) Die Spannungen senkrecht zur Mittelfläche werden vernachlässigt. Ihr Einfluß wäre ähnlich wie bei c) nur in örtlich sehr begrenzten Bereichen vorhanden.

e) Die Querdehnzahl ist nur bei einigen Schnittgrößen von Einfluß und wird zu $\nu = 1/6$ angenommen.

Die *positiven Belastungen, Verformungen und Schnittgrößen* werden nach Dischinger [23] bzw. Aas-Jakobsen [24] angenommen und sind in Bild 1 dargestellt.

2. Fundamentals of the Analysis of Circular Cylindrical Shells

Because it will be always necessary to satisfy the boundary conditions at the transverse and at the longitudinal boundaries later, the most important fundamentals shall be given here.

The *assumptions* for the determination of the condition or influence surfaces, respectively, are those as in Flügge's Theory of Shells:

a) The material shall be homogeneous and isotropic, and follows Hooke's Law. This is not true, in particular, for reinforced concrete. However, the method of computation of the section stress resultants, derived with this assumption, is sufficiently accurate for the working load condition.

b) The deformations shall be small in relation to the constant shell thickness which, in turn, shall be small in relation to the other dimensions. Thus, stability cases and so-called thick-walled shells shall be excluded.

c) The normal on the middle surface of the shell shall remain as such after the deformation. This condition is not satisfied in the close vicinity of concentrated loads and at the shell boundaries only, and will affect only a range of the size of the shell thickness.

d) The stresses perpendicular to the middle surface shall be disregarded. Their influence would exist only, similar to c) in very limited ranges.

e) Poisson's ratio is of importance only for a few section stress resultants, and shall be assumed $\nu = 1/6$.

The *positive loads, deformations and section stress resultants* shall be assumed in accordance with Dischinger [23] or Aas-Jakobsen [24], respectively, and are shown in Fig. 1.

Bild 1. Positive Belastungen, Verformungen und Schnittgrößen

Figure 1. Positive loads, deformations and stress resultants.

Schreibt man für die Differentialquotienten $\partial f/\partial\varphi = f^{\cdot}$ und $a\,(\partial f/\partial x) = f'$, so folgen die von Flügge [8] aufgestellten Zusammenhänge zwischen den Verschiebungen und Schnittgrößen zu:

When denoting the differential quotients by $\partial f/\partial\varphi = f^{\cdot}$ and $a\,(\partial f/\partial x) = f'$, the relations between displacements and stress resultants are, in accordance with Flügge [8], as follows:

$$
\begin{aligned}
n_x &= \frac{D}{a}(u' + \nu v^{\cdot} + \nu w) - \frac{K}{a^3} w'' \\
n_\varphi &= \frac{D}{a}(v^{\cdot} + w + \nu u') + \frac{K}{a^3}(w + w^{\cdot\cdot}) \\
n_{x\varphi} &= \frac{D}{a}\frac{1-\nu}{2}(u^{\cdot} + v') + \frac{K}{a^3}\frac{1-\nu}{2}(v' - w'^{\cdot}) \\
n_{\varphi x} &= \frac{D}{a}\frac{1-\nu}{2}(u^{\cdot} + v') + \frac{K}{a^3}\frac{1-\nu}{2}(u^{\cdot} + w'^{\cdot}) \\
q_x &= \frac{K}{a^3}\left(w''' + w'^{\cdot\cdot} - \frac{1+\nu}{2} v'^{\cdot} - u'' + \frac{1-\nu}{2} u^{\cdot\cdot}\right)
\end{aligned}
\qquad
\begin{aligned}
m_x &= \frac{K}{a^2}(w'' + \nu w^{\cdot\cdot} - u' - \nu v^{\cdot}) \\
m_\varphi &= \frac{K}{a^2}(w^{\cdot\cdot} + \nu w'' + w) \\
m_{x\varphi} &= \frac{K}{a^2}(1-\nu)(w'^{\cdot} - v') \\
m_{\varphi x} &= \frac{K}{a^2}(1-\nu)\left(w'^{\cdot} + \frac{u^{\cdot}}{2} - \frac{v'}{2}\right) \\
q_\varphi &= \frac{K}{a^3}\{w^{\cdot} + w^{\cdot\cdot\cdot} + w''^{\cdot} - (1-\nu)\,v''\}
\end{aligned}
\qquad (1)
$$

Zur Erfüllung der Randbedingungen werden noch die Randkräfte benötigt:

To satisfy the boundary conditions, the boundary forces are required also:

$$
\begin{aligned}
\bar{n}_{x\varphi} &= n_{x\varphi} - \frac{m_{x\varphi}}{a} = \frac{D}{a}\frac{1-\nu}{2}(u^{\cdot} + \nu') + \frac{K}{a^3}\,3\,\frac{1-\nu}{2}(v' - w'^{\cdot}) \\
\bar{q}_x &= q_x + \frac{m^{\cdot}_{x\varphi}}{a} = \frac{K}{a^3}\left\{w''' + (2-\nu)\,w'^{\cdot\cdot} - \frac{3-\nu}{2} v'^{\cdot} - u'' + \frac{1-\nu}{2} u^{\cdot\cdot}\right\} \\
\bar{q}_\varphi &= q_\varphi + \frac{m'_{\varphi x}}{a} = \frac{K}{a^3}\left\{w^{\cdot} + w^{\cdot\cdot\cdot} + (2-\nu)\,w''^{\cdot} - 3\,\frac{1-\nu}{2} v'' + \frac{1-\nu}{2} u'^{\cdot}\right\}
\end{aligned}
\qquad (2)
$$

Hierin ist $D = Et/(1-\nu^2)$ die Dehnsteifigkeit und $K = Et^3/[12\,(1-\nu^2)]$ die Biegesteifigkeit der isotropen Kreiszylinderschale. Im folgenden können die beiden Ausdrücke zum Steifigkeitsparameter $k = K/Da^2 = t^2/12a^2$ zusammengefaßt werden. Bei den im Bauwesen üblichen Schalen liegt k etwa zwischen 10^{-6} und 10^{-4}. Bei kleineren k-Werten müssen die Schalen meist durch Ringrippen ausgesteift werden, und bei größeren nähert man sich den dickwandigen Behältern.

Aus den sechs Gleichgewichtsbedingungen des Schalenelementes erhält man bei Elimination der Querkräfte und Einsetzen der Beziehungen (1) die drei gekoppelten partiellen Differentialgleichungen von Flügge für die drei Verschiebungskomponenten u, v und w:

Herein denote, $D = Et/(1-\nu^2)$, the strain rigidity, and, $K = Et^3/[12\,(1-\nu^2)]$, the bending rigidy of the isotropic circular cylindrical shell. Both expressions can be combined in the stiffness parameter $k = K/Da^2 = t^2/12a^2$ for the following. For shells used in building constructions, k will be approximately between 10^{-6} and 10^{-4}. For smaller k-values, the shells will have to be stiffened, in most cases, by ring ribs, and one approaches the thick-walled containers for larger values.

From the six equilibrium conditions of the shell element, one obtains, by elimination of the transverse forces and by substitution of the relations (1), the three coupled partial differential equations for the three displacement components u, v and w, given by Flügge:

$$
\begin{aligned}
u'' + \frac{1-\nu}{2} u^{\cdot\cdot} + \nu w' + \frac{1+\nu}{2} v'^{\cdot} + k\left(\frac{1-\nu}{2} u^{\cdot\cdot} - w''' + \frac{1-\nu}{2} w'^{\cdot\cdot}\right) &= -X\frac{a^2}{D} \\
\frac{1+\nu}{2} u'^{\cdot} + v^{\cdot\cdot} + \frac{1-\nu}{2} v'' + w^{\cdot} + k\left(3\,\frac{1-\nu}{2} v'' - \frac{3-\nu}{2} w''^{\cdot}\right) &= -Y\frac{a^2}{D} \\
\nu u' + v^{\cdot} + w + k\left(\frac{1-\nu}{2} u'^{\cdot\cdot} - u''' - \frac{3-\nu}{2} v''^{\cdot} + w'''' + 2w''^{\cdot\cdot} + w^{\cdot\cdot\cdot\cdot} + 2w^{\cdot\cdot} + w\right) &= -Z\frac{a^2}{D}
\end{aligned}
\qquad (3)
$$

3. Lösungsansatz für radiale Belastungen

Bei der hier allein interessierenden radialen Belastung wird $X = Y = 0$, und die drei Gleichungen (3) lassen sich durch Einführung der Lösungsfunktion $F = F(x, \varphi)$ auf eine einzige partielle Differentialgleichung achter Ordnung zurückführen. Wird der Laplace-Operator zweiter und enstprechend höherer Ordnung

3. Solution Assumption for Radial Loadings

For the radial loading, considered here, it is $X = Y = 0$, and the three equations (3) can be reduced to a single partial differential equation of the eighth order by introduction of the solution function $F = F(x, \varphi)$. When introducing the Laplace operator of the second, and of correspondingly higher order,

$$\nabla^2 f = a^2 \frac{\partial^2 f}{\partial x^2} + \frac{\partial^2 f}{\partial \varphi^2} \qquad \nabla^4 f = a^4 \frac{\partial^4 f}{\partial x^4} + 2a^2 \frac{\partial^4 f}{\partial x^2\, \partial \varphi^2} + \frac{\partial^4 f}{\partial \varphi^4} \quad \text{usw.}$$

eingeführt, so ergibt sich diese wegen $k \ll 1$ nach [6] praktisch genau zu:

the equation becomes, in accordance with [6] and because of $k \ll 1$, practically exact:

$$\nabla^8 F + 2\nabla^6 F + \nabla^4 F - 2(1-\nu)\left(F'''''' - F''{::} - F''{:}\right) + \frac{1-\nu^2}{k} F'''' = -\frac{a^4}{K} Z \tag{4}$$

Die Verschiebungen folgen mit der hieraus zu bestimmenden Lösungsfunktion F durch einfache Differentiation:

The displacements follow from simple differentiation, with the solution function F to be determined herefrom:

$$\begin{aligned} u &= k\left(F''''' - F'{::} - 3k\frac{1-\nu}{2} F'''{:}\right) - \nu F''' + F'{:} \\ v &= 2k\left(F''''{\cdot} + F''{:\cdot}\right) - (2+\nu) F''{\cdot} - F{:\cdot} \\ w &= \nabla^4 F \end{aligned} \tag{5}$$

Entsprechend (1) und (2) können somit alle interessierenden Größen der Kreiszylinderschale berechnet werden.

Zur Erfüllung von Randbedingungen ist die homogene Differentialgleichung (4) mit $Z = 0$ zu lösen. Bei gleichem Ansatz für F, der im vorliegenden Fall auch für die Randstörungen an den Binderscheiben geeignet ist, können also die mitgeteilten Ergebnisse für den Einzellastangriff bis auf die Konstanten zur Erfüllung der Randbedingungen an den Querrändern übernommen werden.

Thus, all quantities of the circular cylindrical shell of interest can be computed in accordance with (1) and (2).

To satisfy boundary conditions, the homogeneous differential equation (4), with $Z = 0$, has to be solved. With the same assumption for F which, for the considered case, is suitable also for the boundary disturbances at the frame supports, the results, shown for the single load application, can be used, except for the constants, to satisfy the boundary conditions at the transverse boundaries.

4. Das geschlossene Rohr unter einer Einzellastgruppe

4.1 Lösungsweg

Die im Gleichgewicht stehenden Einzellasten $P = 1$ werden entsprechend Bild 2 am Ring $\xi = x/a = 0$ in $\varphi = 0$ und $\varphi = \pi$ wirkend angenommen. Wird die Einzel-

4. The Closed Pipe under a Group of Concentrated Loads

4.1 Method of Solution

The concentrated loads $P = 1$, being in equilibrium, are considered acting at the ring, $\xi = x/a = 0$, at $\varphi = 0$, and $\varphi = \pi$, in accordance with Fig. 2. When ex-

lastgruppe in Ringrichtung φ in eine Fourier-Reihe mit der Periode 2π und in Längsrichtung x/a in ein Fourier-Integral entwickelt, dann folgt für die radiale Belastung Z der Gleichung (4):

pressing the load group in the hoop direction φ by a Fourier series, with the period 2π, and in the longitudinal direction x/a by a Fourier integral, for the radial load Z, of equation (4), is obtained:

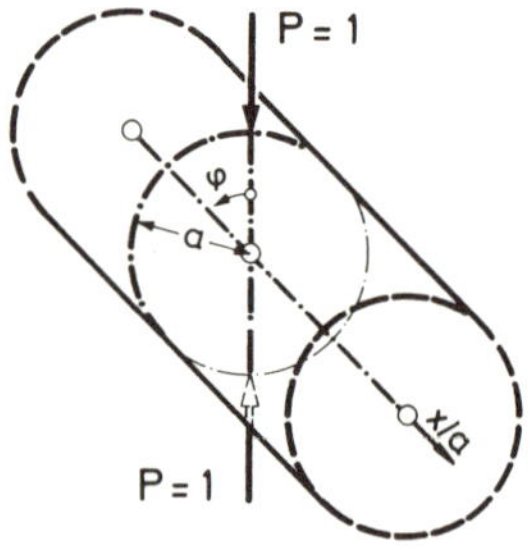

Bild 2. Radiale Einzellastgruppe

Figure 2. Group of radial concentrated loads.

$$Z = Z(\xi,\varphi) = \frac{2}{a^2\pi^2}\left\{\frac{1}{2}\int_{\alpha=0}^{\infty}\cos\alpha\xi\,\mathrm{d}\alpha + \sum_{n=2,4,6\ldots}^{\infty}\cos n\varphi\int_{\alpha=0}^{\infty}\cos\alpha\xi\,\mathrm{d}\alpha\right\} \tag{6}$$

Diese Reihe ist divergent. Jedoch folgt das gleiche Ergebnis, wenn der Grenzübergang aus der konzentrierten, gleichmäßig verteilten Belastung zur Einzellastgruppe erst bei der Lösungsfunktion durchgeführt wird.

Für die Lösungsfunktion F wird ein in $\xi = x/a$ und φ formal gleicher Ansatz mit den noch freien Konstanten $C_{\alpha,n}$ gewählt:

This series is divergent. However, the same result is obtained when the limit transition from the concentrated, uniformly distributed load to the concentrated load group is made later on the solution function.

For the solution function F, an assumption is chosen which is formally equal in $\xi = x/a$ and φ with the, still free, constants $C_{\alpha,n}$:

$$F = F(\xi,\varphi) = -\frac{2a^2}{\pi^2 K}\left\{\frac{1}{2}\int_{\alpha=0}^{\infty}C_{\alpha,0}\cos\alpha\xi\,\mathrm{d}\alpha + \sum_{n=2,4,6\ldots}^{\infty}\cos n\varphi\int_{\alpha=0}^{\infty}C_{\alpha,n}\cos\alpha\xi\,\mathrm{d}\alpha\right\} \tag{7}$$

Werden die nach (4) erforderlichen Differentiationen ausgeführt und diese zusammen mit der Belastung (6) in die Differentialgleichung (4) eingesetzt, so entsteht eine Gleichung, die für jede Harmonische n und jedes $\xi = x/a$ bzw. φ nur dann erfüllt sein kann, wenn $C_{\alpha,n} = 1/\mathrm{Ch}(\alpha,n)$ wird. Hierin ist

When executing the differentiations, as required by (4), and introducing these and the loading (6) into the differential equation (4), an equation is obtained which can be satisfied only for every harmonic n and each $\xi = x/a$ or φ, respectively, if $C_{\alpha,n} = 1/Ch(\alpha, n)$, where

$$\begin{aligned}\mathrm{Ch}(\alpha,n) &= (\alpha^2+n^2)^4 - 2(\alpha^2+n^2)^3 + (\alpha^2+n^2)^2 - 2(1-\nu)(-\alpha^6+\alpha^2n^4-\alpha^2n^2) + \frac{1-\nu^2}{k}\alpha^4\\ &= \alpha^8 + 2(2n^2-\nu)\alpha^6 + \left(6n^4-6n^2+\frac{1-\nu^2}{k}\right)\alpha^4 + 2n^2(n^2-1)(2n^2-2+\nu)\alpha^2 + (n^2-1)^2n^4\end{aligned} \tag{8}$$

die charakteristische Gleichung der Kreiszylinderschalen.

is the characteristic equation of the circular cylindrical shell.

Wird die so gefundene Lösung für die Konstanten $C_{\alpha,0}$ bzw. $C_{\alpha,n}$ in den Lösungsansatz (7) eingesetzt, dann sind mit

When substituting the solution for the constants $C_{\alpha,0}$ and $C_{\alpha,n}$, thus found, in the solution assumption (7), one obtains with

$$F = F(\xi,\varphi) = -\frac{2a^2}{\pi^2 K}\left\{\frac{1}{2}\int_{\alpha=0}^{\infty}\frac{\cos\alpha\xi}{\mathrm{Ch}\,(\alpha,0)}\,\mathrm{d}\alpha + \sum_{n=2,4,6\ldots}^{\infty}\cos n\varphi\int_{\alpha=0}^{\infty}\frac{\cos\alpha\xi}{\mathrm{Ch}\,(\alpha,n)}\,\mathrm{d}\alpha\right\} \tag{9}$$

über (5) die Verschiebungen und über (1) bzw. (2) sämtliche Schnittgrößen für das mit einer Einzellastgruppe belastete Rohr bestimmt.

Zur Lösung der Fourier-Integrale werden noch die Wurzeln der charakteristischen Gleichung (8) benötigt. Diese sind sowohl für die Größe als auch für den Verlauf aller interessierenden Verschiebungen und somit Schnittlasten der Kreiszylinderschale maßgebend.

the displacements by (5), and by (1) or (2), respectively, all of the section stress resultants for the pipe, subjected to a group of concentrated loads.

For the solution of the Fourier integrals, the roots of the characteristic equation (8) are still needed. They are controlling for the magnitude as well as for the functional behavior of all of the displacements of interest, and thus for the section stress resultants of the circular cylindrical shell.

4.2 Die Wurzeln der charakteristischen Gleichung

Entsprechend dem Aufbau der charakteristischen Gleichung (8) ist keine ihrer Wurzeln ganz reell oder imaginär. Sie sind immer paarweise konjugiert komplex und paarweise entgegengesetzt gleich, so daß sie mit den Real- und Imaginärteilen $\varkappa$ und μ in der Form

4.2 The Roots of the Characteristic Equation

According to the structure of the characteristic equation (8), none of its roots is entirely real or imaginary. They are always in pairs conjugate complex and opposite equal, thus, can be written, with the coefficients $\varkappa$ and μ, for real and imaginary part, in the form

$$\alpha_{1,2 \atop 3,4} = \pm\varkappa_1 \pm \mathrm{i}\mu_1 \quad \text{und} \quad \alpha_{5,6 \atop 7,8} = \pm\varkappa_2 \pm \mathrm{i}\mu_2 \tag{10}$$

geschrieben werden können.

Die Bestimmung von $\varkappa$ und μ erfolgt entweder mit Hilfe der kubischen Resolvente [23] oder iterativ mit einer bekannten Näherung unter Anwendung des Newtonschen Verfahrens für zwei Veränderliche [25]. Bei Benutzung von programmgesteuerten Rechenanlagen bietet sich das letzte Verfahren an, da von Holand [26] eine recht gute Näherungslösung nach dem verbesserten Donnellschen Verfahren angegeben wird. Die Real- bzw. Imaginärteile der Wurzeln weichen um weniger als 1% von der genauen Lösung ab und können somit auch bei der Berechnung des Randstörungsproblems (vgl. Abschnitt 5.1) benutzt werden.

Die Flüggeschen Werte der Harmonischen $n = 2$ finden sich für die in der Arbeit gewählten Steifigkeitsparameter k in der Tafelübersicht. Für höhere Harmonische und abweichende Steifigkeitsparameter können die folgenden Ausdrücke benutzt werden:

The determination of $\varkappa$ and μ is done either by the cubic resolvent [23], or by iteration with a known approximation, using Newton's method for two variables [25]. When using computers, the latter method is preferable, because a good approximate solution, with the improved Donnell method, was given by Holand [26]. The real and imaginary parts of the roots differ, herein, less than 1% from the exact solution, thus, can be used also for the computation of the boundary disturbances (see Section 5.1).

Flügge's values of the harmonic $n = 2$, for the stiffness parameters k chosen in this work, can be found under Chart Classification. For higher harmonics, and different stiffness parameters, the following expressions can be used:

$$\varkappa_{1 \atop 2} = +\frac{1}{2\sqrt{2}}\sqrt[4]{\frac{1-\nu^2}{k}}\left\{1 \mp \sqrt{\sqrt{\varepsilon^2+1}-\varepsilon}\right\} \qquad \mu_{1 \atop 2} = \mp\frac{1}{2\sqrt{2}}\sqrt[4]{\frac{1-\nu^2}{k}}\left\{1 \mp \sqrt{\sqrt{\varepsilon^2+1}+\varepsilon}\right\}$$

$$\text{mit/with} \qquad \varepsilon = 4n\sqrt{n^2-1}\sqrt{\frac{k}{1-\nu^2}} \tag{11}$$

Für die Harmonische $n = 0$ ergeben sich nur vier von Null verschiedene Wurzeln

For the harmonic $n = 0$, only four roots, different from zero, are obtained,

$$\alpha_{1,2 \atop 3,4} = \pm \varkappa_0 (1 \pm \mathrm{i}), \tag{10a}$$

deren Real- und Imaginärteile bei Annahme von $k \ll (1 - \nu^2)/\nu^2$ den gleichen Wert

whose real and imaginary parts, with the assumption $k \ll (1 - \nu^2)/\nu^2$, assume the same value:

$$\varkappa_0 = \sqrt[4]{\frac{1 - \nu^2}{4k}} \tag{11a}$$

annehmen.

4.3 Koeffizienten der Lösungsfunktion

Die Berechnung der Fourier-Integrale in der Lösungsfunktion (9) kann entweder mit der Partialbruchzerlegung [6] oder mit Cauchys Integralformel [9, 25] erfolgen. Die Integration führt auf eine Summe von vier gedämpften Schwingungen. Das eine Paar klingt mit wachsender Entfernung von der Last ab, das andere Paar mit abnehmendem $\xi = x/a$. Die letzten beiden Schwingungen müssen bei dem unendlich langen Rohr verschwinden, so daß die Lösungsfunktion folgende Form erhält:

4.3 Coefficients of the Solution Function

The computation of the Fourier integrals of the solution function (9) can be done either by resolution into partial fractions [6], or with Cauchy's integral formula [9, 25]. The integration leads to a sum of four dampened oscillations. The one pair decreases with increasing distance from the load, the other pair with decreasing $\xi = x/a$. The last two oscillations have to vanish for the infinite pipe, thus, the solution function takes the following form:

$$\begin{aligned} F(\xi,\varphi) &= -\frac{a^2}{\pi K}\left[\frac{1}{2}\,2A_{F0}\,\mathrm{e}^{-\varkappa_0\xi}(\cos\varkappa_0\xi + \sin\varkappa_0\xi) + \sum_{n=2,4,6\ldots}^{\infty}\cos n\varphi\left\{\mathrm{e}^{-\mu_1\xi}(A_{Fn}\cos\varkappa_1\xi + B_{Fn}\sin\varkappa_1\xi) + \mathrm{e}^{-\mu_2\xi}(C_{Fn}\cos\varkappa_2\xi + D_{Fn}\sin\varkappa_2\xi)\right\}\right] \\ F(\xi,\varphi) &= -\frac{a^2}{\pi K}\sum_{n=0,2,4\ldots}^{\infty}\cos n\varphi\left\{\mathrm{e}^{-\mu_1\xi}(A_{Fn}\cos\varkappa_1\xi + B_{Fn}\sin\varkappa_1\xi) + \mathrm{e}^{-\mu_2\xi}(C_{Fn}\cos\varkappa_2\xi + D_{Fn}\sin\varkappa_2\xi)\right\} \end{aligned} \tag{12}$$

Für $n = 0$ ist dann $\mu_1 = \varkappa_1 = \varkappa_0$, $C_{F0} = D_{F0} = 0$ und

$$A_{F0} = B_{F0} = \frac{1}{2}\,2\,A_{F0} = -\frac{1}{32\,\varkappa_0^7}$$

For $n = 0$ is $\mu_1 = \varkappa_1 = \varkappa_0$, $C_{F0} = 0$ and

$$A_{F0} = B_{F0} = \frac{1}{2}\,2\,A_{F0} = -\frac{2}{32\,\varkappa_0^7}$$

Für $n = 2, 4, 6\ldots$ ergeben sich alle Konstanten aus den Real- bzw. Imaginärteilen der Wurzeln der charakteristischen Gleichung zu:

For $n = 2, 4, 6,\ldots$, all constants are obtained from the real and imaginary parts of the roots of the characteristic equation:

$$A_{Fn} = \frac{2\delta}{\gamma^2 + \delta^2} \qquad B_{Fn} = -\frac{2\gamma}{\gamma^2 + \delta^2} \qquad C_{Fn} = \frac{2\psi}{\zeta^2 + \psi^2} \qquad D_{Fn} = -\frac{2\zeta}{\zeta^2 + \psi^2} \tag{13}$$

mit

with

$$
\begin{aligned}
\gamma &= -4\varkappa_1\,\mu_1^2\,(5\varkappa_1^4 + \mu_1^4 - 10\varkappa_1^2\,\mu_1^2 + \varkappa_2^4 + \mu_2^4 + 2\varkappa_2^2\,\mu_2^2 - 6\varkappa_1^2\,\varkappa_2^2 - 2\mu_1^2\,\mu_2^2 + 2\varkappa_2^2\,\mu_1^2 + 6\varkappa_1^2\,\mu_2^2)\\
\delta &= +4\mu_1\,\varkappa_1^2\,(5\mu_1^4 + \varkappa_1^4 - 10\mu_1^2\,\varkappa_1^2 + \mu_2^4 + \varkappa_2^4 + 2\mu_2^2\,\varkappa_2^2 - 6\mu_1^2\,\mu_2^2 - 2\varkappa_1^2\,\varkappa_2^2 + 2\mu_2^2\,\varkappa_1^2 + 6\mu_1^2\,\varkappa_2^2)\\
\zeta &= -4\varkappa_2\,\mu_2^2\,(5\varkappa_2^4 + \mu_2^4 - 10\varkappa_2^2\,\mu_2^2 + \varkappa_1^4 + \mu_1^4 + 2\varkappa_1^2\,\mu_1^2 - 6\varkappa_2^2\,\varkappa_1^2 - 2\mu_2^2\,\mu_1^2 + 2\varkappa_1^2\,\mu_2^2 + 6\varkappa_2^2\,\mu_1^2)\\
\psi &= +4\mu_2\,\varkappa_2^2\,(5\mu_2^4 + \varkappa_2^4 - 10\mu_2^2\,\varkappa_2^2 + \mu_1^4 + \varkappa_1^4 + 2\mu_1^2\,\varkappa_1^2 - 6\mu_2^2\,\mu_1^2 - 2\varkappa_2^2\,\varkappa_1^2 + 2\mu_1^2\,\varkappa_2^2 + 6\mu_2^2\,\varkappa_1^2)
\end{aligned}
\qquad (14)
$$

4.4 Verformungen und Schnittgrößen

Alle Ableitungen der Lösungsfunktion haben die gleiche Form (12), lediglich die Konstanten ändern sich. Für Differentiationen nach $\xi = x/a$ können diese aus einer einfachen Rekursionsformel bestimmt werden. Sind A_m bis D_m die Konstanten der m-ten Ableitung, so folgen diese aus der $(m-1)$-ten Ableitung nach

4.4 Deformations and Section Stress Resultants

All derivations of the solution function have the same form (12), only the constants change. For the differentiations with respect to $\xi = x/a$, they can be obtained by a simple recursion formula. If A_m to D_m are the constants of the *m-th* derivation, they are obtained from the $(m-1)$-*th* derivation by

$$A_m = -\mu_1 A_{m-1} + \varkappa_1 B_{m-1} \qquad B_m = -\varkappa_1 A_{m-1} - \mu_1 B_{m-1} \qquad C_m = -\mu_2 C_{m-1} + \varkappa_2 D_{m-1} \qquad D_m = -\varkappa_2 C_{m-1} - \mu_2 D_{m-1} \qquad (15)$$

Es ist also ratsam, alle interessierenden Größen in Abhängigkeit von der mit $\xi = x/a$ veränderlichen Funktion

Thus, it is advisible to express all quantities of interest in dependence of the function

$$f_n = f_n(\xi) = e^{-\mu_1\xi}(A_{Fn}\cos\varkappa_1\xi + B_{Fn}\sin\varkappa_1\xi) + e^{-\mu_2\xi}(C_{Fn}\cos\varkappa_2\xi + D_{Fn}\sin\varkappa_2\xi) \qquad (16)$$

anzugeben.

Um den Tafelanhang nicht zu umfangreich werden zu lassen, muß eine gewisse Auswahl der Größen von (1), (2) und (5) getroffen werden. Bei der Erfüllung der Randbedingungen werden im allgemeinen zusätzlich noch die Verdrehungen in Längs- (χ_x) und Ringrichtung (χ_φ) benötigt. Eine Vertafelung der ersteren ist nicht nötig, da sie leicht über die Durchbiegungsdifferenz zweier benachbarter Punkte ermittelt werden kann:

which varies with $\xi = x/a$.

In order to keep the Chart Appendix short, a certain selection of the quantities (1), (2) and (5) has to be made. To satisfy the boundary condition, in general, the rotations in longitudinal (χ_x) and in hoop direction (χ_φ) are required additionally. A tabulation of the former is not necessary, because they can be obtained easily from the deflection difference of two adjacent points:

$$\chi_x = \frac{\partial w}{\partial x} = \frac{1}{a}\,\frac{w(\xi + \Delta\xi) - w(\xi)}{\Delta\xi} = \frac{w(x/a + \Delta x/a) - w(x/a)}{a\cdot\Delta x/a}$$

Bei der Verdrehung in Ringrichtung muß noch die Verschiebung v berücksichtigt werden. Daher wurde $\chi_\varphi = (v - w^{\cdot})/a$ zusätzlich vertafelt.

Die Schub- ($n_{x\varphi}$, $n_{\varphi x}$) und die Querkräfte (q_x, q_φ) werden vor allem zur Erfüllung der Randbedingungen gebraucht. Zur Ermittlung der Hauptspannungen und deren Richtung reicht meist ein Näherungswert aus. Da sowieso die Schubkräfte und Drillmomente in den beiden Richtungen nahezu gleich sind (Donnellsche Näherung) und der Einfluß des Drillmomentes bei der Randschubkraft klein, scheint es ausreichend, nur $m_{x\varphi} \approx m_{\varphi x}$ und $\bar{n}_{x\varphi} \approx n_{x\varphi} \approx n_{\varphi x}$ zu vertafeln. Bei den Randquerkräften ist der Einfluß der Drillmomente größer, trotzdem kann auch

For the rotation in the hoop direction, the displacement v has to be considered. Therefore, $\chi_\varphi = (v - w^{\cdot})/a$ was tabulated in addition.

The shearing ($n_{x\varphi}$, $n_{\varphi x}$) and transverse forces (q_x, q_φ) are needed mainly to satisfy the boundary conditions. For determination of the principal stresses and their directions, an approximate value is sufficient in most cases. Because shearing forces and twisting moments in the two directions are almost equal (Donnell's approximation), and the influence of the twisting moment on the boundary shearing force is small, it seems to be sufficient to tabulate only $m_{x\varphi} \approx m_{\varphi x}$ and $\bar{n}_{x\varphi} \approx n_{x\varphi} \approx n_{\varphi x}$. The influence of the twisting moments on the transverse boun-

die Schubspannungsermittlung näherungsweise mit den dargestellten Werten $\bar{q}_x \approx q_x$ und $\bar{q}_\varphi \approx q_\varphi$ erfolgen.

Werden die Differentiationen nach φ ausgeführt und die gedämpften Schwingungen in Längsrichtung nach (16) eingesetzt, dann ergeben sich für die vertafelten Flächen mit der Lösungsfunktion (12) aus (1), (2) und (5):

dary forces is larger. However, the computation of the shear stresses can be made, approximately, with the presented values $\bar{q}_x \approx q_x$ and $\bar{q}_\varphi \approx q_\varphi$.

When executing the differentiations with respect to φ, and substituting the dampened oscillations (16) of the longitudinal direction, the tabulated surfaces are obtained from (1), (2) and (5) with the solution function (12):

$$\begin{aligned}
u &= \frac{a^2}{\pi K} \sum_{n=0,2,4\ldots}^{\infty} \cos n\varphi \left\{ -k f_n''''' + \left[\nu - \frac{3}{2}(1-\nu)\, n^2 k^2\right] f_n''' + [n^2 + n^4 k]\, f_n' \right\} \\
v &= \frac{a^2}{\pi K} \sum_{n=2,4,6\ldots}^{\infty} \sin n\varphi \, \{2\, nk f_n'''' - [(2+\nu)\, n + 2\, n^3 k]\, f_n'' + n^3 f_n\} \\
w &= \frac{a^2}{\pi K} \sum_{n=0,2,4\ldots}^{\infty} \cos n\varphi \, \{-f_n'''' + 2n^2 f_n'' - n^4 f_n\} \\
\chi_\varphi &= \frac{a}{\pi K} \sum_{n=2,4,6\ldots}^{\infty} \sin n\varphi \, \{-n f_n'''' + [2n^3 - (2+\nu)\, n]\, f_n'' - [n^5 - n^3]\, f_n\} \\
n_x &= \frac{1}{\pi a} \sum_{n=0,2,4\ldots}^{\infty} \cos n\varphi \left\{ -2\,(1-\nu)\, n^2 f_n'''' + \left[\frac{1-\nu^2}{k} n^2 + 2\,(1-\nu)\, n^4\right] f_n'' \right\} \\
n_\varphi &= \frac{1}{\pi a} \sum_{n=0,2,4\ldots}^{\infty} \cos n\varphi \left\{ -\nu f_n'''''' - \left[\frac{1-\nu^2}{k} - 3n^2\right] f_n'''' - [(4-\nu)\, n^4 - 2n^2]\, f_n'' + [n^6 - n^4]\, f_n \right\} \\
\bar{n}_{x\varphi} &= \frac{1}{\pi a} \sum_{n=2,4,6\ldots}^{\infty} \sin n\varphi \, \frac{1-\nu}{2} \left\{ 6nk f_n''''' - \left[2\, \frac{1+\nu}{k} n - 4n^3\right] f_n''' - [4n^5 - 3n^3]\, f_n' \right\} \\
m_x &= \frac{1}{\pi} \sum_{n=0,2,4\ldots}^{\infty} \cos n\varphi \, \{-f_n'''''' + [(2+\nu)\, n^2 - \nu]\, f_n'''' - [(1+2\nu)\, n^4 + (1-2\nu-\nu^2)\, n^2]\, f_n'' + \nu\, [n^6 - n^4]\, f_n\} \\
m_\varphi &= \frac{1}{\pi} \sum_{n=0,2,4\ldots}^{\infty} \cos n\varphi \, \{-\nu f_n'''''' + [(1+2\nu)\, n^2 - 1]\, f_n'''' - [(2+\nu)\, n^4 - 2n^2]\, f_n'' + [n^6 - n^4]\, f_n\} \\
m_{x\varphi} &= \frac{1}{\pi} \sum_{n=2,4,6\ldots}^{\infty} \sin n\varphi \, (1-\nu) \, \{n f_n''''' - [2n^3 - (2+\nu)\, n]\, f_n''' + [n^5 - n^3]\, f_n'\} \\
\bar{q}_x &= \frac{1}{\pi a} \sum_{n=0,2,4\ldots}^{\infty} \cos n\varphi \, \{-f_n''''''' + [(4-\nu)\, n^2 - \nu]\, f_n''''' - [(5-2\nu)\, n^4 - 2n^2]\, f_n''' + [(2-\nu)\, n^6 - (2-\nu)\, n^4]\, f_n'\} \\
\bar{q}_\varphi &= \frac{1}{\pi a} \sum_{n=2,4,6\ldots}^{\infty} \sin n\varphi \, \{(2-\nu)\, n f_n'''''' - [(5-2\nu)\, n^3 - (4-2\nu-\nu^2)\, n]\, f_n'''' + [(4-\nu)\, n^5 - 2\,(2-\nu)\, n^3]\, f_n'' - [n^7 - n^5]\, f_n\}
\end{aligned} \tag{17}$$

Die hier mitgeteilten Ausdrücke für die statischen Größen können auch für die Randstörungsrechnung an den Binderscheiben benutzt werden. Es sind im allgemeinen Fall lediglich die mit abnehmendem x/a abklingenden Schwingungen hinzuzufügen und die neuen Konstanten A_0 bis H_0 als Freiwerte zu betrachten. Die Wurzeln der charakteristischen Gleichung (11) bleiben genauso erhalten wie die Rekursionsformeln (15) zur Bestimmung der Konstanten bei den Ableitungen nach $\xi = x/a$.

The expressions of the static quantities, given here, can also be used in the computation of the boundary disturbance at the frame supports. For the general case, only the oscillations, which decrease with decreasing x/a have to be added, and the new constants A_0 to H_0 have to be considered free values. The roots of the characteristic equation (11) as well as the recursion formulas (15) for the determination of the constants are preserved in the derivations with respect to x/a.

5. Erfüllung der Randbedingungen

5.1 Der Querrand

5.1.1 Randstörungsrechnung

Wird aus dem unendlich langen Rohr entsprechend der gegebenen Laststellung x_0 ein einfeldriger Bereich von der Länge l herausgeschnitten, so sind an den Rändern $x = 0$ und $x = l$ aus der Einzelkraftbelastung folgende Verformungen und Schnittgrößen vorhanden (vgl. Bild 3):

5. Satisfying the Boundary Conditions

5.1 The Transverse Boundary

5.1.1 Computation of the Boundary Disturbance

When cutting from the infinite pipe, in accordance with the load position x_0, a single span of length l, the following deformations and section stress resultants, due to the single load, occur at the boundaries $x = 0$ and $x = l$ (see Fig. 3):

$$u^0, u^l;\; v^0, v^l;\; w^0, w^l;\; \chi_x^0, \chi_x^l;\; n_x^0, n_x^l;\; \bar{n}_{x\varphi}^0, \bar{n}_{x\varphi}^l;\; m_x^0, m_x^l;\; \bar{q}_x^0, \bar{q}_x^l. \tag{18}$$

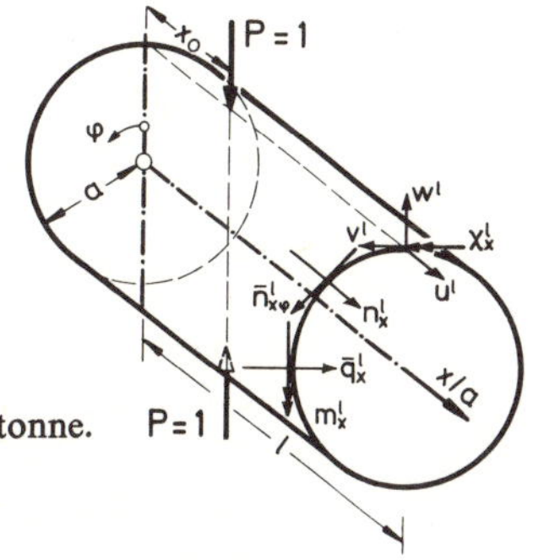

Bild 3. Einfeldtonne.

Figure 3. Single-span barrel shell.

Je nach Auflagerung werden nun an den beiden Rändern jeweils vier statische Größen entweder zu Null, oder sie müssen aus Verformungsbedingungen bestimmt werden. Über Binderscheiben durchlaufende Schalen werden dabei meist näherungsweise aus der gelenkig gelagerten Einfeldschale abgeleitet ([27, 28], vgl. auch [29, 30]).

Häufig sind die Binderscheiben in ihrer Ebene sehr steif und senkrecht dazu weich. Es ergeben sich dann für die Einfeldtonne an beiden Querrändern Naviersche Randbedingungen:

According to the support, at both boundaries four static quantities either vanish, or have to be determined from deformation conditions. Herein, shells, continuous at transverse frame supports, are, in most of the cases, deduced approximately from the hinge supported single-span shell ([27, 28] compare also [29, 30]).

The transverse support frames are, in most cases, very rigid in their plane, but soft perpendicular to the plane, resulting in Navier's boundary conditions at both transverse boundaries for the single-span barrel shell:

$$v = w = n_x = m_x = 0, \tag{19}$$

die auch bei der Berechnung der Längsränder mit Tafelwerken meist vorausgesetzt werden.

Die aus dem partikulären Integral der Differentialgleichung herrührenden statischen Größen (18) müssen für die geforderten Randbedingungen (z.B. (19)) durch eine reguläre Fläche aus der Lösung der homogenen Differentialgleichung (4) zum Verschwinden gebracht werden (oder eine vorgegebene Größe annehmen). Man kann nun entweder die acht Konstanten A_0 bis H_0 der erweiterten Gleichung (16) oder die Maximalgrößen X_1 bis X_8 von zwei mal vier sogenannten Randlastfällen als Unbekannte einführen. Die erste Methode erfordert die Bestimmung aller Konstanten der benötigten Größen nach (17) mit Hilfe der Rekursionsformel (15). Gute Näherungen für diese sogenannten charakteristischen Koeffizienten haben Aas-Jakobsen in [24] und Holand in [26] angegeben.

which are also assumed, in most cases, in the computation of the longitudinal boundaries with tables.

The static quantities (18), resulting from the particular integral of the differential equation, have to be reduced to zero, for the existing, prescribed boundary conditions (for instance (19)), by a regular surface from the solution of the homogeneous differential equation (4) (or have to be of prescribed magnitude). One can introduce, as unknown, either the eight constants A_0 to H_0 of the expanded equation (16), or the maximum quantities X_1 to X_8 of two times four so-called boundary load cases. The first method necessitates the determination of all constants, according to (17), by using the recursion formula (15). Good approximations for these, so-called, characteristic coefficients have been given by Aas-Jakobsen in [24], and by Holand in [26].

Der zweite Lösungsweg ist bei Holand [26] für die Randlastfälle

$$w^0 = X_1 \frac{a}{Et}\frac{1}{n^2}\sqrt{\frac{1-\nu^2}{k}}\sin n\varphi$$

$$\chi_x^0 = X_2 \frac{1}{Et}\frac{1}{n^2}\sqrt{\frac{1-\nu^2}{2k}}\sqrt[4]{\frac{1-\nu^2}{k}}\sin n\varphi$$

$$n_x^0 = X_3 \sin n\varphi$$

$$n_{x\varphi}^0 = X_4 \frac{1}{\sqrt{2}}\frac{1}{n}\sqrt[4]{\frac{1-\nu^2}{k}}\cos n\varphi$$

vertafelt. Er empfiehlt sich dann, wenn die elastische Nachgiebigkeit der Randträger berücksichtigt werden soll.

In beiden Fällen muß, wenn nicht in symmetrische und antimetrische Lastanteile aufgespalten wird, ein achtgliedriges Gleichungssystem gelöst werden. Die rechten Seiten ergeben sich durch eine Fourier-Entwicklung der betrachteten Randgrößen. Im folgenden soll nun der Rechengang am Beispiel der Randbedingung (19) mittels der Lösungsfunktion erläutert werden.

Ratsamerweise werden die Verformungen bzw. Schnittgrößen (z. B. v, w, n_x und m_x) an den beiden Rändern aus den Zustandsflächen abgegriffen und aufgetragen. Für die hier gewählte Form der regulären Fläche

The second solution method is tabulated by Holand [26] for the boundary load cases

$$\chi_x^0 = n_x^0 = n_{x\varphi}^0 = 0$$

$$w^0 = n_x^0 = n_{x\varphi}^0 = 0$$

$$w^0 = \chi_x^0 = n_{x\varphi}^0 = 0$$

$$w^0 = \chi_x^0 = n_x^0 = 0$$

This is recommended when the elastic settlement at the boundary trusses is to be considered.

An eight-element equation system has to be solved in both cases when not resolving into symmetric and anti-symmetric loadings. The right sides are obtained from a Fourier development of the considered boundary quantities. The computation steps, with the solution function, shall be explained, in the following, for the example of boundary condition (19).

It is advisable to scale and plot the deformations or section stress resultants (for instance v, w, n_x and m_x) at both boundaries from the conditions surfaces. For the form of the regular surface, chosen herein,

$$F(\xi,\varphi)_R = \frac{a^2}{\pi K}\sum_{n=0,2,4\ldots}^{\infty}\cos n\varphi \quad f_n(\xi)_R$$

muß entsprechend (17) für den Rand $\xi = x/a = 0$

it has to be assumed, at the boundary $\xi = x/a = 0$ according to (17),

$$v^0 = \frac{a^2}{\pi K}\sum_{n=2,4,6\ldots}^{\infty} V_n^0 \sin n\varphi \qquad w^0 = \frac{a^2}{\pi K}\left\{\frac{W_0^0}{2} + \sum_{n=2,4,6\ldots}^{\infty} W_n^0\cos n\varphi\right\}$$

$$n_x^0 = \frac{1}{\pi a}\left\{\frac{N_{x_0}^0}{2} + \sum_{n=2,4,6\ldots}^{\infty} N_{x_n}^0\cos n\varphi\right\} \qquad m_x^0 = \frac{1}{\pi}\left\{\frac{M_{x_0}^0}{2} + \sum_{n=2,4,6\ldots}^{\infty} M_{x_n}^0\cos n\varphi\right\} \tag{20}$$

angesetzt werden. Für den zweiten Rand, $\xi = x/a = l/a$ wird nur jeweils 0 durch l ersetzt.

Die Beiwerte V_n^0, W_0^0, W_n^0 etc. ergeben sich bei Benutzung der Simpsonschen Regel und den aus den Zustandsflächen an einer ungeraden Anzahl von äquidistanten Punkten abgegriffenen Werten $v^0(\varphi_k)$, $w^0(\varphi_k)$ etc. zu:

For the second boundary $\xi = x/a = l/a$, 0 has to be substituted only by l.

The coefficients V_n^0, W_0^0, W_n^0 etc. are obtained by using Simpson's rule and with the values $v^0(\varphi_k)$, $w^0(\varphi_k)$ etc., scaled from the condition surfaces at an odd number of equidistant points, from:

$$V_n^0 = \frac{4}{\pi}\int_0^{\pi/2} v^0(\varphi)\sin n\varphi\,d\varphi = \frac{4}{3}\frac{\Delta\varphi}{\pi}\sum_k \varkappa_k\, v^0(\varphi_k)\sin n\varphi_k = \frac{4}{3}\frac{\Delta\varphi}{\pi}(0 + 4\,v^0(\Delta\varphi)\sin n\Delta\varphi + 2\,v^0(2\Delta\varphi)\sin n2\Delta\varphi + \ldots + 4\,v^0(\pi/2-\Delta\varphi)\sin n(\pi/2-\Delta\varphi) + 0)$$

$$W_0^0 = \frac{4}{\pi}\int_0^{\pi/2} w^0(\varphi)\,d\varphi = \frac{4}{3}\frac{\Delta\varphi}{\pi}\sum_k \varkappa_k\, w^0(\varphi_k) = \frac{4}{3}\frac{\Delta\varphi}{\pi}(w^0(0) + 4\,w^0(\Delta\varphi) + 2\,w^0(2\Delta\varphi) + 4\,w^0(3\Delta\varphi) + \ldots + 4\,w^0(\pi/2-\Delta\varphi) + w^0(\pi/2)) \tag{21}$$

$$W_n^0 = \frac{4}{\pi}\int_0^{\pi/2} w^0(\varphi)\ \cos n\varphi\, d\varphi = \frac{4}{3}\frac{\Delta\varphi}{\pi}\sum_k \varkappa_k\ w^0(\varphi_k)\ \cos n\varphi_k$$

$$= \frac{4}{3}\frac{\Delta\varphi}{\pi}(w^0(0) + 4\ w^0(\Delta\varphi)\ \cos n\Delta\varphi + 2\ w^0(2\Delta\varphi)\ \cos n2\Delta\varphi + \ldots + 4\ w^0(\pi/2 - \Delta\varphi)\ \cos n(\pi/2 - \Delta\varphi) + w^0(\pi/2))$$

In gleicher Weise finden sich die Beiwerte

$$V_n^l;\ W_0^l, W_n^l;\ N_{x_0}^0, N_{x_n}^0;\ N_{x_0}^l, N_{x_n}^l;\ M_{x_0}^0, M_{x_n}^0;\ M_{x_0}^l, M_{x_n}^l$$

aus den entsprechenden Randschnitten.

Die Randgrößen der regulären Fläche werden ratsamerweise mit den gleichen Multiplikatoren S_c wie in (17) bzw. im Anhang (ohne Zehnerpotenzen) behaftet. Die vervollständigte Gleichung (16)

$$f_n(\xi)_R = e^{-\mu_1\xi}(A_0 \cos \varkappa_1\xi + B_0 \sin \varkappa_1\xi) + e^{-\mu_2\xi}(C_0 \cos \varkappa_2\xi + D_0 \sin \varkappa_2\xi)$$
$$+ e^{+\mu_1\xi}(E_0 \cos \varkappa_1\xi + F_0 \sin \varkappa_1\xi) + e^{+\mu_2\xi}(G_0 \cos \varkappa_2\xi + H_0 \sin \varkappa_2\xi) \quad (22)$$

muß entsprechend (17) nach $\xi = x/a$ abgeleitet und mit den von k, ν und n abhängigen Konstanten multipliziert werden. Da alle statischen Größen auf A_0, B_0 usw. zurückgeführt werden müssen, ist es ratsam, mit den aus (15) zu bestimmenden Rekursionsformeln für die m-te Ableitung

$$a_m = -\mu_1 a_{m-1} - \varkappa_1 b_{m-1} \qquad b_m = -\mu_1 b_{m-1} + \varkappa_1 a_{m-1} \quad (23)$$

mit $a_0 = +1$ und $b_0 = 0$, sowie

$$c_m = -\mu_2 c_{m-1} - \varkappa_2 d_{m-1} \qquad d_m = -\mu_2 d_{m-1} + \varkappa_2 c_{m-1}$$

mit $c_0 = +1$ und $d_0 = 0$

die Konstanten A_m, B_m usw. anzugeben:

$$\begin{aligned} A_m &= a_m A_0 + b_m B_0 & E_m &= (-1)^m (a_m E_0 - b_m F_0)\\ B_m &= -b_m A_0 + a_m B_0 & F_m &= (-1)^m (b_m E_0 + a_m F_0)\\ C_m &= c_m C_0 + d_m D_0 & G_m &= (-1)^m (c_m G_0 - d_m H_0)\\ D_m &= -d_m C_0 + c_m D_0 & H_m &= (-1)^m (d_m G_0 + c_m H_0) \end{aligned} \quad (24)$$

Jede Verformung oder Schnittgröße S hat somit folgende Form:

$$S(\xi,\varphi)_R = S_c \sum_n \begin{matrix}\sin n\varphi\\ \cos n\varphi\end{matrix} \Big\{ e^{-\mu_1\xi}[(+\bar a_s A_0 + \bar b_s B_0)\cos\varkappa_1\xi + (-\bar b_s A_0 + \bar a_s B_0)\sin\varkappa_1\xi] + e^{-\mu_2\xi}[(+\bar c_s C_0 + \bar d_s D_0)\cos\varkappa_2\xi + (-\bar d_s C_0 + \bar c_s D_0)\sin\varkappa_2\xi]$$
$$+ e^{+\mu_1\xi}[(\pm\bar a_s E_0 \mp \bar b_s F_0)\cos\varkappa_1\xi + (\pm\bar b_s E_0 \pm \bar a_s F_0)\sin\varkappa_1\xi] + e^{+\mu_2\xi}[(\pm\bar c_s G_0 \mp \bar d_s H_0)\cos\varkappa_2\xi + (\pm\bar d_s G_0 \pm\bar c_s H_0)\sin\varkappa_2\xi]\Big\} \quad (25)$$

Die oberen Vorzeichen gelten für gerade m, also nach (17) für v, w, χ_φ, n_x, n_φ, m_x, m_φ und $\bar q_\varphi$. Für die unteren Vorzeichen verbleiben u, $\bar n_{x\varphi}$, $m_{x\varphi}$ und $\bar q_x$.

Similarily, the coefficients $V_n^l;\ W_0^l, W_n^l;\ N_{x_0}^0, N_{x_n}^0;\ N_{x_0}^l, N_{x_n}^l;\ M_{x_0}^0, M_{x_n}^0;\ M_{x_0}^l, M_{x_n}^l$ are obtained from the corresponding boundary sections.

Preferably, the boundary quantities of the regular surface should be provided with the same multipliers S_c as in (17) or in the Appendix (without the powers of ten). The supplemented equation (16) (22) has to be, according to (17), differentiated with respect to $\xi = x/a$, and multiplied by the constants which are depending on k, ν and n. Because all static quantities have to be deduced to A_0, B_0 etc., it is advisable to denote the constants A_m, B_m etc. with the recursion formulas for the m-th derivation, obtained from (15), (23)

with $a_0 = +1$ and $b_0 = 0$, and furthermore

with $c_0 = +1$ and $d_0 = 0$.

The constants A_m, B_m etc. become (24)

Thus, every deformation or section quantity has the following form: (25)

The upper signs apply for even m, i.e., according to (17) for v, w, χ_φ, n_x, n_φ, m_x, m_φ and $\bar q_\varphi$. For the lower signs remain u, $\bar n_{x\varphi}$, $m_{x\varphi}$ and $\bar q_x$.

Die allein noch vorhandenen neuen Koeffizienten $\bar{a}_s$, $\bar{b}_s$, $\bar{c}_s$ und $\bar{d}_s$ sind schnell nach (17) mit (23) zu berechnen. So wird z.B. für die Verschiebung v oder das Biegemoment m_x:

The remaining new coefficients $\bar{a}_s$, $\bar{b}_s$, $\bar{c}_s$ and $\bar{d}_s$ are obtained immediately from (17) with (23). Thus, for the displacement v or the bending moment m_x are, for example:

$$\begin{aligned}
\bar{a}_v &= 2nk\, a_4 - \left[(2+\nu)\, n + 2n^3k\right] a_2 + n^3 \\
\bar{b}_v &= 2nk\, b_4 - \left[(2+\nu)\, n + 2n^3k\right] b_2 + 0 \\
\bar{c}_{mx} &= -\, c_6 + \left[(2+\nu)\, n^2 - \nu\right] c_4 - \left[(1+2\nu)\, n^4 + (1-2\nu-\nu^2)\, n^2\right] c_2 + \nu\left[n^6 - n^4\right] \\
\bar{d}_{mx} &= -\, d_6 + \left[(2+\nu)\, n^2 - \nu\right] d_4 - \left[(1+2\nu)\, n^4 + (1-2\nu-\nu^2)\, n^2\right] d_2
\end{aligned} \tag{26}$$

Für das Gleichungssystem benötigt man die zwei mal vier statischen Größen an den Rändern $\xi = x/a = 0$ und $\xi = x/a = l/a$. Im ersten Fall wird sofort z.B.

Required are, for the equation system, the two times four static quantities at the boundaries $\xi = x/a = 0$ and $\xi = x/a = l/a$. For the first case, for instance, it becomes immediately

$$v_R^0 = \frac{a^2}{\pi K} \sum_n \sin n\varphi\, (\bar{a}_v A_0 + \bar{b}_v B_0) + (\bar{c}_v C_0 + \bar{d}_v D_0) + (\bar{a}_v E_0 - \bar{b}_v F_0) + (\bar{c}_v G_0 - \bar{d}_v H_0) \tag{27}$$

Am Rand $x/a = l/a$ dagegen müssen erst die trigonometrischen bzw. logarithmischen Funktionen errechnet und in (25) eingesetzt werden. Somit wird z.B.

However, at the boundary $x/a = l/a$, the trigonometric or logarithmic functions, respectively, have to be computed first, and substituted in (25). Thus, it becomes for instance

$$\begin{aligned}
m_{x_R}^l = \frac{1}{\pi} \sum_n \cos n\varphi \Big\{ & e^{-\mu_1 l/a} \left[(\bar{a}_{m_x} \cos \varkappa_1\, l/a - \bar{b}_{m_x} \sin \varkappa_1\, l/a)\, A_0 + (\ \bar{b}_{m_x} \cos \varkappa_1\, l/a + \bar{a}_{m_x} \sin \varkappa_1\, l/a)\, B_0\right] \\
& + e^{-\mu_2 l/a} \left[(\bar{c}_{m_x} \cos \varkappa_2\, l/a - \bar{d}_{m_x} \sin \varkappa_2\, l/a)\, C_0 + (\ \bar{d}_{m_x} \cos \varkappa_2\, l/a + \bar{c}_{m_x} \sin \varkappa_2\, l/a)\, D_0\right] \\
& + e^{+\mu_1 l/a} \left[(\bar{a}_{m_x} \cos \varkappa_1\, l/a + \bar{b}_{m_x} \sin \varkappa_1\, l/a)\, E_0 + (-\bar{b}_{m_x} \cos \varkappa_1\, l/a + \bar{a}_{m_x} \sin \varkappa_1\, l/a)\, F_0\right] \\
& + e^{+\mu_2 l/a} \left[(\bar{c}_{m_x} \cos \varkappa_2\, l/a + \bar{d}_{m_x} \sin \varkappa_2\, l/a)\, G_0 + (-\bar{d}_{m_x} \cos \varkappa_2\, l/a + \bar{c}_{m_x} \sin \varkappa_2\, l/a)\, H_0\right] \Big\} \\
= \frac{1}{\pi} \sum_n \cos n\varphi \Big\{ & \bar{a}^l_{m_x} A_0 + \bar{b}^l_{m_x} B_0 + \bar{c}^l_{m_x} C_0 + \bar{d}^l_{m_x} D_0 + \bar{e}^l_{m_x} E_0 + \bar{f}^l_{m_x} F_0 + \bar{g}^l_{m_x} G_0 + \bar{h}^l_{m_x} H_0 \Big\}
\end{aligned} \tag{28}$$

worin

where:

$$\bar{a}^l_{m_x} = (\bar{a}_{m_x} \cos \varkappa_1\, l/a - \bar{b}_{m_x} \sin \varkappa_1\, l/a)\, e^{-\mu_1 l/a} \qquad \bar{b}^l_{m_x} = (\bar{b}_{m_x} \cos \varkappa_1\, l/a + \bar{a}_{m_x} \sin \varkappa_1\, l/a)\, e^{-\mu_1 l/a} \text{ etc.} \tag{29}$$

In den Gleichungen (20) und (27) bzw. (28) treten dieselben trigonometrischen Funktionen in φ und Multiplikatoren S_c auf, so daß für jedes Reihenglied $n = 0, 2, 4, \ldots$ getrennt ein Gleichungssystem (30) für die zwei mal vier Randbedingungen aufgestellt werden kann.

In equations (20) and (27) or (28), respectively, the same trigonometric functions of φ and multipliers S_c occur, thus, for each element $n = 0, 2, 4 \ldots$ of the series, an equation system (30) can be set up separately for the two times four boundary conditions.

	A_0	B_0	C_0	D_0	E_0	F_0	G_0	H_0	$\otimes$		
$v^0 =$	$\bar{a}_v$	$\bar{b}_v$	$\bar{c}_v$	$\bar{d}_v$	$\bar{a}_v$	$-\bar{b}_v$	$\bar{c}_v$	$-\bar{d}_v$	V_n^0	$= 0$	Rand / Boundary
$w^0 =$	$\bar{a}_w$	$\bar{b}_w$	$\bar{c}_w$	$\bar{d}_w$	$\bar{a}_w$	$-\bar{b}_w$	$\bar{c}_w$	$-\bar{d}_w$	W_n^0	$= 0$	$\frac{x}{a} = 0$
$n_x^0 =$	$\bar{a}_{n_x}$	$\bar{b}_{n_x}$	$\bar{c}_{n_x}$	$\bar{d}_{n_x}$	$\bar{a}_{n_x}$	$-\bar{b}_{n_x}$	$\bar{c}_{n_x}$	$-\bar{d}_{n_x}$	$N_{x_n}^0$	$= 0$	
$m_x^0 =$	$\bar{a}_{m_x}$	$\bar{b}_{m_x}$	$\bar{c}_{m_x}$	$\bar{d}_{m_x}$	$\bar{a}_{m_x}$	$-\bar{b}_{m_x}$	$\bar{c}_{m_x}$	$-\bar{d}_{m_x}$	$M_{x_n}^0$	$= 0$	
$v^l =$	$\bar{a}_v^l$	$\bar{b}_v^l$	$\bar{c}_v^l$	$\bar{d}_v^l$	$\bar{e}_v^l$	$\bar{f}_v^l$	$\bar{g}_v^l$	$\bar{h}_v^l$	V_n^l	$= 0$	Rand / Boundary
$w^l =$	$\bar{a}_w^l$	$\bar{b}_w^l$	$\bar{c}_w^l$	$\bar{d}_w^l$	$\bar{e}_w^l$	$\bar{f}_w^l$	$\bar{g}_w^l$	$\bar{h}_w^l$	W_n^l	$= 0$	$\frac{x}{a} = \frac{l}{a}$
$n_x^l =$	$\bar{a}_{n_x}^l$	$\bar{b}_{n_x}^l$	$\bar{c}_{n_x}^l$	$\bar{d}_{n_x}^l$	$\bar{e}_{n_x}^l$	$\bar{f}_{n_x}^l$	$\bar{g}_{n_x}^l$	$\bar{h}_{n_x}^l$	$N_{x_n}^l$	$= 0$	
$m_x^l =$	$\bar{a}_{m_x}^l$	$\bar{b}_{m_x}^l$	$\bar{c}_{m_x}^l$	$\bar{d}_{m_x}^l$	$\bar{e}_{m_x}^l$	$\bar{f}_{m_x}^l$	$\bar{g}_{m_x}^l$	$\bar{h}_{m_x}^l$	$M_{x_n}^l$	$= 0$	

(30)

Die Auflösung des Gleichungssystems (30) liefert die Konstanten der Spannungsfunktion A_0 bis H_0 und mit den bekannten Koeffizienten $\bar{a}_s$ bis $\bar{d}_s$ auch den Verlauf jeder statischen Größe, wenn zu dieser regulären Fläche noch die Zustandsfläche aus dem Anhang addiert wird.

Steht die Einzellast nicht in unmittelbarer Nähe eines Randes, dann ist meist – vor allem bei kleinem Steifigkeitsparameter k – die Ermittlung der Harmonischen $n = 2$ ausreichend. Der in Ringrichtung konstante Anteil der Last ($n = 0$) ist ohnehin klein. Wünscht man eine größere Genauigkeit, dann sollte jedoch noch $n = 4$ und evtl. $n = 6$ zusätzlich berechnet werden.

Stehen mehrere Einzellasten auf der Schale, dann addiert man ratsamerweise deren Wirkung vor der Erfüllung der Randbedingungen. Lediglich bei gleicher oder spiegelbildlicher Anordnung von Lasten oder Lastgruppen in Längsrichtung kann die Superposition nachträglich vorgenommen werden, da dann kein neuer Lastfall zu ermitteln ist. Es wird in diesen Fällen lediglich $\varphi = 0$ neu festgelegt.

5.1.2 Lastumordnungsverfahren

Für den häufig vorkommenden Fall von zwei gelenkig gelagerten Querrändern entsprechend den Randbedingungen (19) kann man mit Erfolg vom Lastumordnungsverfahren Gebrauch machen, wie es Huber [31] bereits 1926 bei kreuzweise gespannten Platten vorgeschlagen hat.

The solution of the equation system (30) produces the constants A_0 to H_0 of the stress function, and with the known coefficients $\bar{a}_s$ to $\bar{d}_s$ the contour of every static quantity, when adding to this regular surface the condition surface from the Appendix.

If the load is not located close to a boundary, the determination of the harmonic $n = 2$ is sufficient, in most cases, —in particular, for a small stiffness parameter k –. The share of the load ($n = 0$), constant in the hoop direction, is small anyhow. However, if one desires a greater accuracy, $n = 4$ and, perhaps, $n = 6$ should be computed in addition.

If several concentrated loads are applied to the shell, it is advisable to add their effects before satisfying the boundary conditions. Only for equal or mirrored arrangement of loads or load groups with respect to the longitudinal direction, the superposition can be made afterwards because no new load case has to be analyzed. In such cases, only $\varphi = 0$ has to be established anew.

5.1.2 Load Rearrangement Method

For the frequent case of two hinge supported transverse boundaries, according to the boundary conditions (19), the load rearrangement method can be applied with advantage, as was suggested already by Huber [31] in 1926 for two-way slabs.

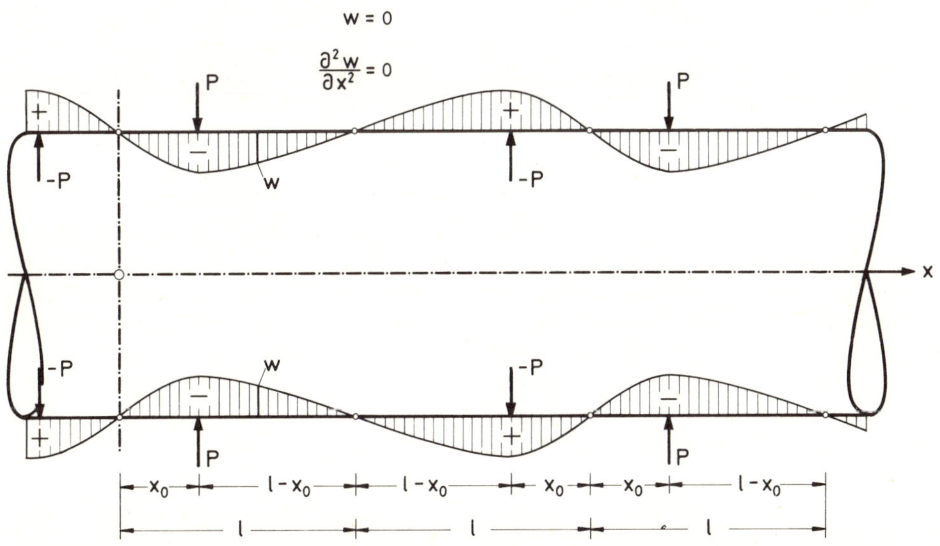

Bild 4. Einfeldtonne mit beidseitig gelenkiger Lagerung.

Figure 4. Single-span barrel, both sides hinge supported.

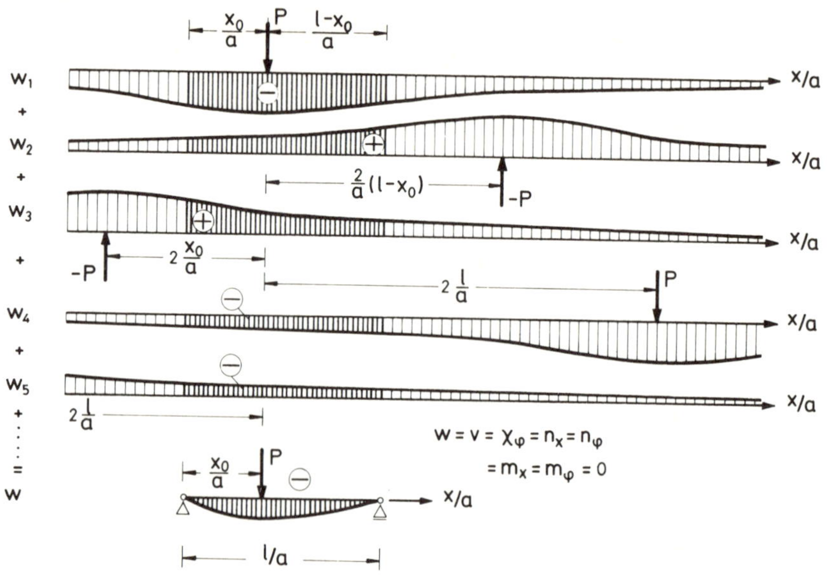

Bild 5. Summation der Zustandsflächen.

Figure 5. Summation of the condition surfaces.

Werden am unendlich langen Rohr entsprechend Bild 4 abwechselnd positive und negative Einzellasten oder Einzellastgruppen angebracht, so ergeben sich bei $x = 0$ und $x = l$ Wende- und Nullpunkte der Biegelinie. Gleichzeitig verschwinden hier am ganzen Umfang die Verschiebung v, die Verdrehung χ_φ, die Normalkräfte n_x und n_φ und die Biegemomente m_x und m_φ. Die Randbedingungen (19) sind also erfüllt. Die interessierende statische Größe bestimmt man entweder durch Summation der nach rechts und nach links verschobenen positiven bzw. negativen Zustandsflächen-Ordinaten entsprechend Bild 5 oder für einen vorgegebenen Punkt durch Addition der Einflußflächen-Ordinaten für die angegebene Laststellung mit $\pm P$. Analog verfährt man bei Lastgruppen.

Da die Zustands- bzw. Einflußflächen nur bis $x/a = \pm 3{,}5$ aufgetragen sind, die Werte jedoch bis $x/a = \pm 10$ und bei den Verformungen sogar bis $x/a = \pm 20$ für eine einigermaßen genaue Ermittlung benötigt werden, muß der Verlauf für $x/a > 3{,}5$ vorgegeben sein. Nun zeigt sich aber, daß die mit $\xi = x/a$ schnell abklingenden Schwingungen, für die $\varkappa_2$ und μ_2 (vgl. (11)) verantwortlich sind, bereits keinen Einfluß mehr haben. Ferner sind die Werte der höheren Harmonischen $n \geqslant 4$ (neben $n = 0$) – außer bei n_φ, $m_{x\varphi}$, $\bar{q}_x$ und $\bar{q}_\varphi$ – vernachlässigbar klein. Schließlich werden noch die Koeffizienten vor $\cos \varkappa_1 \xi$ und $\sin \varkappa_1 \xi$ bis auf das Vorzeichen nahezu gleich – außer bei m_x – oder einer von beiden sehr klein. Da jedoch alle Schnittgrößen dieser Ausnahmen – vor allem bei kleinem Steifigkeitsparameter k – bereits bei $\xi = x/a = 3{,}5$ stark abgeklungen sind und somit der Fehler wieder gering wird, kann mit folgenden Näherungen gerechnet werden.

If alternate positive and negative concentrated loads or groups of concentrated loads are applied to the infinite pipe, according to Fig. 4, inflection or zero points of the deflection line are obtained at $x = 0$ and $x = l$. At the same time, the displacements v, the rotation χ_φ, the normal forces n_x and n_φ, and the bending moments m_x and m_φ vanish along the entire periphery. Thus, the boundary conditions (19) are satisfied. The static quantities of interest are determined either by summation of the positive or negative condition surface ordinates, shifted to the right or left according to Fig. 5, or, for a given point, by addition of the influence surface ordinates for the considered load position with $\pm P$. One proceeds similarily for groups of loads.

Because the condition and influence surfaces, respectively, are plotted only up to $x/a = \pm 3{,}5$, and for a somewhat accurate determination the values are required to $x/a = \pm 10$, and for the displacements even up to $x/a = \pm 20$, the graph has to be given for $x/a > 3{,}5$. However, it is evident that the oscillations which decrease with x/a rapidly and which are being governed by $\varkappa_2$ and μ_2 (see (11)), have already no effect. Furthermore, the values of the higher harmonics, $n \geqslant 4$ (besides $n = 0$), are negligible small—except for n_φ, $m_{x\varphi}$, $\bar{q}_x$ and $\bar{q}_\varphi$—. Finally, the coefficients in front of $\cos \varkappa_1 \xi$ and $\sin \varkappa_1 \xi$ become, except for the sign, almost equal or one of the two becomes very small—excepting m_x—. Because all of the section stress resultants of these exceptions have decreased very much at $x/a = 3{,}5$— in particular for small stiffness parameters k—, and, therefore, the error becomes small again, one can compute with the following approximations.

Für $\xi = x/a > 3{,}5$:

For $\xi = x/a > 3{,}5$:

$$u \approx C_u \sin \varkappa_1 \xi \; e^{-\mu_1 \xi}$$
$$v \approx C_v (\cos \varkappa_1 \xi + \sin \varkappa_1 \xi)\, e^{-\mu_1 \xi}$$
$$w \approx C_w (\cos \varkappa_1 \xi + \sin \varkappa_1 \xi)\, e^{-\mu_1 \xi}$$
$$\chi_\varphi \approx C_{\chi_\varphi} (\cos \varkappa_1 \xi + \sin \varkappa_1 \xi)\, e^{-\mu_1 \xi}$$
$$n_x \approx C_{n_x} (-\cos \varkappa_1 \xi + \sin \varkappa_1 \xi)\, e^{-\mu_1 \xi}$$
$$n_\varphi \approx C_{n_\varphi} (\cos \varkappa_1 \xi + \sin \varkappa_1 \xi)\, e^{-\mu_1 \xi}$$

($n = 4$ noch nicht ganz abgeklungen)

($n = 4$ not vanished completely) (31)

$$\bar{n}_{x\varphi} \approx C_{n_{x\varphi}} \cos \varkappa_1 \xi \; e^{-\mu_1 \xi}$$
$$m_x \approx C_{m_x} (\cos \varkappa_1 \xi + \sin \varkappa_1 \xi)\, e^{-\mu_1 \xi}$$

(ungenau, da $A_{m_x} \neq B_{m_x}$)

(inaccurate because $A_{mx} \neq B_{mx}$)

$$m_\varphi \approx C_{m_\varphi} (\cos \varkappa_1 \xi + \sin \varkappa_1 \xi)\, e^{-\mu_1 \xi}$$
$$m_{x\varphi} \approx C_{m_{x\varphi}} \sin \varkappa_1 \xi \; e^{-\mu_1 \xi}$$

($n = 4$ noch nicht abgeklungen)

($n = 4$ not vanished)

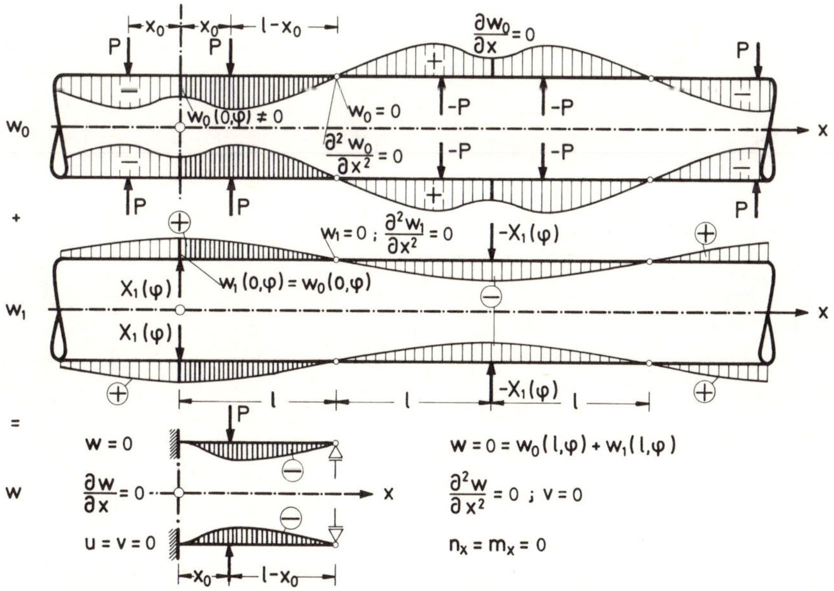

Bild 6. Lastumordnungsverfahren bei einer einseitig eingespannten Einfeldschale.

Figure 6. Load rearrangement method for a single-span shell, fixed at one side.

$$\bar{q}_x \approx C_{q_x} \sin \varkappa_1 \xi \;\; e^{-\mu_1 \xi}$$

($n = 4$ noch nicht abgeklungen) ($n = 4$ not vanished)

$$\bar{q}_\varphi \approx C_{q_\varphi} (\cos \varkappa_1 \xi + \sin \varkappa_1 \xi) \, e^{-\mu_1 \xi}$$

($n = 4$ noch nicht ganz abgeklungen) ($n = 4$ not vanished completely).

Die Konstanten C_s werden aus der Bedingung hergeleitet, daß bei $\xi = x/a = 3{,}5$ die aufgetragenen Größen (z. B. $w(3{,}5)$) und diejenigen aus der Näherung gleich sein müssen. So wird z. B.

The constants C_s are derived from the condition that, at $\xi = x/a = 3{,}5$, the plotted quantities (for instance w (3,5)) and those of the approximation have to be equal. Thus, for instance,

$$C_w = \frac{w\,(3{,}5)}{(\cos 3{,}5\,\varkappa_1 + \sin 3{,}5\,\varkappa_1)\, e^{-3{,}5\mu_1}} \qquad \text{oder/or} \qquad C_{n_{x\varphi}} = \frac{\bar{n}_{x\varphi}\,(3{,}5)}{\cos 3{,}5\,\varkappa_1 \;\; e^{-3{,}5\mu_1}} \tag{32}$$

Da die Ordinaten der aufgetragenen Größen (z. B. $w(3{,}5)$) für den gewünschten Längsschnitt $\varphi = \text{const}$ entnommen werden, erübrigt sich auch eine Abhängigkeit der Näherungslösung von φ. Die hier benötigten Real- bzw. Imaginärteile der Wurzeln $\varkappa_1$ und μ_1 sind für $n = 2$ in der Tafelübersicht zusammengestellt.

Es sollte erwähnt werden, daß diese Methode grundsätzlich auch bei einseitiger oder beidseitiger fester Einspannung der Rohrenden benutzt werden kann. Entsprechend Bild 6 muß jedoch bei einseitiger Einspannung eine zusätzliche Ringlast $X_1(\varphi)$ so bestimmt werden, daß für alle φ die Verschiebung $w_0(0, \varphi)$ bei $x = 0$, $2l$, $4l \dots$ infolge der Gruppenlasten P verschwindet.

A dependence of the approximate solution on φ is unnecessary, because the ordinates of the plotted quantity (for instance w (3,5)) are taken for the desired longitudinal section $\varphi = \text{const}$. The real and imaginary parts of the roots, $\varkappa_1$ and μ_1, required herein, are recorded in the Chart Classification.

It shall be noted that, in principle, this method can be used also when one or both pipe ends are fixed. However, for one-sided fixity, according to Fig. 6, an additional ring load $X_1(\varphi)$ has to be determined such that, for all φ, the displacements $w_0(0, \varphi)$ at $x = 0$, $2l$, $4l\dots$, due to the group loads P, vanish.

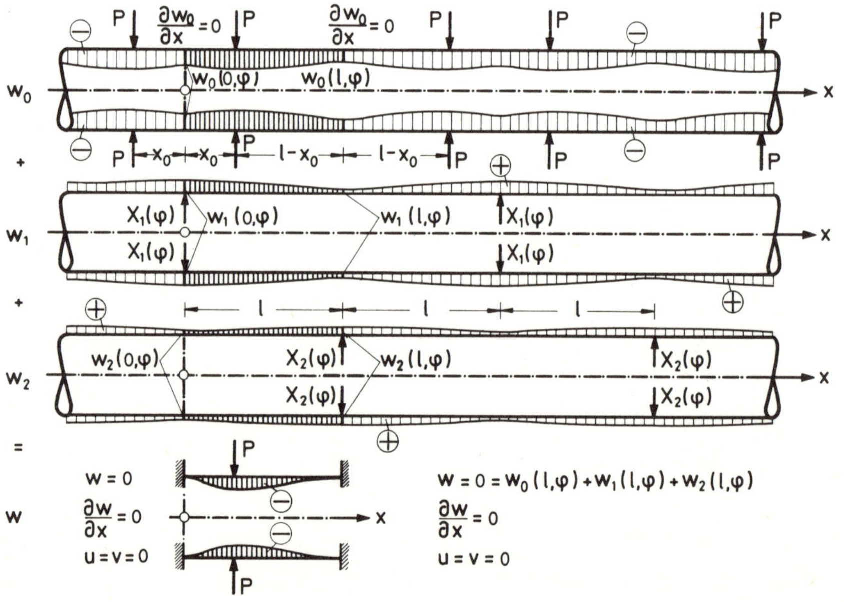

Bild 7. Lastumordnungsverfahren bei einer Einfeldschale mit zwei eingespannten Rändern.

Figure 7. Load rearrangement method for a single-span shell, fixed at two sides.

Sind beide Ränder $x = 0$ und $x = l$ unverschieblich und unverdrehbar gelagert, müssen sogar zwei Ringlasten $X_1(\varphi)$ und $X_2(\varphi)$ aus den Durchbiegungsbedingungen bei $x = 0, 2l, 4l, \ldots$ und $x = l, 3l, 5l, \ldots$ ermittelt werden (vgl. Bild 7). Dieser Rechengang ist wegen der Differenzen großer Zahlen sehr empfindlich und zudem mit erheblichem numerischen Aufwand verbunden, weil die Ermittlung von $X_1(\varphi)$ ratsamerweise wieder mit orthogonalen Funktionen erfolgen muß.

Selbstverständlich können auch alle anderen in der Schalenstatik üblichen Methoden benutzt werden, um die Randbedingungen an den Querrändern zu erfüllen. Es seien nur das Differenzenverfahren [32, 33] und die Finiten Elemente (z. B. [34]) erwähnt. Meist jedoch wird man mit diesen Hilfsmitteln gleichzeitig die Quer- und Längsränder befriedigen.

Eine detaillierte Beschreibung dieser Verfahren würde den Umfang der vorliegenden Arbeit überschreiten.

5.2 Der Längsrand

Zur Erfüllung der Randbedingungen an den Längsrändern gibt es mehrere ausgezeichnete Tabellenwerke [24, 26, 35–41]. Hier sollen jedoch nur die zwei beschrieben und bei den Beispielen benutzt werden, die in deutscher und englischer Sprache zur Verfügung stehen.

Das ASCE-Manual No. 31 [28] (deutsche Fassung in [35]) hat für die Randstörungsrechnung zwei Sätze von Tabellen und zwar einmal einen (Tabellen 2A und 2B) für lange Schalen mit $a/l \leqslant 0{,}6$ (vgl. Bild 3) und zum anderen einen für kurze Schalen (Tabellen 3A und 3B). Im ersten Fall sind alle erforderlichen Randwerte und die wesentlichen Schnittgrößen im Abstand von 10° Öffnungswinkel für vier ausgesuchte symmetrische Randlastfälle angegeben. Auch hier wird von der Fourierschen Analyse Gebrauch gemacht, jedoch nur das erste Reihenglied $n = 1$ (entsprechend $\sin \pi x/l$ bzw. $\cos \pi x/l$) berücksichtigt. Die Handhabung ist sehr übersichtlich, einfach und schnell, lediglich die meist erforderlich werdende zweifache Interpolation lästig und bei linearem Ansatz etwas ungenau.

Bei den Tabellen für die kurze Schale (Tabellen 3A und 3B) wird nur einfach interpoliert, jedoch muß bei einem Öffnungswinkel von etwa $2\varphi_E < 2 \cdot 30°$ der Einfluß des gegenüberliegenden Längsrandes berücksichtigt werden. Hierbei sind aber die Werte für zwei Reihenglieder $n = 1$ und $n = 3$ vertafelt.

Die Berechnung mit diesem Tabellenwerk ist in dem ersten Anwendungsbeispiel gezeigt, auch die unterschiedliche Bezeichnung und die positiven Schnitt- bzw. Randgrößen. Es soll lediglich noch auf die Ermittlung der Fourier-Koeffizienten aus dem Einzellastangriff eingegangen werden. Der Koordinatenursprung muß entgegen den Einfluß- bzw. Zustandsflächen an einen Schalenrand verschoben werden. Da nur $n = 1$ und evtl. $n = 3$ berücksichtigt werden kann, sind mit diesem Tabellenwerk auch nur zu $x = l/2$ symmetrische – oder nahezu symmetrische – Lastfälle berechenbar.

If both boundaries, $x = 0$ and $x = l$, are prevented from displacement and rotation, two ring loads, $X_1(\varphi)$ and $X_2(\varphi)$, have to be derived from the deflection conditions at $x = 0, 2l, 4l\ldots$ and $x = l, 3l, 4l\ldots$ (comp. Fig. 7). This method of analysis is very sensitive, because of differences of large numbers, and requires considerable computational effort, because, the determination of $X_1(\varphi)$ should preferably be done, again, with orthogonal functions.

Of course, all other usual methods of the analysis of shells can be used to satisfy the boundary conditions at the transverse boundaries. Only the difference method [32, 33], and the method of finite elements (for instance [34]) shall be named. In most cases, however, one will satisfy by these methods both, the transverse and the longitudinal boundaries.

A detailed description of these methods would go beyond the scope of this work.

5.2 The Longitudinal Boundary

For satisfying the boundary conditions at the longitudinal boundaries, several excellent table volumes exist [24, 26, 35–41]. However, only the two shall be described and used in the examples which are available in the German and the English language.

The ASCE-Manual No. 31 [28] (German version in [35]) has two sets of tables for the computation of the boundary disturbances, i.e. one set (Tables 2A and 2B) for long shells, with $a/l \leqslant 0{,}6$ (see Fig. 3), and another set (Tables 3A and 3B) for short shells. For the first case, all required boundary quantities and the more important stress resultants are presented, for four selected symmetrical boundary load cases, at 10° spacings of the central angle. Herein, the Fourier analysis was applied also, however, only the first term of the series, $n = 1$ (corresponding to $\sin \pi x/l$ or $\cos \pi x/l$, respectively), was considered. The application is very clear, simple and expedient, only the double interpolation, necessary in most cases, is inconvenient, and, with a linear assumption, somewhat inaccurate.

With the tables for short shells (Tables 3A and 3B), only one interpolation is required. However, the influence of the opposite longitudinal boundary has to be considered for a central angle of, approximately, $2\varphi_E < 2 \cdot 30°$. Hence, the values for two terms of the series, $n = 1$ and $n = 3$, were tabulated.

The analysis with this table volume is demonstrated in the first application example, and the different notations and positive stress resultants and boundary quantities are pointed out. Only the determination of the Fourier ocefficients, for single concentrated load, shall be explained further. Contrary to the influence and condition surfaces, the coordinate origin has to be moved to a shell boundary. Because only $n = 1$ and, perhaps, $n = 3$ can be considered, only load cases symmetric – or almost symmetric – to $l/2$ can be analyzed with this table volume.

Somit ergeben sich die Fourier-Koeffizienten z. B. für

Hence, the Fourier coefficients, for instance, for

$$\chi_\varphi(x) = \sum_{n=1,3\ldots} \Theta_n \sin n\pi x/l \qquad \text{und/and} \qquad \bar{n}_{x\varphi}(x) = \sum_{n=1,3\ldots} S_{L_n} \cos n\pi x/l$$

in ähnlicher Weise wie nach (21), wenn die Werte $\chi_\varphi(x)$ und $\bar{n}_{x\varphi}(x)$ wieder an einer ungeraden Anzahl von äquidistanten Punkten abgegriffen werden:

are obtained in a manner similar to (21), if the values $\chi_\varphi(x)$ and $\bar{n}_{x\varphi}(x)$ are scaled, again, at an odd number of equidistant points:

$$\begin{aligned}
\Theta_n &= \frac{2}{l}\int_0^l \chi_\varphi(x)\ \sin n\pi x/l\ \mathrm{d}x = \frac{2}{l}\frac{\Delta x}{3}\sum_k \varkappa_k\ \chi_\varphi(x)\ \sin n\pi x/l \\
&= \frac{2}{3}\frac{\Delta x}{l}(0 + 4\ \chi_\varphi(\Delta x)\ \sin n\pi\Delta x/l + 2\ \chi_\varphi(2\Delta x)\ \sin n\pi 2\Delta x/l + \ldots + 4\ \chi_\varphi(l-\Delta x)\ \sin n\pi(l-\Delta x)/l + 0) \\
S_{L_n} &= \frac{2}{l}\int_0^l \bar{n}_{x\varphi}(x)\cos n\pi x/l\ \mathrm{d}x = \frac{2}{l}\frac{\Delta x}{3}\sum_k \varkappa_k\ \bar{n}_{x\varphi}(x)\ \cos n\pi x/l \\
&= \frac{2}{3}\frac{\Delta x}{l}(\bar{n}_{x\varphi}(0) + 4\ \bar{n}_{x\varphi}(\Delta x)\ \cos n\pi\Delta x/l + 2\ \bar{n}_{x\varphi}(2\Delta x)\ \cos n\pi 2\Delta x/l + \ldots + 4\ \bar{n}_{x\varphi}(l-\Delta x)\ \cos n\pi(l-\Delta x)/l - \bar{n}_{x\varphi}(l))
\end{aligned} \tag{33}$$

Bei zu $x = l/2$ voll symmetrischer Fläche braucht auch nur bis zu $x = l/2$ summiert und der Wert verdoppelt zu werden. Die Faktoren S_c (vgl. (25)) entsprechend der Tafelübersicht werden ratsamerweise erst nachträglich berücksichtigt.

Das zweite hier benutzte Tabellenwerk von Rüdiger-Urban [22] arbeitet mit den gleichen vier Randlastfällen eines Längsrandes wie ASCE [28] in den Tabellen 3A und 3B. Wegen der Abhängigkeit des Schalenparameters ($\varepsilon = 2{,}4\, n \sqrt{a\, t}/l$) von n können die Harmonischen $n = 1$ und $n = 3$ mit dem gleichen Tabellensatz berechnet werden. Der Vorteil dieser Tabellen liegt in der größeren Flexibilität. So können hier beliebige Öffnungswinkel φ_E und praktisch alle Randbedingungen an beiden Längsrändern behandelt werden. Leider sind fast alle Bezeichnungen und Koordinaten gegenüber den bisherigen Ausführungen verändert.

Der Rechengang ist im dritten Beispiel vorgeführt und analog demjenigen nach dem ASCE-Manual. Vorerst müssen die zwei mal vier Randbedingungen an den beiden Längsrändern formuliert werden. Außer den Schnittgrößen bzw. Verformungen aus dem Einzellastangriff, die aus dem Tabellenanhang entnommen – mit dem Querrandeinfluß kombiniert – und in eine Fourier-Reihe entwickelt werden, sind noch die zwei mal acht Fourier-Koeffizienten aus den Randlastfällen X_1 bis X_8 zu berücksichtigen (entsprechend V_L, H_L, S_L und M_L/a bei [28]). Bei symmetrischen Systemen und Belastungen reichen vier Überzählige X_1 bis X_4 aus. Diese Fourier-Koeffizienten finden sich mit gewissen Konstanten behaftet in den beiden Tabellenwerken. Aus der achtgliedrigen bzw. bei Symmetrie viergliedrigen Matrix bestimmt man die Überzähligen und mit den in Abhängigkeit von φ angegebenen Koeffizienten schließlich die wichtigsten Schnittgrößen für den Mittelschnitt.

For a surface entirely symmetric to $x = l/2$, one has to summarize only to $x = l/2$, and to double the value. The factors S_c (see (25)), given in the Chart Classification, are preferably considered afterwards.

The second table volume by Rüdiger-Urban [22], applied herein, uses the same four boundary load cases for a longitudinal boundary as ASCE [28] in Tables 3A and 3B. Because of the dependence of the shell parameter ($\varepsilon = 2{,}4\, n \sqrt{a\, t}/l$) on n, the harmonics $n = 1$ and $n = 3$ can be computed with the same set of tables. The advantage of these tables lies in their better adaptability. With it, arbitrary central angles φ_E, and practically all boundary conditions at both longitudinal boundaries can be handled. Unfortunately, almost all notations and coordinates are different from the previous explanations.

The computation steps are demonstrated in the third example, and are analogous to those of the ASCE Manual. At first, the two times four boundary conditions at both longitudinal boundaries need to be formulated. Besides the section stress resultants or deformations, due to the single concentrated load, and taken from the Chart Appendix – combined with the traverse boundary effect – and developed into a Fourier series, the two times eight Fourier coefficients from the boundary load cases X_1 to X_8 have to be considered (corresponding to V_L, H_L, S_L and M_L/a in [28]). For symmetric systems and loadings four redundants, X_1 to X_4, are sufficient. These Fourier coefficients, with certain constants attached, are found in both table volumes. From the eight-term, or for symmetry, from the four-term matrix, respectively, one determines the redundant quantities and with given, and φ-dependent coefficients, the important section stress resultants of the center section.

Da bei dem Tabellenwerk von Rüdiger-Urban die Längsrichtung von der Schalenmitte aus zählt, muß zum Schluß noch die geringfügig geänderte Bestimmung der Fourier-Koeffizienten aus der Einzellastbeanspruchung angegeben werden. Wenn man von vornherein von der Symmetrie (z. B. bei n_φ) bzw. der Antimetrie (z. B. bei u) Gebrauch macht und sogar von einer geraden Anzahl äquidistanter Punkte ausgeht, werden z. B. für

Because the longitudinal direction counts from the shell center in Rüdiger-Urban's table volume, the slightly different determination of the Fourier coefficients for a single concentrated load application has to be noted, finally. When utilizing, from the start, the symmetry (for instance for n_φ) or the anti-symmetry (for instance for u), respectively, and even proceeds from an even number of equidistant points, one obtains, for instance, for

$$u = \sum_{n=1,3\ldots} U_n \sin n\pi x/l \qquad \text{oder/or} \qquad n_\varphi = \sum_{n=1,3\ldots} N_{\varphi n} \cos n\pi x/l$$

die Fourier-Koeffizienten

the Fourier coefficients

$$\begin{aligned}
U_n &= 2\frac{2}{l}\int_0^{l/2} u(x)\ \sin n\pi x/l\ \mathrm{d}x = \frac{4}{l}\frac{\Delta x}{3}\sum_k \varkappa_k\ u(x)\ \sin n\pi x/l \\
&= \frac{4}{3}\frac{\Delta x}{l}(0 + 4\,u(\Delta x)\ \sin n\pi\Delta x/l + 2\,u(2\Delta x)\ \sin n\pi 2\Delta x/l + \ldots + 2\,u(l/2-\Delta x)\ \sin n\pi\,(l/2-\Delta x)/l + 2\,u(l/2)\ \sin n\pi/2) \\
N_{\varphi n} &= 2\frac{2}{l}\int_0^{l/2} n_\varphi(x)\ \cos n\pi x/l\ \mathrm{d}x = \frac{4}{l}\frac{\Delta x}{3}\sum_k \varkappa_k\ n_\varphi(x)\ \cos n\pi x/l \\
&= \frac{4}{3}\frac{\Delta x}{l}(n_\varphi(0) + 4\,n_\varphi(\Delta x)\ \cos n\pi\Delta x/l + 2\,n_\varphi(2\Delta x)\ \cos n\pi 2\Delta x/l + \ldots + 2\,n_\varphi(l/2-\Delta x)\ \cos n\pi(l/2-\Delta x)/l + 0)
\end{aligned} \tag{34}$$

6. Konzentrierte Flächenlasten

Wie bereits in der Einleitung erwähnt, kann man bei den Kreiszylinderschalen nicht wie bei den Platten das Volumen des unendlich langen Schlauches im Aufpunkt der Momenten- und Querkrafteinflußflächen vernachlässigen. Das liegt einfach daran, daß wegen der Krümmung der Mittelfläche diese Schnittgrößen gegenüber denen der Platten erheblich reduziert werden, die Singularität des Aufpunktes jedoch erhalten bleibt. Es scheint ratsam, entgegen der bisherigen Berechnungsweise (vgl. [6]) hier von den letzten in den Einflußflächen gezeichneten Höhenschichtlinien auszugehen. Dabei kann entsprechend Bild 8 angenommen werden, daß das abgeschnittene Volumen ΔV aus der Schalensingularität gleich dem der Plattensingularität ist, wenn die betrachtete Höhenschichtlinie gleiche Form hat. Für $\nu = 1/6$ folgt nach [6] die in Bild 8 gezeichnete Kontur für m_x (bzw. m_φ bei Vertauschen von x und φ) sowie für $\bar{q}_\varphi$ (bzw. $\bar{q}_x$).

Hierzu wird vorerst die Ordinate (Höhenschichtlinie) gesucht, die für ein vorgegebenes R_0 (vgl. Bild 8) nach der klassischen Plattensingularität erhalten wird. Nach Pucher (vgl. z. B. [42]) wird diese entsprechend [6]* für das *Biegemoment*:

*) Die in [6] auf Seite 52 angegebene Gleichung für m_x gilt für $R(\eta)/2$, somit liefert dort auch die Gleichung (41) den Wert $4\,\Delta V$

6. Concentrated Surface Loads

As mentioned in the Introduction, the volume of the infinite trunk at the reference point of the moment and transverse force influence surfaces for circular cylindrical shells cannot be neglected as for plates. The reason is simply, that, because of the curvature of the middle surface, these stress resultants, as compared to those in plates, will be reduced considerably, the singularity at the reference, however, remains. It seems advisable, contrary to the usual method of computation (cf. [6]), to proceed, herein, from the last contour line, plotted in the influence surface. Corresponding to Fig. 8, it can be assumed, herein, that the cut-off volume ΔV, due to the shell singularity, is equal to the plate singularity if the considered contour line is of the same shape. For $\nu = 1/6$ follows from [6], as plotted in Fig. 8, the contour line for m_x (or m_φ, respectively, by exchanging x and φ), and for $\bar{q}_\varphi$ (or $\bar{q}_x$, respectively).

First, the ordinate (contour line) is determined, herein, which is obtained from the classic plate singularity (see Fig. 8) for a given R_0. By Pucher (see for instance [42]) this becomes, corresponding to [6]*, for the *bending moment*:

*) The equation for m_x, on page 52 of [6], is valid for $R(\eta)/2$, hence, the equation (41) produces there the value $4\,\Delta V$ also.

$$\pi\, m_x = -\frac{1}{8}\left\{2\,(1+\nu)\ln\frac{4\,R(\eta)}{\pi\, a} + (1+\nu) + 2\cos^2\eta + 2\nu\sin^2\eta\right\}$$

oder für $\eta = \pi/2$, d.h. $R(\eta) = R_0$, und mit $\nu = 1/6$:

or for $\eta = \pi/2$, i.e. $R(\eta) = R_0$, and with $\nu = 1/6$:

$$\pi\, m_x = -\frac{1}{8}\left\{\frac{7}{3}\ln\frac{4\,R_0}{\pi\, a} + \frac{3}{2}\right\}.$$

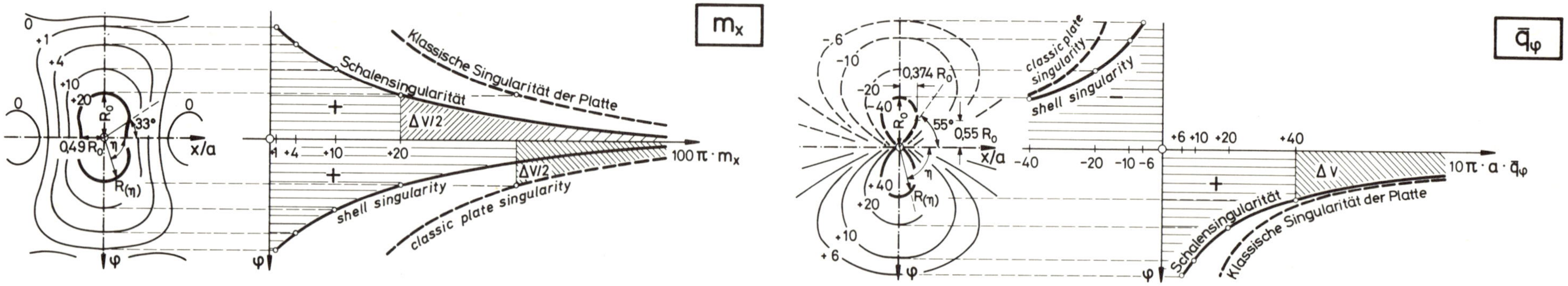

Bild 8. Genäherte Volumenbestimmung der abgeschnittenen Schläuche der Momenten- (m_x) bzw. Randquerkraft-Singularität ($\bar{q}_\varphi$).

Figure 8. Approximate volume determination of the cut-off trunks of the moment (m_x) and the transverse boundary force singularity ($\bar{q}_\varphi$), respectively.

Für die *Randquerkraft* folgt in gleicher Weise:

For the *transverse boundary force* is obtained, similarily:

$$\pi\,\bar{q}_\varphi = +\frac{1}{2}\frac{\cos\eta}{R(\eta)}\left(1 + \frac{1-\nu}{2}\cos 2\eta\right)$$

oder für $\eta = 0$, d.h. $R(\eta) = R_0$, und mit $\nu = 1/6$:

or for $\eta = 0$, i.e. $R(\eta) = R_0$, and with $\nu = 1/6$:

$$\pi\,\bar{q}_\varphi = \frac{1}{2\,R_0}\,\frac{17}{12} = \frac{0{,}7083}{R_0}.$$

Die Volumen der abgeschnittenen Teile über den letzten angegebenen Ordinaten – z.B. $\pi\, m_x = 20/100$ bzw. $\pi\,\bar{q}_\varphi = 40/10\, a$ – werden wieder nach Pucher (vgl. auch [6]):
für die *Biegemomente*:

The volumes of the cut-off parts over the last ordinates shown – for instance, $\pi\, m_x = 20/100$ or $\pi\,\bar{q}_\varphi = 40/10\, a$, respectively – again by Pucher (see also [6]):
for *the bending moments*:

$$\Delta V = 0{,}013778\, a^2 \exp\,(-\,6{,}857\,\pi\, m_x) = 0{,}0138\, a^2 \exp\left[+\,0{,}857\left(1{,}5 + 2{,}33\ln\frac{4\,R_0}{\pi\, a}\right)\right] \tag{35}$$

und für die *Randquerkräfte*:

and for the *transverse boundary forces*:

$$\Delta V = \frac{0{,}09397}{\pi\, q_\varphi} = 0{,}133\, R_0.$$

Bei $k = 10^{-4}$ greift man z.B. aus Tafel 104 für $100\,\pi\, m_x = 20$ einen Wert $R_0 = 0{,}12\, a$ ab, der nach Gleichung (35) ein vernachlässigtes Volumen von

One scales, for instance, from Chart 104 with $k = 10^{-4}$ for $100\,\pi\, m_x = 20$ a value $R_0 = 0{,}12\, a$ which, according to equation (35), would produce a neglected volume of

$$\Delta V = 0{,}0138\, a^2 \exp\left[+\,0{,}857\left(1{,}5 + 2{,}33\ln\frac{0{,}48}{\pi}\right)\right] = 0{,}0012\, a^2$$

liefern würde.

7. Anwendungsbeispiele

7.1 Tonnenreihe mit einer Einzellast

7.1.1 Aufgabenstellung

Es soll die Mittelschale einer Tonnenreihe entsprechend Bild 9 für eine Einzellast *P* in Feldmitte untersucht werden. Die Randbedingungen der Navierschen Lagerung an den Binderscheiben werden mit Hilfe des Lastumordnungsverfahrens (vgl. Abschnitt 5.1.2) erfüllt. Die Längsränder bei $\varphi_E = \pm 30°$ können näherungsweise nach dem ASCE-Manual [28] befriedigt werden, da hier nur folgende Randbedingungen zu erfüllen sind:

7. Application Examples

7.1 Shell Row under a Concentrated Load

7.1.1 Problem Definition

The interior shell under a concentrated load *P* at the span center, according to Fig. 9, shall be investigated. The boundary conditions for a "Navier support" at the transverse frames are being satisfied by the load rearrangement method (see Section 5.1.2). The longitudinal boundaries, at $\varphi_E = \pm 30°$, can be satisfied approximately by the method of the ASCE-Manual [28], because only the following boundary conditions have to be satisfied here:

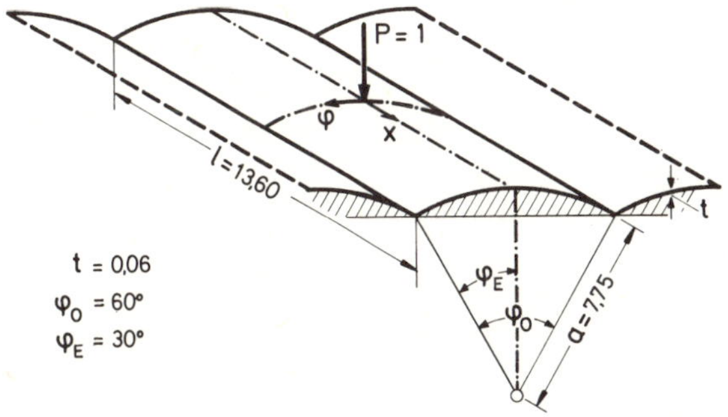

Bild 9. Mittelschale einer Tonnenreihe.

Figure 9. Interior shell of a barrel row.

Bei der Mitteltonne müssen die horizontale Verschiebung und die Verdrehung sowie die senkrechte Auflagerkomponente und die Schubkraft verschwinden (vgl. Bild 10):

For the interior shell, the horizontal displacement and the rotation as well as the vertical reaction component and the shearing force have to vanish (see Fig. 10):

$$\begin{aligned} \Delta H - w \sin \varphi_E - v \cos \varphi_E &= 0 \\ \Theta + \chi_\varphi &= 0 \\ V_L + n_\varphi \sin \varphi_E + \bar{q}_\varphi \cos \varphi_E &= 0 \\ - S_L + \bar{n}_{x\varphi} &= 0 \end{aligned} \qquad (36)$$

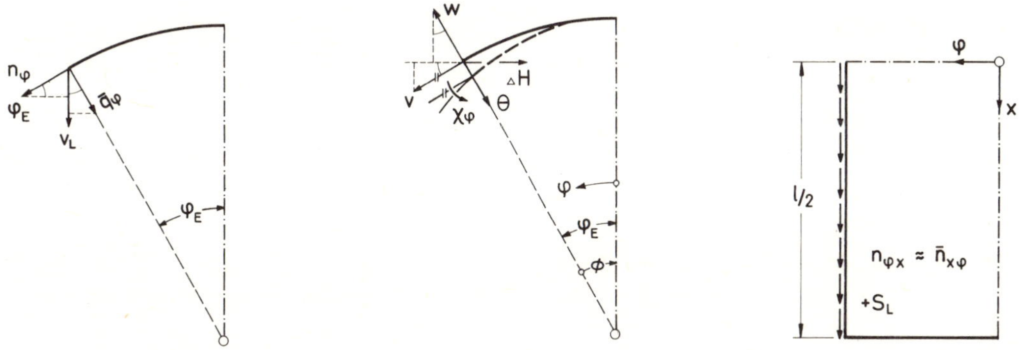

Bild 10. Randbedingungen an den Längsrändern.

Figure 10. Boundary conditions at the longitudinal boundaries.

Die ersten Terme (einschließlich der Bezeichnungen) jeder Gleichung folgen aus der Randstörungsrechnung nach [28] und die weiteren aus der Lastumordnung entsprechend Abschnitt 5.1.2.

7.1.2 Erfüllung der Randbedingungen an den Querrändern

Für die Einzellast findet man in den anliegenden Zustandsflächen (Tafeln 30, 26, 27, 28, 36 und 31) bei $\varphi_E = +30°$ für den Schalenparameter:

$$k = \frac{t^2}{12\,a^2} = \frac{0{,}06^2}{12 \cdot 7{,}75^2} = 5 \cdot 10^{-6}$$

die in Bild 11 dargestellten Schnittgrößen n_φ, v, w, χ_φ, $\bar{q}_\varphi$ und $\bar{n}_{x\varphi}$. Der Fehler bei der Schubkraft $n_{\varphi x} \approx \bar{n}_{x\varphi}$ ist kleiner ($< 5\%$) als die Ungenauigkeiten durch das Lastumordnungsverfahren (vgl. auch Abschnitt 4.4).

Da die Einzellast in Feldmitte steht, also $x_0/a = l/2a = 0{,}877$ (vgl. Bild 5) wird, braucht nur die Summation für die $\pm P$-Lasten rechts des betrachteten Feldes vorgenommen zu werden. Der Anteil der übrigen Lasten folgt durch verschwenken – und bei antimetrischen Flächen zusätzlichem Vorzeichenwechsel – der so gewonnenen Summenlinien.

Die einfachen aber umfangreichen Summationen sollen nur an Hand der Mittelordinate der n_φ-Linie gezeigt werden. Bis zu $x/a = 3{,}5$ – d.h. für die drei Einzelanteile $+P(x/a = 0)$, $-P(x/a = 1{,}75)$ und $+P(x/a = 3{,}5)$ – werden die Ordinaten direkt aus Tafel 30 entnommen. Für $x/a > 3{,}5$ wird von der Näherung nach (31) Gebrauch gemacht.

Mit $n_\varphi(3{,}5) = +0{,}06/\pi a$ und $\varkappa_1 = 0{,}1158$, $\mu_1 = 0{,}1174$ (vgl. Tafelübersicht) liefert die angepaßte Gleichung (32):

The first terms (including the notations) of each equation follow from the boundary disturbance computation, according to [28], and the additional ones from the load rearrangement, according to Section 5.1.2.

7.1.2 Satisfying the Boundary Conditions at the Transverse Boundaries

For the concentrated load, one obtains from the condition surfaces in the Appendix (Charts 30, 26, 27, 28, 36 and 31) at $\varphi_E = +30°$, and the shell parameter

the sections values n_φ, v, w, χ_φ, $\bar{q}_\varphi$ and $\bar{n}_{x\varphi}$, shown in Fig. 11. The error for the shearing force $n_{\varphi x} \approx \bar{n}_{x\varphi}$ is smaller ($< 5\%$) than the inaccuracies from the load rearrangement method (see also Section 4.4).

Because the concentrated load is located at the span center, i.e., $x_0/a = l/2a = 0{,}877$ (see Fig. 5), the summation has to be performed only for the $\pm P$-loads at the right of the considered span. The contribution of the remaining loads follows from mirroring—and additional sign change for anti-symmetric areas—of the summation lines, thus obtained.

The simple but voluminous summations shall be demonstrated for the middle ordinate of the n_φ-diagram only. Up to $x/a = 3{,}5$—i.e., for the three individual contributions $+P(x/a = 0)$, $-P(x/a = 1{,}75)$ and $+P(x/a = 3{,}5)$ — the ordinates are taken directly from Chart 30. For $x/a > 3{,}5$, the approximation according to (31) is applied.

With $n_\varphi(3{,}5) = +0{,}06/\pi a$ and $\varkappa_1 = 0{,}1158$, $\mu_1 = 0{,}1174$ (see Chart Classification), the adapted equation (32) produces:

$$\pi\,a\,C_{n\varphi} = \frac{0{,}06}{(\cos 3{,}5 \cdot 0{,}1158 + \sin 3{,}5 \cdot 0{,}1158)\,e^{-3{,}5 \cdot 0{,}1174}} = 0{,}069$$

und (31) z.B. die Mittelordinate für $-P(x/a = 5{,}26)$:

and (31), for instance, the middle ordinate for $-P(x/a = 5{,}26)$:

$$\pi\,a\,n_\varphi = -0{,}069\,(\cos 0{,}1158 \cdot 5{,}26 + \sin 0{,}1158 \cdot 5{,}26)\,e^{-0{,}1174 \cdot 5{,}26} = -0{,}052$$

Die Restanteile können leicht abgeschätzt werden. Da jeweils positive und negative Ordinaten abwechseln, ergibt sich in Feldmitte nur etwa ein Wert von $\pi a\,n_\varphi = +0{,}02$.

Für alle rechts von dem betrachteten Feld stehenden Lasten folgt die in Bild 11 bei n_φ ausgezogene Kurve mit der Mittelordinate:

The remaining contributions can be estimated easily. Because positive and negative ordinates alternate, at the span center is obtained only a value of circa $\pi a n_\varphi = +0{,}02$.

For all loads, located to the right of the considered span, the curve shown solidly in Fig. 11, is obtained for n_φ, with the middle ordinate:

$$\pi\,a\,n_\varphi = +0{,}10 + 0{,}06 - 0{,}05 + 0{,}02 = +0{,}13$$

Die Addition mit der verschwenkten Kurve liefert den Randeinfluß der Querränder und ist hier negativ eingetragen, um die endgültigen Ordinaten direkt abgreifen zu können.

The summation with the mirrored curve produces the boundary effect of the transverse boundaries, and is shown negative, herein, to be able to scale the final ordinates directly.

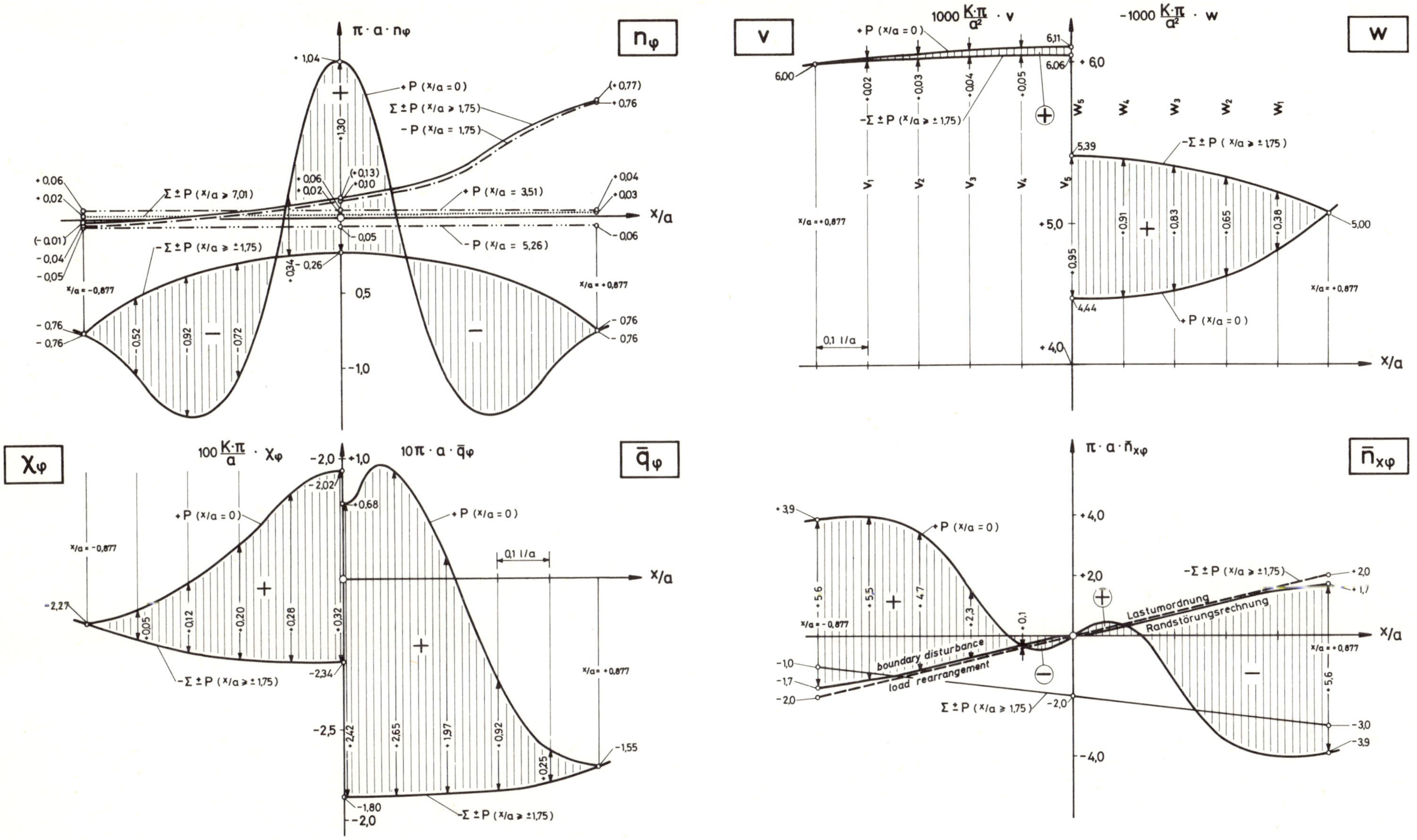

Bild 11. Lastumordnung für $\varphi_E = 30°$.

Figure 11. Load rearrangement for $\varphi_E = 30°$.

In genau gleicher Weise werden die Ordinaten für die übrigen Schnittgrößen ermittelt, wobei darauf zu achten ist, daß bei den Verschiebungen wegen der langsamen Abnahme der Ordinaten mit x/a mehrere Lastanteile mitzunehmen sind. Bei $\bar{n}_{x\varphi}$ in Bild 11 wird schließlich noch der Vergleich mit einer Randstörungs-

The ordinates for the remaining section values are obtained in exactly the same way; however, one has to take note, that for the displacements several load terms have to be considered because of the slow decrease of the ordinates with x/a. Finally, the comparison with a boundary disturbance calculation (according to

rechnung (entsprechend Abschnitt 5.1.1) gezeigt, bei der insgesamt fünf Reihenglieder mitgenommen wurden. Diese genaueren Ergebnisse sind mit ähnlich kleinen Abweichungen auch bei den anderen Schnittgrößen der weiteren Rechnung zugrunde gelegt.

Section 5.1.1) for $n_{x\varphi}$ is shown in Fig. 11, where a total of five terms of the series were considered. These more accurate results, with similarily small deviations, were also the basis of the subsequent calculation for the other section quantities.

7.1.3 Erfüllung der Randbedingungen an den Längsrändern

Da im ASCE-Manual [28] nur der Einfluß des ersten Gliedes einer Fourier-Reihe bei einem symmetrischen System berücksichtigt ist, brauchen auch nur diese vier Fourier-Koeffizienten aus dem Einzellastangriff nach Bild 11 bestimmt zu werden. Dabei werden entsprechend Gleichung (36) sogleich die beiden Verschiebungen (v und w) und die Schnittkräfte n_φ und q_φ zusammengefaßt und $x = 0$ an den linken Rand gelegt. Bei Benutzung der Simpsonschen Regel ergibt sich nach (33) unter Berücksichtigung der Symmetrie zu $x = l/2$ und mit $\Delta x = l/10$:

7.1.3 Satisfying the Boundary Conditions at the Longitudinal Boundaries

Because the influence of only the first term of the Fourier series is considered in the ASCE Manual [28] for a symmetrical system, only these four Fourier coefficients for the single concentrated load, according to Fig. 11, need to be determined. Herein, the two displacements (v and w), and the stress resultants n_φ and q_φ are immediately combined, according to equation (36), and $x = 0$ is shifted to the left boundary. By use of Simpson's rule, considering the symmetry to $x = l/2$ and with $\Delta x = l/10$ is obtained, in accordance with (33):

$$- w \sin \varphi_E - v \cos \varphi_E = \Delta H_{\mathrm{Rand}} \approx \Delta H_1 \sin \frac{\pi x}{l}$$

$$\frac{1000\,K\pi}{a^2} \Delta H_1 = \frac{2}{15} \Big\{ 0 - 4 (w_1 \sin \varphi_E + v_1 \cos \varphi_E) \sin 0{,}1\pi - 2 (w_2 \sin \varphi_E + v_2 \cos \varphi_E) \sin 0{,}2\pi$$
$$- 4 (w_3 \sin \varphi_E + v_3 \cos \varphi_E) \sin 0{,}3\pi - 2 (w_4 \sin \varphi_E + v_4 \cos \varphi_E) \sin 0{,}4\pi - 2 (w_5 \sin \varphi_E + v_5 \cos \varphi_E) \sin 0{,}5\pi \Big\}$$

Mit $\sin \varphi_E = 0{,}50$ und $\cos \varphi_E = 0{,}866$ und den Ordinaten von Bild 11 folgen die ersten Fourier-Koeffizienten nach Tabelle 1.

With $\sin \varphi_E = 0{,}50$ and $\cos \varphi_E = 0{,}866$, and the ordinates of Fig. 11, the first Fourier coefficients become as in Table 1.

Tabelle 1. Ermittlung der ersten Fourier-Koeffizienten

Table 1. Determination of the first Fourier coefficients

x/l	$\varkappa_k$	$\Delta H_1 = - w \sin\varphi_E - v \cos\varphi_E$	$\Delta H_1 \varkappa_k \times \sin \pi x/l$	$\Theta_1 = \chi_\varphi$	$\Theta_1 \varkappa_k \times \sin \pi x/l$	$V_{L1} = + n_\varphi \sin\varphi_E + \bar{q}_\varphi \cos\varphi_E$	$V_{L1} \varkappa_k \times \sin \pi x/l$	$S_{L1} \approx - \bar{n}_{x\varphi}$	$S_{L1} \varkappa_k \times \cos \pi x/l$
0	1	0	0	0	0	0	0	− 5,6	− 5,6
0,1	4	− 0,21	− 0,26	+ 0,05	+ 0,06	− 0,24	− 0,30	− 5,5	− 20,9
0,2	2	− 0,35	− 0,41	+ 0,12	+ 0,14	− 0,38	− 0,45	− 4,7	− 7,6
0,3	4	− 0,45	− 1,46	+ 0,20	+ 0,65	− 0,19	− 0,61	− 2,3	− 5,4
0,4	2	− 0,50	− 0,95	+ 0,28	+ 0,53	+ 0,40	+ 0,76	− 0,1	− 0,1
0,5	2	− 0,52	− 1,04	+ 0,32	+ 0,64	+ 0,86	+ 1,72	0	0
Σ	=		− 4,12		+ 2,02		+ 1,12		− 39,6
$\left(\frac{2}{15}\Sigma\right)$	=	$\frac{a^2}{1000\,K\pi}$	(− 0,549)	$\frac{a}{100\,K\pi}$	(+ 0,269)	$\frac{1}{\pi a}$	(+ 0,149)	$\frac{1}{\pi a}$	(− 5.28)

Interessant sind noch die Fourier-Koeffizienten für $n = 3$:

Also of interest are the Fourier coefficients for $n = 3$:

$$\Delta H_3 = \frac{a^2}{1000\,K\pi}(-0{,}037), \qquad \Theta_3 = \frac{2}{100\,K\pi}(-0{,}044), \qquad V_{L_3} = \frac{1}{\pi a}(-0{,}523), \qquad S_{L_3} = \frac{1}{\pi a}(-0{,}90).$$

Sie sind bis auf V_L kleiner als 20% der Werte für $n = 1$. Trotz des überwiegenden Anteils der dritten Harmonischen bei V_L sind die Ergebnisse bereits für $n = 1$ recht gut (vgl. Bild 12).

Die Ermittlung der Fourier-Koeffizienten aus den Randlastfällen erfolgt nach [28], Tabelle 2A und 2B, für die Parameter

Except for V_L, they are less than 20% of the values for $n = 1$. In spite of the governing contribution of the third harmonic to V_L, the results are already quite good for $n = 1$ (see Fig. 12).

The determination of the Fourier coefficients for the boundary load cases is done according to [28], Tables 2A and 2B, with the parameters

$$r/t \triangleq a/t = 7{,}75/0{,}06 = 129 > 100; \qquad r/l \triangleq a/l = 7{,}75/13{,}60 = 0{,}57 < 0{,}60; \qquad \phi_k \triangleq \varphi_E = 30°.$$

Die doppelte geradlinige Interpolation liefert mit den jeweils vier Randgrößen infolge der symmetrischen Randlastfälle V_L, H_L, S_L und M_L/a die folgenden Verformungen bzw. Randlasten für $x = 0$ bzw. $x = l/2$ bei S_L:

The double, linear interpolation produces, with the four boundary quantities each due to the symmetrical boundary load cases V_L, H_L, S_L and M_L/a, the following deformations and boundary loads, respectively, at $x = 0$ or $x = l/2$ for S_L, respectively:

$$\Delta H = \frac{l^4}{a^3 t E}\Big(\underset{\mathit{129}}{136}\, V_L - \underset{\mathit{54,9}}{59{,}3}\, H_L + \underset{\mathit{1,59}}{1{,}73}\, S_L - \underset{\mathit{528}}{576}\,\frac{M_L}{a}\Big)$$

$$\Theta = \frac{a^2}{EI}\Big(\underset{\mathit{0,0569}}{0{,}0568}\, V_L - \underset{\mathit{0,0250}}{0{,}0250}\, H_L + \underset{\mathit{0,000599}}{0{,}000598}\, S_L - \underset{\mathit{0,328}}{0{,}328}\,\frac{M_L}{a}\Big)$$

$$V_L = (1{,}0\, V_L + 0 + 0 + 0)$$

$$S_L = -\,(0 + 0 + 1{,}0\, S_L + 0)$$

Die kursiv geschriebenen Zahlen ergeben sich bei Benutzung der Flüggeschen Differentialgleichung und $\nu = 0$.

Unter Berücksichtigung von $EI = E t^3/12 \approx K$ nehmen nach Multiplikation der Fourier-Koeffizienten aus Tabelle 1 mit

The values printed in italics are obtained by application of Flügge's differential equation, and $\nu = 0$.

After multiplication of the Fourier coefficients in Table 1 by

$$\frac{12\,a^5}{1000\,\pi\, l^4\, t^2} = 0{,}867 \quad \text{für/for } \Delta H, \qquad \frac{1}{100\,\pi\, a} = 0{,}000411 \quad \text{für/for } \Theta \quad \text{bzw./resp.} \quad \frac{1}{\pi a} = 0{,}0411$$

für V_L und S_L die Gleichungen (36) die folgende Form an:

for V_L and S_L, and considering that $EI = E t^3/12 \approx K$, the equations (36) assume the following form:

$$\begin{aligned}
136\, V_L - 59{,}3\, H_L + 1{,}73\, S_L - 576\, M_L/a - 0{,}476 &= 0\\
56{,}8\, V_L - 25{,}0\, H_L + 0{,}598\, S_L - 328\, M_L/a + 0{,}111 &= 0\\
1{,}0\, V_L \qquad\qquad\qquad\qquad + 0{,}00612 &= 0\\
1{,}0\, S_L \qquad\qquad - 0{,}217 &= 0
\end{aligned}$$

Hieraus ergeben sich die Überzähligen zu

Herefrom, the redundant quantities

$$V_L = -0{,}00612 \qquad H_L = -0{,}0484 \qquad S_L = +0{,}217 \qquad M_L/a = +0{,}00336$$

und mit den Einheitswerten nach [28], Tabelle 2A, schließlich die Schnittgrößen aus der Störung am Längsrand nach Tabelle 2.

are obtained, and finally, with the unit values according to [28], Table 2A, the stress resultants, due to longitudinal boundary disturbance, in Table 2.

Tabelle 2. Schnittgrößen für Längsrandstörung bei $x/a = 0$ bzw. $x/a = 0{,}877$ für $\bar{n}_{x\varphi}$

Table 2. Section stress resultants, due to the longitudinal boundary disturbance, at $x/a = 0$ and $x/a = 0{,}877$, respectively, for $\bar{n}_{x\varphi}$ ($\varphi = \varphi_E - \phi$).

$$\pi\, a\, n_x \mathrel{\hat{=}} \pi\, a\, T_x = \frac{\pi\, l^2}{a} \sum_{i=1}^{4} X_i\, T_x^i \qquad (T_x^i = \text{Tafelwert / Table value [28]})$$

	$\varphi = 30°$		$\varphi = 20°$		$\varphi = 10°$		$\varphi = 0°$	
X_i	T_x^i	$X_i\, T_x^i$	T_x^i	$X_i\, T_x^i$	T_x^i	$X_i\, T_x^i$	T_x^i	$X_i\, T_x^i$
$V_L = -0{,}00612$	+ 33,9	− 0,208	− 8,1	+ 0,050	− 3,2	+ 0,020	+ 3,3	− 0,020
$H_L = -0{,}0484$	− 9,36	+ 0,453	+ 3.97	− 0,192	− 0,31	+ 0,015	− 3,43	+ 0,166
$S_L = +0{,}217$	+ 2,46	+ 0,534	+ 0,46	+ 0,100	− 0,22	− 0,048	− 0,35	− 0,076
$M_L/a = +0{,}00336$	− 74,7	− 0,251	+ 28,3	+ 0,095	− 1,6	− 0,005	− 25,0	− 0,084
$\sum$ =		+ 0,528		+ 0,053		− 0,018		− 0,014
$\pi\, a\, n_x$	+ 39,588		+ 3,974		− 1,350		− 1,050	

$$\pi\, a\, \bar{n}_{x\varphi} \mathrel{\hat{=}} \pi\, a\, S = \pi\, l \sum_{i=1}^{4} X_i\, S^i$$

	$\varphi = 30°$		$\varphi = 20°$		$\varphi = 10°$		
X_i	S^i	$X_i\, S^i$	S^i	$X^i\, S^i$	S^i	$X_i\, S^i$	
$V_L = -0{,}00612$	0	0	− 3,50	+ 0,021	+ 0,50	− 0,003	
$H_L = -0{,}0484$	0	0	+ 0,08	− 0,004	− 1,24	+ 0,060	
$S_L = +0{,}217$	+ 0,570	+ 0,124	− 0,147	− 0,032	− 0,146	− 0,032	
$M_L/a = +0{,}00336$	0	0	+ 0,40	+ 0,001	− 9,21	− 0,031	
$\sum$ =		+ 0,124		− 0,014		− 0,006	
$\pi\, a\, \bar{n}_{x\varphi}$	+ 5,298		− 0,598		− 0,256		

7.1.4 Überlagerung der Einzelanteile

Die Überlagerung der drei Anteile aus der jeweiligen Zustandsfläche (Tafeln 29 bis 33), der Lastumordnung (vgl. Bild 11 für $\varphi_E = 30°$) und aus der ersten Harmonischen für die Störung am Längsrand (vgl. Tabelle 2) liefert schließlich die im Bild 12 jeweils auf der rechten Seite eingetragenen Schnittgrößen. Zum Vergleich sind noch die genaueren Werte bei Berechnung der Längsrandstörung mittels der Flüggeschen Differentialgleichung, einmal mit einem Reihenglied und $\nu = 0$ sowie mit vier Reihengliedern und $\nu = 1/6$ eingetragen.

Die Vergleiche zeigen, daß die Ergebnisse in den Schnittgrößen bei Berechnung nach dem ASCE-Manual von denen der genaueren Theorie kaum abweichen, obwohl die Verschiebungen bis zu 10% (ΔH) verschieden sind. Der Einfluß von ν ist sehr gering. Wird eine größere Genauigkeit gefordert, sollten jedoch mindestens

7.1.4 Superposition of the Individual Contributions

The superposition of the three contributions, from the corresponding condition surface (Charts 29 to 33), the load rearrangement (see Fig. 11 for $\varphi_E = 30°$) and from the first harmonic due to the disturbance at the longitudinal boundary (see Table 2), produces finally the stress resultants, shown on the right sides in Fig. 12. For comparison, the more exact values are plotted also, obtained from a computation of the boundary disturbance with Flügge's differential equation, once with one term of the series and $\nu = 0$, and furthermore, with four terms of the series and $\nu = 1/6$.

The comparison indicates that the results for the section stress resultants from computation by the ASCE Manual differ very little from those of the more accurate theory, although the displacements differ up to 10% (ΔH). The influence of ν is very small. If more accuracy is required, at least two terms of the series

Tabelle 2 (Fortsetzung) Table 2 (continued)

	$\pi a n_\varphi \triangleq \pi a T_\phi = \pi a \sum_{i=1}^{4} X_i T_\phi^i$							
	$\varphi = 30°$		$\varphi = 20°$		$\varphi = 10°$		$\varphi = 0°$	
X_i	T_ϕ^i	$X_i T_\phi^i$	T_ϕ^i	$X_i T_\phi^i$	T_ϕ^i	$X_i T_\phi^i$	T_ϕ^i	$X_i T_\phi^i$
$V_L = -0{,}00612$	+ 0,50	− 0,0031	− 1,60	+ 0,0098	− 2,38	+ 0,0146	− 2,09	+ 0,0128
$H_L = -0{,}0484$	+ 0,866	− 0,0419	+ 1,268	− 0,0614	+ 0,869	− 0,0421	+ 0,451	− 0,0218
$S_L = +0{,}217$	0	0	+ 0,0662	+ 0,0144	− 0,0400	− 0,0087	− 0,0913	− 0,0198
$M_L/a - +0{,}00336$	0	0	+ 2,47	+ 0,0083	− 0,91	− 0,0031	− 4,10	− 0,0138
$\sum =$		− 0,045		− 0,0289		− 0,0393		− 0,0426
$\pi a n_\varphi$	− 1,096		− 0,704		− 0,957		− 1,037	
	$100 \pi m_\varphi \triangleq 100 \pi M_\phi = 100 \pi a \sum_{i=1}^{4} X_i M_\phi^i$							
	$\varphi = 30°$		$\varphi = 20°$		$\varphi = 10°$		$\varphi = 0°$	
X_i	M_ϕ^i	$X_i M_\phi^i$	M_ϕ^i	$X_i M_\phi^i$	M_ϕ^i	$X_i M_\phi^i$	M_ϕ^i	$X_i M_\phi^i$
$V_L = -0{,}00612$	0	0	− 0,088	+ 0,00054	− 0,147	+ 0,00090	− 0,164	+ 0,00100
$H_L = -0{,}0484$	0	0	+ 0,0451	− 0,00218	+ 0,0617	− 0,00299	+ 0,0634	− 0,00307
$S_L = +0{,}217$	0	0	− 0,0001	− 0,00002	− 0,0014	− 0,00030	− 0,0029	− 0,00063
$M_L/a = +0{,}0036$	+ 1,000	+ 0,00336	+ 0,706	+ 0,00237	+ 0,488	+ 0,00164	+ 0,394	+ 0,00132
$100 \sum =$		+ 0,336		+ 0,071		− 0,075		− 0,138
$100 \pi m_\varphi$	+ 8,181		+ 1,729		− 1,826		− 3,360	

zwei Reihenglieder mitgenommen werden, zumal wenn – wie hier bei V_L – der zweite Fourier-Koeffizient größer ist als der erste. Die Gleichgewichtskontrollen $\int n_x \, ds = 0$ und $\int n_x \bar{z} \, ds = 1 \cdot l/4$ ergaben maximal 4% Fehler.

should be considered this, in particular, when the second Fourier coefficient is larger than the first—as herein for V_L—. The equilibrium checks, $\int n_x \, ds = 0$ and $\int n_x \bar{z} \, ds = 1 \cdot l/4$, produced a maximum error of 4%.

7.2 Kühlturm mit Seilabspannung

7.2.1 Aufgabenstellung

Um die Berechnung beim Angriff mehrerer Einzellasten und gleichzeitig die Erfüllung beliebiger Randbedingungen an den Querrändern zu zeigen, soll die Beanspruchung eines Naturzugkühlers infolge der Radialkomponenten der Seilvorspannung aus dem zentralen Turmdrehkran entsprechend [4] näherungsweise bestimmt werden. Als grobe Vereinfachung wird statt des Rotationshyperboloides mit variabler Wandstärke eine zylindrische Schalenfläche mit konstanter Dicke

7.2 Cooling Tower with Cable Stays

7.2.1 Problem Definition

To show the computation for a loading by several concentrated loads, and at the same time the satisfying of arbitrary boundary conditions at the transverse boundaries, the stresses in a natural-draft cooling tower, due to the radial components from rope stays of a center tower crane according to [4], shall be determined approximately. Instead of a hyperboloid of revolution with variable wall thickness, a cylindrical shell surface with constant thickness shall be investigated as a coarse

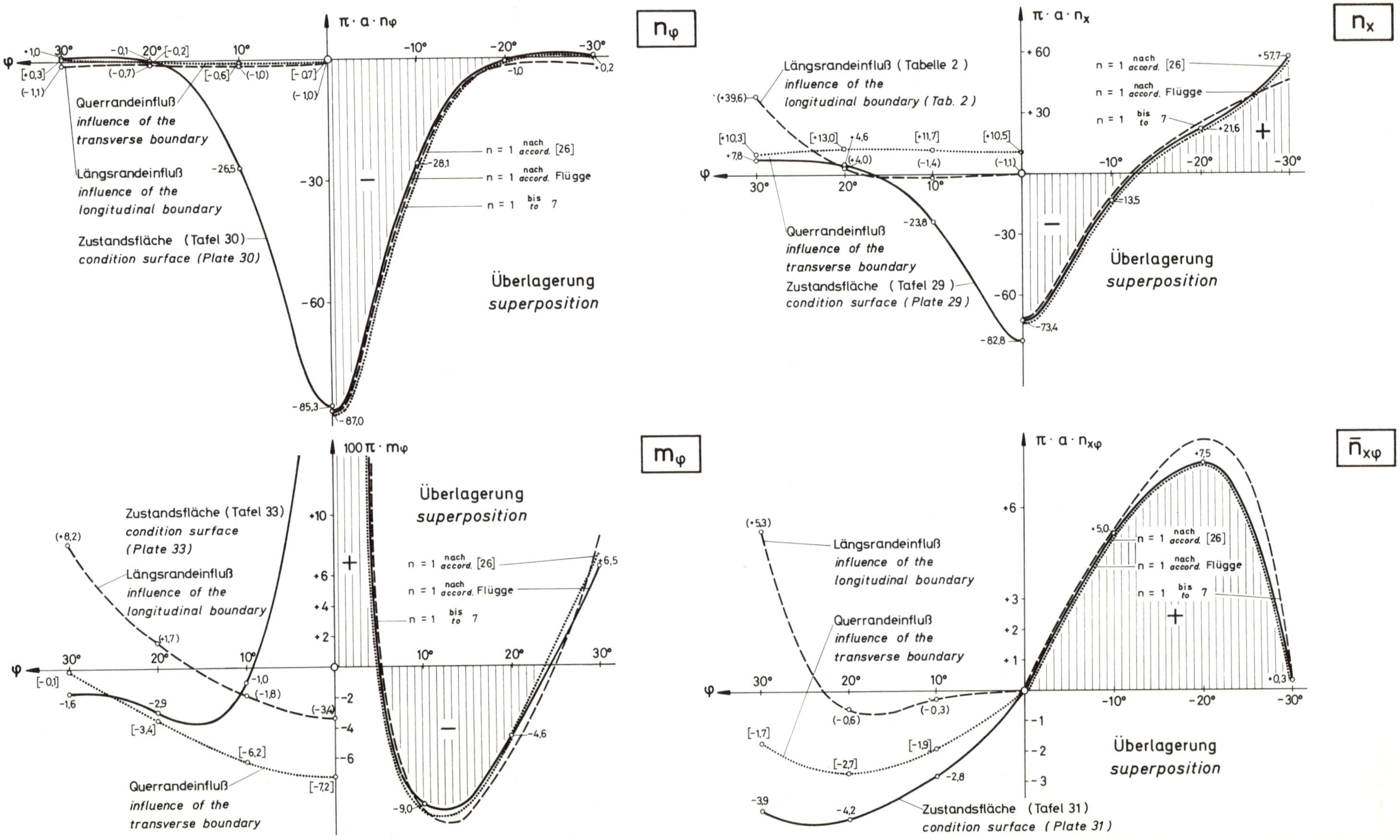

Bild 12. Überlagerung der Einzelanteile für Schalenmitte (n_x, n_φ, m_φ) bzw. Schalenrand ($\bar{n}_{x\varphi}$).

Figure 12. Superposition of individual contributions at shell center (n_x, n_φ, m_φ), and shell boundary ($\bar{n}_{x\varphi}$), respectively.

untersucht. Da auch am oberen Rand keine gelenkige Lagerung entstehen soll, wird die konstruktiv falsche Lösung ohne obere Randverstärkung gewählt. Es ergeben sich dann die Belastungen, Randbedingungen und Abmessungen entsprechend Bild 13, die zu einem Schalenparameter

$$k = t^2/12\,a^2 = 0{,}14^2/12 \cdot 25^2 = 2{,}6 \cdot 10^{-6} \approx 2{,}5 \cdot 10^{-6} \text{ führen.}$$

simplification. Because no hinge support shall be created at the upper edge, the structurally wrong assumption of no upper-edge reinforcement was chosen. Hence, the loadings, boundary conditions and dimensions were those according to Fig. 13, producing a shell parameter

$$k = t^2/12a^2 = 0{,}14^2/12 \cdot 25^2 = 2{,}6 \cdot 10^{-6} \approx 2{,}5 \cdot 10^{-6}$$

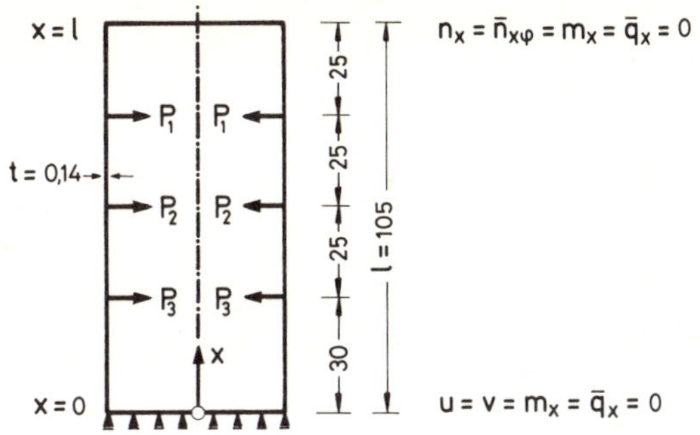

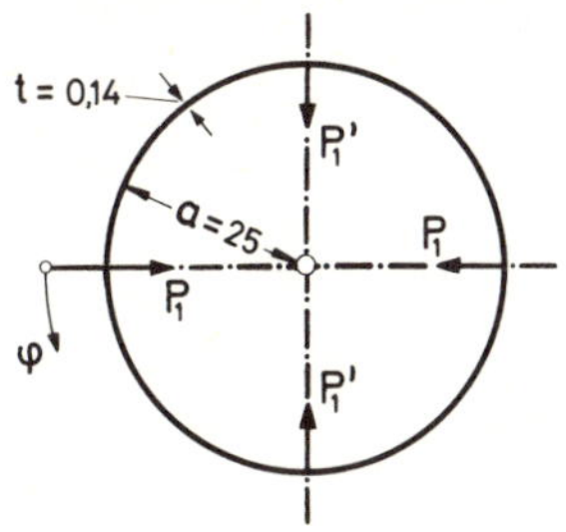

Bild 13. Belastungen, Randbedingungen und Abmessungen für die Ersatzschale.

Figure 13. Loads, boundary conditions and dimensions of the ersatz shell.

7.2.2 Ermittlung der Verformungen und Schnittgrößen an den Querrändern

7.2.2 Determination of the Deformations and the Section Stress Resultants at the Transverse Boundaries

Die Verschiebungen und Schnittgrößen an den Rändern $x = 0$ und $x = l$ werden vorerst nur für die Lastgruppen P_1, P_2, und P_3 überlagert. In den Zustandsflächen der Tafeln 13, 14, 20 und 23 sind somit für den unteren Rand folgende Schnitte zu superponieren (vgl. Bild 13):

Initially, the displacements and stress resultants at the boundaries $x = 0$ and $x = l$ are superposed only for the load groups P_1, P_2 and P_3. Thus, in the condition surfaces of Charts 13, 14, 20 and 23, the following sections have to be superposed for the lower boundary (see Fig. 13):

$$\frac{x}{a} = -\frac{30,00}{25,00} = -1,20; \qquad \frac{x}{a} = -\frac{55,00}{25,00} = -2,20 \qquad \text{und/and} \qquad \frac{x}{a} = -\frac{80,00}{25,00} = -3,20$$

Für den oberen Rand werden die Ordinaten aus den Flächen der Tafeln 17, 19, 20 und 23 entsprechend addiert für:

For the upper boundary, the ordinates from the surfaces of Charts 17, 19, 20 and 23 have to be added, correspondingly, for

$$\frac{x}{a} = \frac{25,00}{25,00} = +1,00; \qquad \frac{x}{a} = \frac{50,00}{25,00} = +2,00 \qquad \text{und/and} \qquad \frac{x}{a} = \frac{75,00}{25,00} = +3,00$$

Die Lastgruppen P_1', P_2' und P_3' liefern wegen der Symmetrie (bzw. Antimetrie bei v und $\bar{n}_{x\varphi}$) zu $\varphi = \pi/2$ gleiche (bzw. negativ gleiche) Ordinaten spiegelbildlich zu $\varphi = \pi/4$. Sie sind in Bild 14 negativ aufgetragen, um die endgültigen Werte direkt abgreifen zu können.

Gegenüber Abschnitt 5.1.1 werden hier nur die Reihenglieder $n = 0, 4, 8, 12, \ldots$ benötigt, da wegen der vier Einzellasten bei gleichem Ansatz die Werte der übrigen Harmonischen verschwinden. Wie die Ergebnisse zeigen, reicht es aus, nur das Reihenglied $n = 4$ zu berücksichtigen. Die Fourier-Koeffizienten für $n = 8$ sind zusätzlich in Klammern angegeben. Mit $\Delta\varphi = \pi/16$ und den Ordinaten aus Bild 14 wird somit nach (21), wenn noch von der Symmetrie bzw. Antimetrie zu $\varphi = \pi/4$ Gebrauch gemacht wird:

Because of the symmetry (anti-symmetry, respectively, for v and $\bar{n}_{x\varphi}$) to $\varphi = \pi/2$, the load groups P_1', P_2' and P_3' produce mirrored equal ordinates (negative equal, respectively) with respect to $\varphi = \pi/4$. They are plotted negatively in Fig. 14 for a direct scaling of the final values.

Contrary to Section 5.1.1, only the series terms $n = 0, 4, 8, 12\ldots$ are required herein, because the values of the remaining harmonics vanish due to the four concentrated loads with equal assumption. As the results show, it is sufficient to consider only the series term $n = 4$. The Fourier coefficients for $n = 8$ are shown, additionally, in parantheses. When utilizing also symmetry or anti-symmetry, respectively to $\varphi = \pi/4$, it becomes according to (21) with $\Delta\varphi = \pi/16$ and the ordinates in Fig. 14:

$$1000\,\frac{K\pi}{a^2}\,U_4^0 = 2\,\frac{4}{3}\,\frac{\pi}{16\,\pi}\left(0,179 + 4\cdot 0,124\cos\frac{\pi}{4} - 2\cdot 0,004\cos\frac{\pi}{2} - 4\cdot 0,124\cos\frac{3\pi}{4} - 0,170\cos\pi\right) = +0,175 \qquad (+0,044 \quad \text{bei/for} \quad n = 8)$$

$$1000\,\frac{K\pi}{a^2}\,V_4^0 = \frac{1}{6}\left(0 + 4\cdot 1,10\sin\frac{\pi}{4} + 2\cdot 1,55\sin\frac{\pi}{2} + 4\cdot 1,10\sin\frac{3\pi}{4} + 0\right) = +1,55 \qquad (+0,003 \quad \text{bei/for} \quad n = 8)$$

$$100\,\pi\, M^0_{x_4} = \frac{1}{6}\left(+\,1{,}47 + 4\cdot 1{,}08\cos\frac{\pi}{4} + 2\cdot 0{,}05\cos\frac{\pi}{2} - 4\cdot 1{,}08\cos\frac{3\pi}{4} - 1{,}58\cos\pi\right) = +\,1{,}53 \qquad (-\,0{,}0005 \quad \text{bei/for} \quad n = 8)$$

$$10\,\pi\, a\, \bar{Q}^0_{x_4} = \frac{1}{6}\left(+\,1{,}05 + 4\cdot 0{,}59\cos\frac{\pi}{4} - 2\cdot 0{,}26\cos\frac{\pi}{2} - 4\cdot 0{,}58\cos\frac{3\pi}{4} - 0{,}53\cos\pi\right) = +\,0{,}814 \qquad (+\,0{,}258 \quad \text{bei/for} \quad n = 8)$$

$$\pi\, a\, N^l_{x_4} = \frac{1}{6}\left(+\,4{,}3 - 4\cdot 1{,}6\cos\frac{\pi}{4} - 2\cdot 6{,}1\cos\frac{\pi}{2} + 4\cdot 1{,}7\cos\frac{3\pi}{4} + 7{,}8\cos\pi\right) = -\,2{,}13 \qquad (+\,6{,}05! \quad \text{bei/for} \quad n = 8)$$

$$\pi\, a\, \bar{N}^l_{x\varphi_4} = \frac{1}{6}\left(0 - 4\cdot 4{,}1\sin\frac{\pi}{4} - 2\cdot 7{,}2\sin\frac{\pi}{2} - 4\cdot 5{,}7\sin\frac{3\pi}{4} + 0\right) = -\,7{,}02 \qquad (+\,0{,}77 \quad \text{bei/for} \quad n = 8)$$

$$100\,\pi\, M^l_{x_4} = \frac{1}{6}\left(+\,1{,}67 + 4\cdot 1{,}21\cos\frac{\pi}{4} + 2\cdot 0{,}01\cos\frac{\pi}{2} - 4\cdot 1{,}21\cos\frac{3\pi}{4} - 1{,}69\cos\pi\right) = +\,1{,}70 \qquad (-\,0{,}013 \quad \text{bei/for} \quad n = 8)$$

$$10\,\pi\, a\, \bar{Q}^l_{x_4} = \frac{1}{6}\left(-\,1{,}26 - 4\cdot 0{,}60\cos\frac{\pi}{4} + 2\cdot 0{,}47\cos\frac{\pi}{2} + 4\cdot 0{,}59\cos\frac{3\pi}{4} + 0{,}32\cos\pi\right) = -\,0{,}824 \qquad (-\,0{,}468 \quad \text{bei/for} \quad n = 8)$$

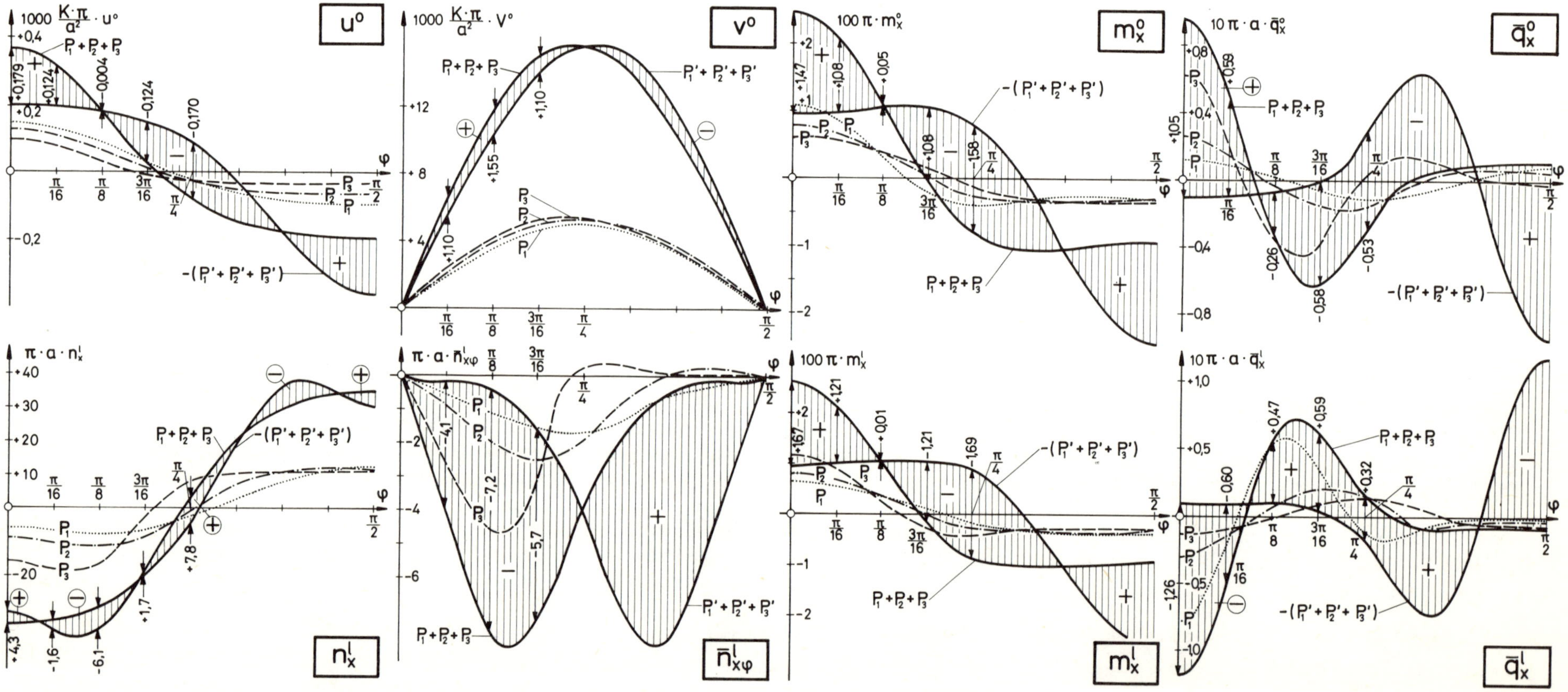

Bild 14. Summation der Randwerte für den unteren ($x = 0$) und den oberen ($x = l$) Rand.

Figure 14. Summation of the boundary values for the lower ($x = 0$) and the upper ($x = l$) boundary.

7.2.3 Randgrößen der regulären Fläche

Um die Rekursionsformel (23) anwenden zu können, müssen erst noch die Wurzeln der charakteristischen Gleichung bestimmt werden, da in der Tafelübersicht diese nur für $n = 2$ angegeben sind:
Für $n = 4$, $k = 2{,}5 \cdot 10^{-6}$ und $\nu = 1/6$ folgt aus (11):

7.2.3 Boundary Values of the Regular Surfaces

To apply the recursion formula (23), the roots of the characteristic equation have to be determined first, because they are given only for $n = 2$ in the Chart Classification:
For $n = 4$, $k = 2{,}5 \cdot 10^{-6}$ and $\nu = 1/6$ follows from (11):

$$\varepsilon = 4 \cdot 4\sqrt{4^2 - 1}\sqrt{\frac{2{,}5 \cdot 10^{-6}}{1 - 1/6^2}} = 0{,}099\,369$$

$$\varkappa_{\frac{1}{2}} = \frac{1}{2\sqrt{2}}\sqrt[4]{\frac{1 - 1/6^2}{2{,}5 \cdot 10^{-6}}}\left\{1 \mp \sqrt{\sqrt{0{,}099\,369^2 + 1} - 0{,}099\,369}\right\} = 8{,}8290\,\{1 \mp 0{,}951\,606\}$$

$$\varkappa_1 = 0{,}42727 \text{ (genau/exact: } 0{,}427\,556); \quad \varkappa_2 = 17{,}231 \text{ (genau/exact: } 17{,}218\,435)$$

$$\mu_{\frac{1}{2}} = \mp\, 8{,}8290\left\{1 \mp \sqrt{\sqrt{0{,}099\,369^2 + 1} + 0{,}099\,369}\right\} = \mp\, 8{,}8290\,\{1 \mp 1{,}050\,854\}$$

$$\mu_1 = 0{,}44899 \text{ (genau/exact: } 0{,}448\,710); \quad \mu_2 = 18{,}107 \text{ (genau/exact: } 18{,}118\,755)$$

Die Rekursionskonstanten werden mit den genauen Werten in Tabelle 3 nach (23) bestimmt.

The recursion constants are being computed in Table 3 with the exact values in accordance with (23).

Tabelle 3. Ermittlung der Rekursionskonstanten

Table 3. Determination of the recursion constants.

q_m \ m	0	1	2	3	4	5	6	7
a_m	+ 1,000	− 0,448710	+ 0,018536	+ 0,155735	− 0,146880	+ 0,071989	− 0,008181	− 0,020313
b_m	0	+ 0,427556	− 0,383698	+ 0,180094	− 0,014225	− 0,056417	+ 0,056094	− 0,028668
c_m	+ 1,000	− 18,11876	+ 31,81480	+ 10167,1	− 388305,4	+ 7719217	− 37126058	− 3477331323
d_m	0	+ 17,21844	− 623,953	+ 11853,1	− 39702,0	− 5966660	+ 241021286	− 5006258288

Die Verschiebungs- und Schnittgrößenkonstanten für $u, v, m_x, \bar{q}_x, n_x$ und $\bar{n}_{x\varphi}$ entsprechend (26) nach (17) sowie die erforderlichen Randgrößenkonstanten für $x = l$ ($\bar{p}^l_{n_x}, \bar{p}^l_{n_{x\varphi}}, \bar{p}^l_{m_x}$ und $\bar{p}^l_{q_x}$ mit $\bar{p} = \bar{a}, \bar{b}, \bar{c}, \bar{d}, \bar{e}, \bar{f}, \bar{g}$ und $\bar{h}$) analog zu (29) werden in Tabelle 4 und 5 errechnet. Die Faktoren zur Ermittlung von p sind mit $n = 4$, $\nu = 1/6$ und $k = 2{,}5 \cdot 10^{-6}$ aus (17) bestimmt. Für die Randgrößenkonstanten $\bar{p}^l$ (oberer Rand $x = l$) benötigt man noch folgende Werte:

The constants for the displacements and stress resultants $u, v, m_x, \bar{q}_x, n_x$ and $\bar{n}_{x\varphi}$, corresponding to (26) and according to (17), and the required constants for the boundary quantities for $x = l$ ($\bar{p}^l_{n_x}, \bar{p}^l_{n_{x\varphi}}, \bar{p}^l_{m_x}$ and $\bar{p}^l_{q_x}$ with $\bar{p} = \bar{a}, \bar{b}, \bar{c}, \bar{d}, \bar{e}, \bar{f}, \bar{g}$ and $\bar{h}$), analogous to (29), are computed in Tables 4 and 5. The factors for the computation of $\bar{p}$ are obtained from (17), with $n = 4$, $\nu = 1/6$ and $k = 2{,}5 \cdot 10^{-6}$. For the constants of the boundary quantities, $\bar{p}^l$ (upper boundary $x = l$), the following values are still required:

$$\begin{Bmatrix} e^{-\bar{\mu}_1} \\ e^{+\bar{\mu}_1} \end{Bmatrix}\begin{Bmatrix} \cos\bar{\varkappa}_1 \\ \sin\bar{\varkappa}_1 \end{Bmatrix} \triangleq \begin{Bmatrix} e^{-\mu_1 l/a} \\ e^{+\mu_1 l/a} \end{Bmatrix}\begin{Bmatrix} \cos\varkappa_1\, l/a \\ \sin\varkappa_1\, l/a \end{Bmatrix} = \begin{Bmatrix} 0{,}151\,893 \\ 6{,}583\,602 \end{Bmatrix}\begin{Bmatrix} -\,0{,}223\,048 \\ +\,0{,}974\,807 \end{Bmatrix} \begin{matrix} = -\,0{,}033\,879 \\ = +\,0{,}148\,066 \\ = -\,1{,}468\,460 \\ = +\,6{,}417\,748 \end{matrix}$$

$$\begin{Bmatrix} e^{-\bar{\mu}_2} \\ e^{+\bar{\mu}_2} \end{Bmatrix}\begin{Bmatrix} \cos\bar{\varkappa}_2 \\ \sin\bar{\varkappa}_2 \end{Bmatrix} \triangleq \begin{Bmatrix} e^{-\mu_2 l/a} \\ e^{+\mu_2 l/a} \end{Bmatrix}\begin{Bmatrix} \cos\varkappa_2\, l/a \\ \sin\varkappa_2\, l/a \end{Bmatrix} = \begin{Bmatrix} 8{,}92737 \cdot 10^{-34} \\ 1{,}12015 \cdot 10^{33} \end{Bmatrix}\begin{Bmatrix} -\,0{,}998\,152 \\ -\,0{,}060\,759 \end{Bmatrix} \begin{matrix} = -\,8{,}91087 \cdot 10^{-34} \\ = -\,5{,}42403 \cdot 10^{-35} \\ = -\,1{,}11808 \cdot 10^{33} \\ = -\,6{,}80574 \cdot 10^{31} \end{matrix}$$

Tabelle 4. Verschiebungskonstanten $\bar{p}_u$ und $\bar{p}_v$.

Table 4. Displacement constants $\bar{p}_u$ and $\bar{p}_v$.

Faktor	$\bar{a}_u$	$\bar{b}_u$	$\bar{c}_u$	$\bar{d}_u$	Faktor	$\bar{a}_v$	$\bar{b}_v$	$\bar{c}_v$	$\bar{d}_v$
$-2{,}5\cdot10^{-6}\,q_5$ $+0{,}1\overline{6}\ldots\,q_3$ $+16{,}0006\,q_1$	– 0,0 + 0,02596 – 7,17965	+ 0,0 + 0,03002 – 6,84117	– 19,30 + 1694,51 – 289,91	+ 14,92 + 1975,51 + 275,50	$+2\cdot10^{-5}\,q_4$ $-8{,}66699\,q_2$ $+64\,q_0$	– 0,0 – 0,1607 + 64,0000	– 0,0 + 3,32550 0	– 7,77 – 275,73 + 64,00	– 0,79 + 5407,79 0
$\bar{p}_u =$	– 7,15369	+ 6,87119	+ 1385,30	+ 2265,93	$\bar{p}_v =$	+ 63,8393	+ 3,32550	– 219,50	+ 5407,00

Die Gleichungen (25) und (29) zeigen, daß die Addition bzw. Subtraktion der beiden Anteile für $\bar{p}^l$ jeweils über zwei Spalten ($\bar{a}$ und $\bar{b}$ bzw. $\bar{c}$ und $\bar{d}$) verläuft. In Tabelle 5 folgen daher die Werte für $\bar{p}^l$ aus seitlich versetzter Summation mit positiven oder negativen Vorzeichen (wie in Klammern angegeben).

The equations (25) and (29) indicate that the addition and subtraction, respectively, of the two parts for $\bar{p}^l$ extend over two columns each ($\bar{a}$ and $\bar{b}$ or $\bar{c}$ and $\bar{d}$, respectively). Hence, the values for $\bar{p}^l$, in Table 5, follow from a staggered summation with positive or negative sign (as shown in parantheses).

Tabelle 5. Schnittgrößenkonstanten $\bar{p}_{n_x}$, $\bar{p}_{n_{x\varphi}}$, $\bar{p}_{m_x}$ und $\bar{p}_{q_x}$ *).

Table 5. Stress resultant constants $\bar{p}_{n_x}$, $\bar{p}_{n_{x\varphi}}$, $\bar{p}_{m_x}$ and $\bar{p}_{q_x}$ *).

Faktor	$\bar{a}_{n_x}$	$\bar{b}_{n_x}$	$10^{-5}\,\bar{c}_{n_x}$	$10^{-5}\,\bar{d}_{n_x}$	Faktor	$\bar{a}_{n_{x\varphi}}$	$\bar{b}_{n_{x\varphi}}$	$10^{-5}\,\bar{c}_{n_{x\varphi}}$	$10^{-5}\,\bar{d}_{n_{x\varphi}}$
$-26{,}6\overline{6}\ldots\,q_4$ $+6222649\,q_2$	+ 4 + 115345	+ 1 – 2387617	+ 103,55 + 1979,72	+ 10,6 – 38826,4	$+25\cdot10^{-6}\,q_5$ $-1555449\,q_3$ $-1626{,}\overline{6}\ldots q_1$	+ 0 – 242238 + 730	– 0 – 280128 – 695	+ 0 – 158143 + 0	– 0 – 184369 – 0
$\bar{p}_{n_x} =$	+ 115349	– 2387616	+ 2083,27	– 38815,8	$\bar{p}_{n_{x\varphi}} =$	– 241508	– 280823	– 158143	– 184369
$e^{-\bar{\mu}}\cos\bar{\varkappa}\cdot\bar{p} =$ $e^{-\bar{\mu}}\sin\bar{\varkappa}\cdot\bar{p} =$	– 3908 + 17079	+ 80891 (±) 353524	– 1,8564(–25) – 0,1130(–25)	+ 34,5883(–25) (∓) 2,1054(–25)		+ 8182,1 – 35759,1	+ 9514,1 (±) 41580,3	+ 140,919(–25) + 8,578(–25)	+ 164,288(–25) (∓) 10,000(–25)
$\bar{a}^l\ldots\bar{d}^l =$	+ 349616	+ 97970	– 3,9618(–25)	+ 34,4753(–25)		+ 49762,4	– 26245,0	+ 130,919(–25)	+ 172,866(–25)
$e^{+\bar{\mu}}\cos\bar{\varkappa}\cdot\bar{p} =$ $e^{+\bar{\mu}}\sin\bar{\varkappa}\cdot\bar{p} =$	– 1,694(+5) + 7,403(+5)	(∓) 35,061(+5) – 153,231(+5)	– 23,293(+40) – 1,418(+40)	(∓) 433,993(+40) + 26,417(+40)		(∓) 3,5464(+5) (±) 15,4993(+5)	+ 4,1238(+5) (±) 18,0225(+5)	(∓) 1768,17(+40) (∓) 107,63(+40)	+ 2061,39(+40) (∓) 125,48(+40)
$\bar{e}^l\ldots\bar{h}^l =$	– 154,925(+5)	– 27,658(+5)	+ 3,124(+40)	– 435,411(+40)		+ 14,4761(+5)	+ 19,6231(+5)	– 1893,65(+40)	+ 1953,76(+40)

*) hierin und im folgenden bedeutet z. B. 490,61 (+ 40) die Zahl $490{,}61\cdot10^{+40}$

*) herein and in the following, denotes for instance, 490,61 (+ 40) the number $490{,}61\cdot10^{+40}$

7.2.4 Lösung des Gleichungssystems

Werden die in den Tabellen 4 und 5 ermittelten Werte für $\bar{p}$ und $\bar{p}^l$ entsprechend dem Schema in das Gleichungssystem (30) eingesetzt, dann ergibt sich schließlich die achtgliedrige Matrix nach Tabelle 6. Die Absolutwerte folgen aus Abschnitt 7.2.2, wenn noch die Zehnerpotenzen berücksichtigt werden. Die weiteren Multiplikatoren S_c (vgl. (25)) bleiben jedoch entsprechend (27) bzw. (28) erhalten. Die

7.2.4 Solution of the Equation System

When substituting the values $\bar{p}$ and $\bar{p}^l$, computed in Tables 4 and 5, in the equation system (30) in accordance with the set-up, the eight-term matrix, in Table 6, is finally obtained. When considering the powers of ten, the absolute values follow from Section 7.2.2. The addtional multipliers S_c (see (25)), however, remain preserved according to (27) or (28), respectively. The solution of the equation system

Tabelle 6. Gleichungssystem der Randstörungsrechnung

Table 6. Equation system for the analysis of the boundary disturbance.

	A_0	B_0	C_0	D_0	E_0	F_0	G_0	H_0	⊗
$u^0 =$	− 7,15369	+ 6,87119	+ 1385,30	+ 2265,93	+ 7,15369	+ 6,87119	− 1385,30	+ 2265,93	+ 17,5 (−5)
$v^0 =$	+ 63,8393	+ 3,32550	− 219,50	+ 5407,00	+ 63,8393	− 3,32550	− 219,50	− 5407,00	+ 155 (−5)
$m_x^0 =$	+ 628,424	+ 134,344	+ 237,199 (+5)	− 2421,66 (+5)	+ 628,424	− 134,344	+ 237,199 (+5)	+ 2421,66 (+5)	+ 0,0153
$\bar{q}_x^0 =$	− 3335,57	+ 2797,19	+ 39374,5 (+5)	+ 46276,4 (+5)	+ 3335,57	+ 2797,19	− 39374,5 (+5)	+ 46276,4 (+5)	+ 0,0814
$n_x^l =$	+ 349616	+ 97970	− 3,9618 (−25)	+ 34,4753 (−25)	− 154,925 (+5)	− 27,658 (+5)	+ 3,124 (+40)	− 435,411 (+40)	− 2,13
$\bar{n}_{x\varphi}^l =$	+ 49762,4	− 26245,0	+ 130,919 (−25)	+ 172,866 (−25)	+ 14,4761 (+5)	+ 19,6231 (+5)	− 1893,65 (+40)	+ 1953,76 (+40)	− 7,02
$m_x^l =$	− 41,1824	+ 88,4967	− 0,34272 (−25)	+ 2,14505 (−25)	− 60,63	+ 4230,35	− 1,0040 (+40)	− 27,2376 (+40)	+ 0,0170
$\bar{q}_x^l =$	− 301,162	− 588,651	− 32,576 (−25)	− 43,372 (−25)	− 22849,8	+ 17299,3	+ 471,73 (+40)	− 490,61 (+40)	− 0,0824

Tabelle 5 (Fortsetzung)

Table 5 (continued)

Faktor	$\bar{a}_{m_x}$	$\bar{b}_{m_x}$	$10^{-5}\,\bar{c}_{m_x}$	$10^{-5}\,\bar{d}_{m_x}$
− 1,0 q_6	+ 0,008	− 0,056	+ 371,261	− 2410,21
+ 34,5 q_4	− 5,067	− 0,491	− 133,956	− 13,64
− 351,5̅ ... q_2	− 6,517	+ 134,891	− 0,112	+ 2,19
+ 640 q_0	+ 640,000	0	+ 0,006	0
$\bar{p}_{m_x} =$	+ 628,424	+ 134,344	+ 237,199	− 2421,66
	− 21,2906 + 93,0482	− 4,5515 (∓) 19,8918	− 0,21137(−25) − 0,01286(−25)	+ 2,15791(−25) (∓) 0,13135(−25)
	− 41,1824	+ 88,4967	− 0,34272(−25)	+ 2,14505(−25)
	− 922,82 + 4033,07	(±) 197,28 + 862,19	− 2,6521(+40) − 0,1614(+40)	(∓) 27,0762(+40) + 1,6481(+40)
	− 60,63	+ 4230,35	− 1,0040(+40)	− 27,2376(+40)

Faktor	$\bar{a}_{q_x}$	$\bar{b}_{q_x}$	$10^{-5}\,\bar{c}_{q_x}$	$10^{-5}\,\bar{d}_{q_x}$
− 1,0 q_7	+ 0,02	+ 0,03	+ 34773,3	+ 50062,6
+ 61,1̅6̅ ... q_5	+ 4,40	− 3,45	+ 4720,7	− 3649,6
− 1162,6̅ ... q_3	− 181,07	− 209,39	− 118,2	− 137,8
+ 7040 q_1	− 3158,92	+ 3010,00	− 1,3	+ 1,2
$\bar{p}_{q_x} =$	− 3335,57	+ 2797,19	+ 39374,5	+ 46276,4
	+ 113,006 − 493,884	− 94,767 (∓) 414,168	− 35,086(−25) − 2,136(−25)	− 41,236(−25) (±) 2,510(−25)
	− 301,162	− 588,651	− 32,576(−25)	− 43,372(−25)
	(∓) 4898,1 (±) 21406,9	− 4107,6 (∓) 17951,7	(±) 440,24(+40) (±) 26,80(+40)	− 517,41(+40) (±) 31,49(+40)
	− 22849,8	+ 17299,3	+ 471,73(+40)	− 490,61(+40)

Lösung des Gleichungssystems liefert die Konstanten A_0 bis H_0 der Spannungsfunktion $f_4\,(\xi)_R$ nach (22):

produces the constants A_0 to H_0 of the stress function $f_4(\xi)_R$, according to (22):

$$A_0 = -1{,}9826 \cdot 10^{-5}; \quad B_0 = -4{,}8895 \cdot 10^{-5}; \quad C_0 = +0{,}1937 \cdot 10^{-10}; \quad D_0 = -0{,}2034 \cdot 10^{-10}$$

$$E_0 = -0{,}1669 \cdot 10^{-5}; \quad F_0 = +0{,}4526 \cdot 10^{-5}; \quad G_0 = +106{,}3 \cdot 10^{-45}; \quad H_0 = +116{,}3 \cdot 10^{-45}$$

7.2.5 Bestimmung der regulären Fläche

Mit der Spannungsfunktion (22) sind alle Verformungen und Schnittgrößen der regulären Fläche aus der Randstörungsrechnung nach (25) bekannt. Die Verschiebungs- und Schnittgrößenkonstanten $\bar{p}$ sind bereits für zwei Verformungen in

7.2.5 Determination of the Regular Surface

With the stress function (22), all of the deformations and stress resultants of the regular surfaces from the analysis of the boundary disturbance, according to (25), are known. The constants $\bar{p}$ of the displacements and the stress resultants are

Tabelle 4 und für vier Schnittgrößen in Tabelle 5 ermittelt. Ratsamerweise berechnet man nun vorab die Konstanten vor den trigonometrischen Funktionen

$$K_{\substack{1\\2}} = \genfrac{}{}{0pt}{}{+\bar a_s}{-\bar b_s} A_0 + \genfrac{}{}{0pt}{}{\bar b_s}{\bar a_s} B_0 \qquad K_{\substack{3\\4}} = \genfrac{}{}{0pt}{}{+\bar c_s}{-\bar d_s} C_0 + \genfrac{}{}{0pt}{}{\bar d_s}{\bar c_s} D_0$$

für die interessierenden statischen Größen und bestimmt dann nach (25) unter Zuhilfenahme von (17) tabellarisch oder mit einem programmierbaren Rechner direkt die Ordinaten an den Stellen x, φ.

Hier sollen nur für $\varphi = 0°$ (bzw. 90°) die Werte der regulären Fläche m_x bei $x = 0$ (als Genauigkeitsvergleich), $x = 0{,}05\,l$, $x = 0{,}3\,l$, $x = 0{,}5\,l$, $x = 0{,}7\,l$, $x = 0{,}95\,l$ und $x = l$ (Kontrolle) berechnet werden.

determined, already, for two deformations in Table 4, and for four stress resultants in Table 5. Preferably, one computes first the constants for the trigonometric functions

$$K_{\substack{5\\6}} = +\genfrac{}{}{0pt}{}{\bar a_s}{\bar b_s} E_0 \genfrac{}{}{0pt}{}{-\bar b_s}{+\bar a_s} F_0 \qquad K_{\substack{7\\8}} = +\genfrac{}{}{0pt}{}{\bar c_s}{\bar d_s} G_0 \genfrac{}{}{0pt}{}{-\bar d_s}{+\bar c_s} H_0$$

of the statical quantities of interest, and hereafter in accordance to (25) and with (17), the ordinates at the loci x, φ by tabulation, or directly with a programmable calculator.

Only the values of the regular surface m_x, for $\varphi = 0°$ (and 90°, respectively) at $x = 0$ (as a check of accuracy), $x = 0{,}05\,l$, $x = 0{,}3\,l$, $x = 0{,}5\,l$, $x = 0{,}7\,l$, $x = 0{,}95\,l$ and $x = l$ (control) shall be computed herein.

$$K_{\substack{1\\2}} = \genfrac{}{}{0pt}{}{+\ 628{,}4}{-\ 134{,}3}(-1{,}9826\cdot 10^{-5}) + \genfrac{}{}{0pt}{}{134{,}3}{628{,}4}(-4{,}8895\cdot 10^{-5}) = \genfrac{}{}{0pt}{}{-1903\cdot 10^{-5}}{-2806\cdot 10^{-5}}$$

$$K_{\substack{3\\4}} = +\genfrac{}{}{0pt}{}{237{,}2}{2421{,}7}\cdot 10^5(+0{,}1937\cdot 10^{-10}) \genfrac{}{}{0pt}{}{-\ 2421{,}7}{+\ 237{,}2}\cdot 10^5(-0{,}2034\cdot 10^{-10}) = \genfrac{}{}{0pt}{}{+\ 539\cdot 10^{-5}}{+\ 421\cdot 10^{-5}}$$

$$K_{\substack{5\\6}} = +\genfrac{}{}{0pt}{}{628{,}4}{134{,}3}(-0{,}1669\cdot 10^{-5}) \genfrac{}{}{0pt}{}{-\ 134{,}3}{+\ 628{,}4}(+0{,}4526\cdot 10^{-5}) = \genfrac{}{}{0pt}{}{-\ 166\cdot 10^{-5}}{+\ 262\cdot 10^{-5}}$$

$$K_{\substack{7\\8}} = \genfrac{}{}{0pt}{}{+\ 237{,}2}{-\ 2421{,}7}\cdot 10^5(+106{,}3\cdot 10^{-45}) + \genfrac{}{}{0pt}{}{2421{,}7}{237{,}2}\cdot 10^5(+116{,}3\cdot 10^{-45}) = \genfrac{}{}{0pt}{}{+3{,}068\cdot 10^{-35}}{-2{,}298\cdot 10^{-35}}$$

Tabelle 7. Bestimmung der Ordinaten für $100\,\pi\, m_x$

$x/l =$	0	0,05	0,30	0,50	0,70	0,95	1,00
$\xi = x/a = 4{,}2\,x/l$	0	0,21	1,26	2,10	2,94	3,99	4,20
$L_1 = e^{-\mu_1\xi}\cos\varkappa_1\xi$	1	+ 0,906	+ 0,488	+ 0,243	+ 0,083	− 0,022	− 0,034
$L_2 = e^{-\mu_1\xi}\sin\varkappa_1\xi$	0	+ 0,082	+ 0,291	+ 0,305	+ 0,254	+ 0,165	+ 0,148
$L_3 = e^{-\mu_2\xi}\cos\varkappa_2\xi$	1	− 0,020	≈ − 0	≈ + 0	≈ + 0	≈ + 0	≈ − 0
$L_4 = e^{-\mu_2\xi}\sin\varkappa_2\xi$	0	− 0,010	≈ + 0	≈ − 0	≈ + 0	≈ − 0	≈ − 0
$L_5 = e^{+\mu_1\xi}\cos\varkappa_1\xi$	1	+ 1,094	+ 1,511	+ 1,599	+ 1,155	− 0,807	− 1,468
$L_6 = e^{+\mu_1\xi}\sin\varkappa_1\xi$	0	+ 0,099	+ 0,903	+ 2,006	+ 3,558	+ 5,937	+ 6,418
$L_7 = e^{+\mu_2\xi}\cos\varkappa_2\xi$	1					+ 23 (+30)	−1118 (+30)
$L_8 = e^{+\mu_2\xi}\sin\varkappa_2\xi$	0					− 10 (+30)	− 68 (+30)

Table 7. Determination of the ordinates for $100\,\pi\, m_x$

$x/l =$	0	0,05	0,30	0,50	0,70	0,95	1,00
$100\,K_1 L_1 =$	− 1,90	− 1,72	− 0,93	− 0,46	− 0,16	+ 0,04	+ 0,06
$100\,K_2 L_2 =$	0	− 0,23	− 0,82	− 0,86	− 0,71	− 0,46	− 0,42
$100\,K_3 L_3 =$	+ 0,54	− 0,01	0	0	0	0	0
$100\,K_4 L_4 =$	0	0	0	0	0	0	0
$100\,K_5 L_5 =$	− 0,17	− 0,18	− 0,25	− 0,27	− 0,19	+ 0,13	+ 0,24
$100\,K_6 L_6 =$	0	+ 0,03	+ 0,24	+ 0,53	+ 0,93	+ 1,56	+ 1,68
$100\,K_7 L_7 =$	0	0	0	0	0	+ 0,07	− 3,43
$100\,K_8 L_8 =$	0	0	0	0	0	+ 0,02	+ 0,16
$100\,\pi\, m_x =$	− 1,53	− 2,11	− 1,76	− 1,06	− 0,13	+ 1,36	− 1,71

In Tabelle 7 sind schließlich die Ordinaten des Momentes $100\,\pi\,m_x$ – um sie direkt mit der singulären Fläche vergleichen zu können – ermittelt. Die Werte der einzelnen Schwingungen im linken Teil der Tabelle können für die anderen statischen Größen direkt benutzt werden.

Finally, the ordinates of the moment $100\,\pi\,m_x$ — to be able to compare directly with the singular surface — are computed in Table 7. The values of the individual oscillations, in the left part of the table, can be used directly for the other statical quantities.

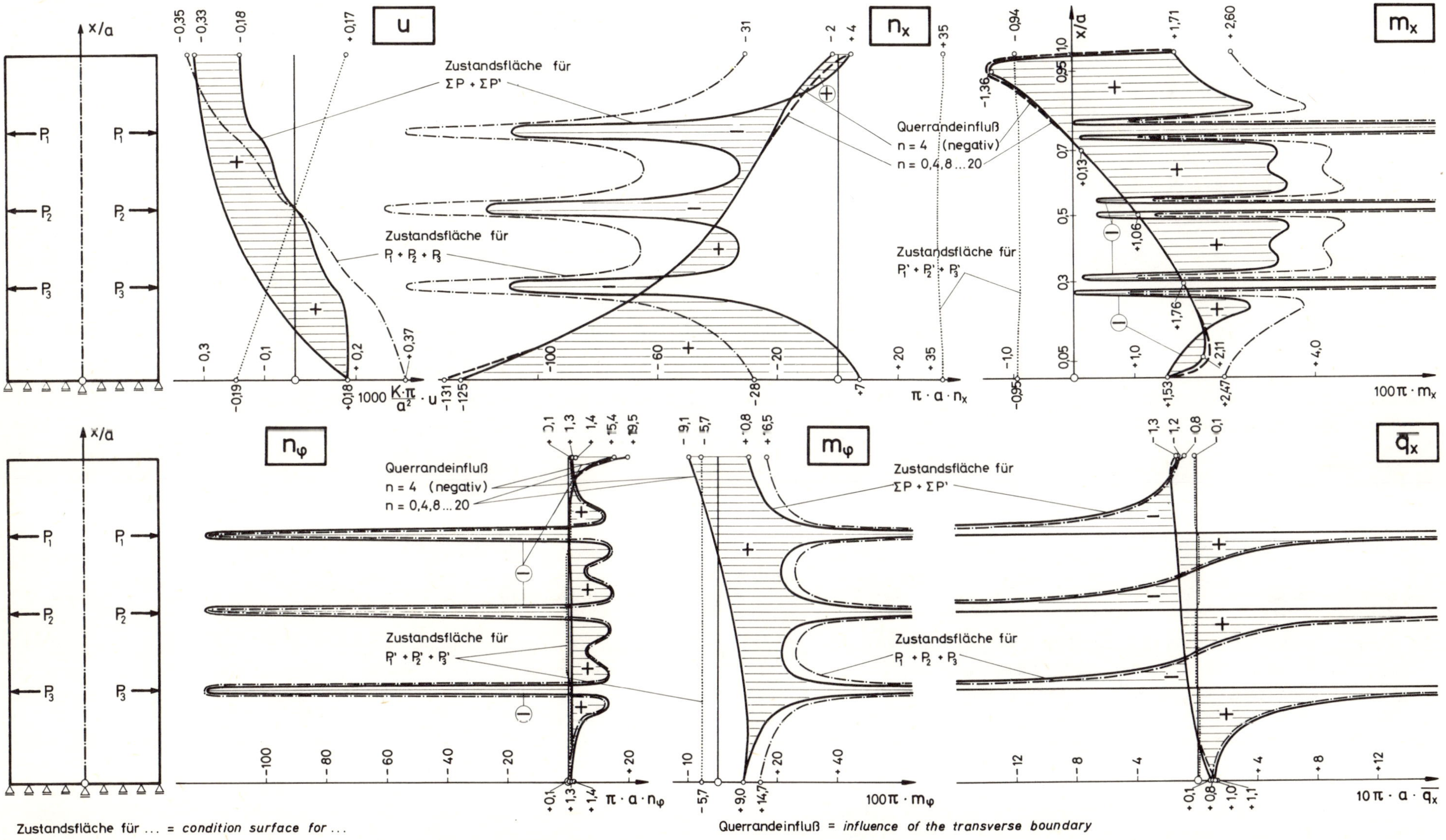

Bild 15. Endgültige statische Größen für $\varphi = 0°$.

Figure 15. Final static quantities for $\varphi = 0°$.

7.2.6 Überlagerung der singulären und regulären Fläche

Entsprechend Abschnitt 7.2.2 müssen die Lastgruppen P_1, P_2 und P_3 überlagert werden. Für $\varphi = 0$ sind diese Werte aus den Tafeln 13, 17, 18, 20, 21 und 23 entnommen und unter Berücksichtigung der Laststellung ($\xi = x/a$) superponiert.

Diese und die gleichermaßen für die Lastgruppen P_1', P_2' und P_3' ($\varphi = \pi/2$) bestimmten Werte sind im Bild 15 aufgetragen. Ihre Summation liefert die singuläre Fläche. Die reguläre Fläche für m_x ist in Tabelle 7 bestimmt. Entsprechend errechnen sich die Werte für die anderen statischen Größen. Diese Ordinaten sind negativ im Bild 15 eingetragen, um mit der Differenz sofort die endgültigen Verschiebungen und Schnittgrößen zeigen zu können.

Aus den zusätzlich angegebenen Kurven für die Summation von $n = 0, 4, 8, 12, 16$ und 20 geht hervor, daß hier bereits bei einem Reihenglied die Genauigkeit ausreichend ist. Selbst bei n_x am oberen Rand, wo $N_{x_8}^l \approx -3\,N_{x_4}^l$ war (vgl. Abschnitt 7.2.2), ist der Fehler bezogen auf die endgültigen Maximalordinaten kleiner als 5%.

Es sollte hier noch erwähnt werden, daß Fehler oder größere Ungenauigkeiten in den Absolutgliedern des Gleichungssystems (vgl. Tabelle 6) – d.h. den Fourier-Koeffizienten nach 7.2.2 – insbesondere die Konstanten vor den schnell abklingenden Funktionen (K_3, K_4 und K_7, K_8) so verändern können, daß die Schnittgrößen in der unmittelbaren Nachbarschaft der Ränder (hier $x/l < 0{,}05$, vgl. Tabelle 7) z.T. stark verfälscht sind. Die hauptsächlich interessierenden Werte in der Nähe der Einzellasten bleiben dagegen richtig und die Randwerte können meist leicht verbessert werden, wenn man berücksichtigt, daß große, kurze Schwingungen der regulären Fläche nicht auftreten werden (vgl. Bild 15).

7.3 Reaktordruckgefäß mit Strahlkraft

7.3.1 Aufgabenstellung

Im vorliegenden Rechenbeispiel soll einmal die Möglichkeit der Extrapolation für $k > 10^{-4}$ und zum anderen die Erfüllung komplexer Längsrandbedingungen gezeigt werden. Die Beanspruchung des Reaktordruckgefäßes nach Bild 16 unter einer Strahlkraft in Behältermitte (vgl. [3]) führt wegen $k = 0{,}35^2/(12 \cdot 7{,}11^2) \approx 2 \cdot 10^{-4}$ und der bei den Einfluß- bzw. Zustandsflächen gewählten Lastgruppe zu dieser Fragestellung.

Die Extrapolation bereitet einige Mehrarbeit, da wegen der großen Änderung der Schnittgrößen mit k eine lineare Extrapolation zu falschen Ergebnissen führen würde. In Bild 17 ist die graphische Lösung vorgestellt, wobei die extrapolierten Werte mit einer numerischen Rechnung entsprechend Abschnitt 5.1.1 verglichen wurden.

7.2.6 Superposition of the Singular and Regular Surface

In accordance with Section 7.2.2, the load groups P_1, P_2 and P_3 have to be superposed. For $\varphi = 0$, these values are taken from Charts 13, 17, 18, 21 and 23, and superposed with consideration for the load position ($\xi = x/a$).

These, and the similarily determined values for the load groups P_1', P_2' and P_3' ($\varphi = \pi/2$) are plotted in Fig. 15. Their summation produces the singular surface. The regular surface for m_x is obtained in Table 7. The values of the other statical quantities are computed accordingly. These ordinates are plotted negatively in Fig. 15, to show immediately the final displacements and stress resultants by the difference.

From the curves, given additionally, for the summation of $n = 0, 4, 8, 12, 16$ and 20, it is evident that the accuracy herein is sufficient already for one term of the series. Even for n_x at the upper boundary, where $N_{x_8}^l \approx -3\,N_{x_4}^l$ (see Section 7.2.2), the error is less than 5% when compared with the final maximum ordinate.

It should be noted here that errors or larger inaccuracies of the absolute terms of the equation system (see Table 6)—i.e., the Fourier coefficients according to Section 7.7.2—can change the constants, in particular those of the rapidly decreasing functions (K_3, K_4 and K_7, K_8), such that section quantities in the immediate vicinity of the boundaries (here $x/l < 0{,}05$, see Table 7) become, partially, very incorrect. However, the values of main interest, close to the concentrated loads, remain correct. The boundary values can be improved, in most cases, when considering that large, short oscillations of the regular surface cannot occur (see Fig. 15).

7.3 Reactor Vessel Subjected to Jet Force

7.3.1 Problem Definition

In the presented example analysis, the possibility of extrapolation for $k > 10^{-4}$, and, furthermore, the satisfying of complex conditions at longitudinal boundaries shall be shown. The subjection of the reactor vessel to a jet force at the container center (see [3]), according to Fig. 16, leads, because of $k = 0{,}35^2/(12 \cdot 7{,}11^2) \approx \approx 2 \cdot 10^{-4}$ and because of the chosen load groups for the influence or condition surfaces, respectively, to this kind of a problem.

The extrapolation causes some additional work but, a linear interpolation would lead to wrong results due to the large changes of the stress resultants with k. The graphical solution is presented in Fig. 17, where the extrapolated values were compared with a numerical analysis, according to Section 5.1.1.

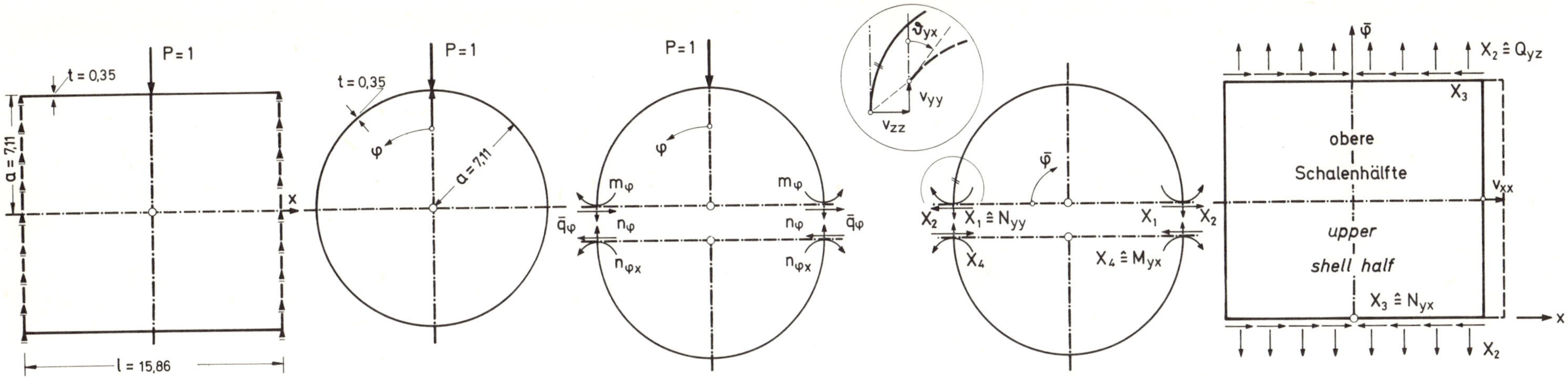

Bild 16. Abmessungen, Belastung und Ersatzsystem der Zylinderschale.

Figure 16. Dimensions, load and ersatz system of the cylindrical shell.

Die für die Ermittlung der Einfluß- und Zustandsflächen angesetzte zweite Last $P = 1$ bei $\varphi = \pi$ wird dadurch entfernt, daß das geschlossene Rohr ratsamerweise bei $\varphi = \pm\pi/2$ aufgetrennt und die untere Hälfte durch eine unbelastete Rohrhälfte ersetzt wird. Diese Hälfte muß die freigewordenen Schnittgrößen der oberen belasteten Hälfte $n_{\varphi x} = 0$, $\bar{q}_\varphi = 0$ (wegen $\varphi = \pm\pi/2$), n_φ und m_φ tragen und erhält somit auch Verformungen (u_{part}, v_{part} etc.) aus der äußeren Last. Die Verformungen der oberen und unteren Schalenhälfte passen nicht zusammen. Diese Klaffung muß durch eine zusätzliche Randstörungsrechnung des Längsrandes geschlossen werden. Hierzu wird das Tabellenwerk [22] benutzt. Mit den dort angegebenen Bezeichnungen lauten die Randbedingungen, die wegen der Symmetrie nur für $\bar{\varphi} = 0$ (vgl. Bild 16) erfüllt sein müssen:

The second load $P = 1$ at $\varphi = \pi$, applied for the determination of the influence and condition surfaces, is removed by cutting the closed pipe, preferably, at $\varphi = \pm\pi/2$, and replacing the lower half by an unloaded pipe half. This half has to carry the freed stress resultants of the loaded upper half, $n_{\varphi x} = 0$, $\bar{q}_\varphi = 0$ (because $\varphi = \pm\pi/2$), n_φ and m_φ, and therefore sustains also deformations (u_{part}, v_{part} etc.) due to the external load. The deformations of the upper and the lower shell halves do not match. This gap has to be closed by an additional boundary disturbance analysis of the longitudinal boundary. The table volume [22] is used herein. The boundary conditions, which have to be satisfied only for $\bar{\varphi} = 0$, because of symmetry (see Fig. 16), read with the notations used in the source:

$$u_{part} + X_1 v_{xx}(N_{yy} = 1) + X_2 v_{xx}(Q_{yz} = 1) + X_3 v_{xx}(N_{yx} = 1) + X_4 v_{xx}(M_{yx} = 1) = 0$$
$$v_{part} + X_1 v_{yy}(N_{yy} = 1) + X_2 v_{yy}(Q_{yz} = 1) + X_3 v_{yy}(N_{yx} = 1) + X_4 v_{yy}(M_{yx} = 1) = 0$$
$$w_{part} + X_1 v_{zz}(N_{yy} = 1) + X_2 v_{zz}(Q_{yz} = 1) + X_3 v_{zz}(N_{yx} = 1) + X_4 v_{zz}(M_{yx} = 1) = 0$$
$$\chi_{\varphi\,part} + X_1 \vartheta_{yx}(N_{yy} = 1) + X_2 \vartheta_{yx}(Q_{yz} = 1) + X_3 \vartheta_{yx}(N_{yx} = 1) + X_4 \vartheta_{yx}(M_{yx} = 1) = 0$$

Bei $\varphi = \pm\pi/2$ (d.h. $\bar{\varphi} = 0$ bzw. π/ω) verschwinden die zu $\varphi = 0$ antimetrischen Verformungen v_{part} und $\chi_{\varphi part}$.

At $\varphi = \pm\pi/2$ (i.e. $\bar{\varphi} = 0$ or π/ω, respectively), the deformations v_{part} and $\chi_{\varphi\,part}$, anti-symmetric to $\varphi = 0$, vanish.

7.3.2 Erfüllung der Randbedingungen an den Querrändern

Ermittelt werden müssen vorerst die Verformungen u_{part} und w_{part} sowie die Schnittgrößen n_φ und m_φ für $\varphi = \pm\pi/2$ (bzw. $\bar{\varphi} = 0$). Die in Bild 17 gezeigte graphische Extrapolation für $x/a = 0$ bis $x/a = 3{,}5$ aus den Werten für $k = 5\cdot10^{-5}$ (Tafeln 73, 75, 78 und 81), $k = 7{,}5\cdot10^{-5}$ (Tafeln 85, 87, 90 und 93) und $k = 10\cdot10^{-5} = 10^{-4}$ (Tafeln 97, 99, 102 und 105) ist nur als grobe Näherung zu betrachten. Auch die Funktionen nach (31) sind wegen des großen Schalenpara-

7.3.2 Satisfying the Boundary Conditions at the Transverse Boundaries

First, the deformations u_{part} and w_{part} as well as the stress resultants n_φ and m_φ for $\varphi = \pm\pi/2$ ($\bar{\varphi} = 0$, respectively) have to be determined. The graphical extrapolation for $x/a = 0$ to $x/a = 3{,}5$ with values for $k = 5\cdot10^{-5}$ (Charts 73, 75, 78 and 81), $k = 7{,}5\cdot10^{-5}$ (Charts 85, 87, 90 and 93) and $k = 10\cdot10^{-5} = 10^{-4}$ (Charts 97, 99, 102 and 105), shown in Fig. 17, is to be considered a coarse approximation only. The functions, according to (31), are also quite inaccurate because

meters k recht ungenau. Das Verfahren soll nur als Hilfsmittel zur angenäherten Ermittlung der Schnittgrößen vorgestellt werden.

Ratsamerweise trägt man noch die Kurven für benachbarte k aus den im Tafelanhang gezeichneten Zustandsflächen ein. Mit der Bedingung, daß bei steigendem k die Krümmungen dieser Kurven geringer werden müssen, ist ein weiteres Hilfsmittel gegeben.

of the large shell parameter k. The method shall be presented only as an auxiliary means for an approximate computation of the stress resultants.

Advisably, one plots also the curves for adjacent k's from the condition surfaces, plotted in the Chart Appendix. By the condition that with increasing k the curvatures of these curves have to decrease, another auxiliary means is given.

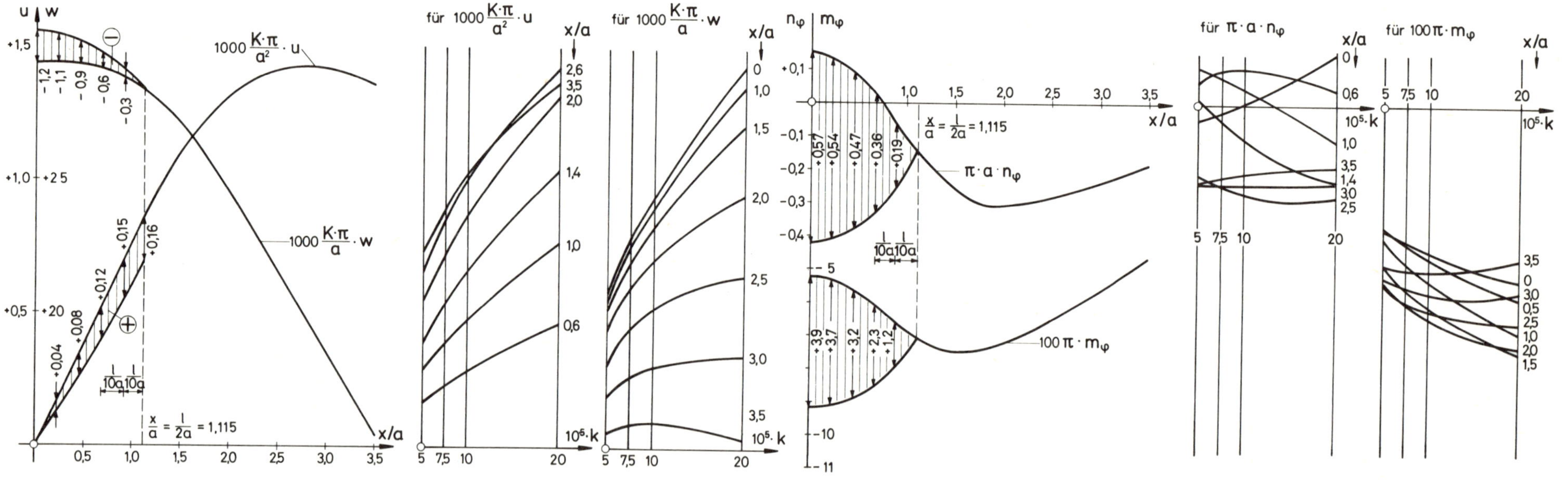

Bild 17. Graphische Extrapolation und Lastumordnung für $\varphi = \pm \pi/2$.

Figure 17. Graphical extrapolation and load rearrangement for $\varphi = \pm \pi/2$.

Die Ermittlung der Zustandslinien bei $\varphi = \pm \pi/2$ für die beidseitig gelenkig gelagerte Einfeldschale mit Hilfe des Lastumordnungsverfahrens nach Abschnitt 5.1.2 ist in gleicher Weise vorgenommen wie in Abschnitt 7.1.2.

Da in [22] (entgegen [28]) $x/a = \xi$ von Schalenmitte aus zählt (vgl. auch Bild 16), sind entsprechend Abschnitt 5.2 die trigonometrischen Funktionen gegenüber Abschnitt 7.1.2 zu vertauschen, und es folgen die ersten Fourier-Koeffizienten ($n = 1$) aus (34) mit $\Delta x = l/10$ zu:

The determination of the condition diagrams at $\varphi = \pm \pi/2$ for the single-span shell, hinge supported at both sides, with the load rearrangement method according to Section 5.1.2, is done the same way as in Section 7.1.2.

Because in [22] (contrary to [28]) $x/a = \xi$ is measured from the shell center (see also Fig. 16), the trigonometric functions, according to Section 5.2, have to be exchanged when compared with Section 7.1.2, and the first Fourier coefficients ($n = 1$) from (34), with $\Delta x = l/10$, are

$$\frac{1000\,K\pi}{a^2}\,U_1 = \frac{2}{15}\{0 + 4\cdot 0{,}04 \sin 0{,}1\,\pi + 2\cdot 0{,}08 \sin 0{,}2\,\pi + 4\cdot 0{,}12 \sin 0{,}3\,\pi + 2\cdot 0{,}15 \sin 0{,}4\,\pi + 2\cdot 0{,}16 \sin 0{,}5\,\pi\} = +\,0{,}15$$

$$\frac{1000\,K\pi}{a^2}\,W_1 = \frac{2}{15}\{-\,1{,}2 - 4\cdot 1{,}1 \cos 0{,}1\,\pi - 2\cdot 0{,}9 \cos 0{,}2\,\pi - 4\cdot 0{,}6 \cos 0{,}3\,\pi - 2\cdot 0{,}3 \cos 0{,}4\,\pi + 0\} = -\,1{,}1$$

$$\pi\,a\,N_{\varphi 1} = \frac{2}{15}\{+\,0{,}57 + 4\cdot 0{,}54 \cos 0{,}1\,\pi + 2\cdot 0{,}47 \cos 0{,}2\,\pi + 4\cdot 0{,}36 \cos 0{,}3\,\pi + 2\cdot 0{,}19 \cos 0{,}4\,\pi + 0\} = +\,0{,}58$$

$$100\,\pi\,M_{\varphi 1} = \frac{2}{15}\{+\,3{,}9 + 4\cdot 3{,}7 \cos 0{,}1\,\pi + 2\cdot 3{,}2 \cos 0{,}2\,\pi + 4\cdot 2{,}3 \cos 0{,}3\,\pi + 2\cdot 1{,}2 \cos 0{,}4\,\pi + 0\} = +\,3{,}9$$

Da die Fourier-Koeffizienten etwa die gleiche Größe wie die Maximalordinaten haben und ihre Kurven annähernd cos- bzw. sin-förmig sind, sieht man bereits, daß der Einfluß von $n = 3, 5, 7 \ldots$ vernachlässigbar klein sein muß.

Because the Fourier coefficients are, approximately, of the same magnitude as the maximum ordinates, and the curves have about the shape of cosine or sine diagrams, one can see already that the influence of $n = 3, 5, 7, \ldots$ has to be negligible small.

7.3.3 Erfüllung der Randbedingungen an den Längsrändern

7.3.3 Satisfying the Boundary Conditions at the Longitudinal Boundaries

Mit den im Tabellenwerk [22] angegebenen Bezeichnungen und positiven Richtungen (vgl. auch Bild 16) ergeben sich die vier interessierenden Verformungen $v_{xx}(\triangleq u)$, $v_{yy}(\triangleq v)$, $v_{zz}(\triangleq w)$ und $\vartheta_{yx}(\triangleq \chi_\varphi)$ am betrachteten ($\bar{\varphi} = 0$) und gegenüberliegenden Rand ($\bar{\varphi} = \pi/\omega$) infolge der Unbekannten X_1 bis X_4 nach Tabelle 8.

Hierzu werden folgende Größen benötigt:

With the notations and positive directions (see also Fig. 16), given in the table volume [22], the four deformations of interest, v_{xx} ($\triangleq u$), v_{yy} ($\triangleq v$), v_{zz} ($\triangleq w$) and ϑ_{yx} ($\triangleq \chi_\varphi$) at the considered ($\bar{\varphi} = 0$) and the opposite boundary ($\bar{\varphi} = \pi/\omega$), due to the unknowns X_1 to X_4, are obtained according to Table 8.

Herein, the following quantities are required:

Schalenparameter ε: Mit / *Shell parameter* ε*:* With

$$k = 2 \cdot 10^{-4}; \qquad \lambda = \frac{n\pi a}{l} = \frac{\pi\, 7{,}11}{15{,}86} = 1{,}41; \qquad \nu = 1/6$$

folgt: / is obtained:

$$\varepsilon = \lambda \sqrt[4]{\frac{4k}{1-\nu^2}} = 1{,}41 \sqrt[4]{\frac{4 \cdot 2 \cdot 10^{-4}}{1 - 1/36}} = 0{,}239 \approx 0{,}24$$

Funktionswerte: Mit / *Function values:* With

$$\omega = \frac{0{,}4899}{1{,}41} = 0{,}348 \quad \text{und/and} \quad \bar{\varphi} = \frac{\pi}{\omega} = \frac{\pi}{0{,}348} = 9{,}03$$

sowie / as well as

$$\varkappa_1 = 1{,}190; \quad \mu_1 = 0{,}4201; \quad \varkappa_2 = 0{,}4980 \quad \text{und/and} \quad \mu_2 = 1{,}004$$

folgt / follows:

$$e^{-\varkappa_1\bar{\varphi}} \sin \mu_1 \bar{\varphi} = -0{,}13 \cdot 10^{-4}; \qquad e^{-\varkappa_1\bar{\varphi}} \cos \mu_1 \bar{\varphi} = -0{,}17 \cdot 10^{-4}$$

$$e^{-\varkappa_2\bar{\varphi}} \sin \mu_2 \bar{\varphi} = +39 \cdot 10^{-4}; \qquad e^{-\varkappa_2\bar{\varphi}} \cos \mu_2 \bar{\varphi} = -104 \cdot 10^{-4}$$

Die Summation der vier Einzelanteile nach Tafel 1 aus [22] führt zur jeweils zweiten Zeile bei jeder Verformungsgröße. Die Werte bei gleichzeitiger Wirkung der entsprechenden Unbekannten $\pm X$ am gegenüberliegenden Rand (d.h. Addition bzw. Subtraktion der ersten und zweiten Zeilen) sind gleicherweise in den dritten und die wirklichen (d.h. mit ψ_w multiplizierten) E-fachen Verformungen in den vierten Zeilen angegeben. Die Verformungen v_{xx} und v_{zz} infolge $X_1 \triangleq N_{yy}$ und $X_4 \triangleq M_{yx}$ an der oberen Rohrhälfte sind gleich denen an der unteren, so daß die Klaffungen von vornherein verschwinden. Das gleiche gilt für v_{yy} und ϑ_{yx} infolge $X_2 \triangleq Q_{yz}$ und $X_3 \triangleq N_{yx}$. Die erstgenannten Werte werden jedoch für die freigewordenen Schnittgrößen n_φ^0 und m_φ^0 benötigt (vgl. Abschnitt 7.3.1).

The summation of the four individual contributions, according to Table 1 in [22], results in the second line of each deformation quantity. The values for simultaneous action of the corresponding unknowns $\pm X$ at the opposite boundary (i.e. addition or subtraction, respectively, of the first and second lines) are shown, similarily, in the third, and the real (i.e., multiplied by ψ_w) E-fold deformations in the fourth lines. The deformations v_{xx} and v_{zz}, due to $X_1 \triangleq N_{yy}$ and $X_4 \triangleq M_{yx}$ at the upper pipe half are equal to those at the lower, thus, the gaps vanish from the outset. The same applies for v_{yy} and ϑ_{yx} due to $X_2 \triangleq Q_{yz}$ and $X_3 \triangleq N_{yx}$. The quantities, mentioned first however, are needed for the freed stress resultants n_φ^0 and m_φ^0 (see Section 7.3.1).

Tabelle 8. Randwerte der Schale / Table 8. Boundary values of the shell

W	ψ_w	$\bar{\varphi}$	$X_1 \triangleq N_{yy} = 1$	$X_2 \triangleq Q_{yz} = 1$	$X_3 \triangleq N_{yx} = 1$	$X_4 \triangleq M_{yx} = 1$
$E\,v_{xx}$	$4{,}167\,\dfrac{7{,}11}{1{,}41 \cdot 0{,}35}$	$\bar{\varphi} = 0$ $\bar{\varphi} = \pi/\omega$	$+\ 3{,}358$ $-\ 0{,}012$	$-\ 1{,}204$ $-\ 0{,}001$	$-\ 2{,}654$ $+\ 0{,}009$	$-\ 0{,}377$ $-\ 0{,}003$
$\triangleq E\,u$	$= 60{,}04$	$\sum =$ $\psi_w \sum =$	$+\ 3{,}346$ $+\ 200{,}1$	$-\ 1{,}205$ $-\ 72{,}3$	$-\ 2{,}645$ $-\ 158{,}8$	$-\ 0{,}380$ $-\ 22{,}8$
$E\,v_{yy}$	$8{,}505\,\dfrac{7{,}11}{1{,}41 \cdot 0{,}35}$	$\bar{\varphi} = 0$ $\bar{\varphi} = \pi/\omega$	$-\ 6{,}962$ $(\mp)\ 0{,}034$			$+\ 1{,}312$ $(\pm)\ 0{,}003$
$\triangleq E\,v$	$= 122{,}5$	$\sum =$ $\psi_w \sum =$	$-\ 6{,}996$ $-\ 857{,}0$			$+\ 1{,}315$ $+\ 161{,}1$
$E\,v_{zz}$	$35{,}71\,\dfrac{7{,}11}{0{,}35}$	$\bar{\varphi} = 0$ $\bar{\varphi} = \pi/\omega$	$+\ 3{,}155$ $-\ 0{,}003$	$-\ 1{,}896$ $+\ 0{,}006$	$-\ 1{,}204$ $-\ 0{,}003$	$-\ 1{,}139$ $+\ 0{,}003$
$\triangleq E\,w$	$= 725{,}4$	$\sum =$ $\psi_w \sum =$	$+\ 3{,}152$ $+\ 2286$	$-\ 1{,}890$ $-\ 1371$	$-\ 1{,}207$ $-\ 876$	$-\ 1{,}136$ $-\ 824$
$E\,\vartheta_{yx}$	$70{,}88\,\dfrac{1{,}41}{0{,}35}$	$\bar{\varphi} = 0$ $\bar{\varphi} = \pi/\omega$	$-\ 1{,}312$ $(\pm)\ 0{,}019$			$+\ 1{,}277$ $(\pm)\ 0{,}001$
$\triangleq E\,\chi_\varphi$	$= 285{,}5$	$\sum =$ $\psi_w \sum =$	$-\ 1{,}293$ $-\ 369{,}2$			$+\ 1{,}278$ $+\ 364{,}9$

Die Lastverformungen (u_{part}, v_{part} etc.) setzen sich entsprechend Abschnitt 7.3.1 aus zwei Anteilen zusammen. Die Verformungen der oberen belasteten Schalenhälfte sind nach Abschnitt 7.3.2 mit $K = E\,t^3/12(1-\nu^2)$:

The deformations (u_{part}, v_{part} etc.), due to the load, are composed, according to Section 7.3.1, of two parts. The deformations of the loaded upper shell half are, according to Section 7.3.2, with $K = E\,t^3/12\,(1-\nu^2)$:

$$E\,u^0 = \frac{E\,a^2}{1000\,K\,\pi}\cdot 0{,}15 = \frac{7{,}11^2 \cdot 12\,(1 - 1/36)}{1000 \cdot 0{,}35^3\,\pi}\cdot 0{,}15 = +\,0{,}66 \qquad \text{und/and} \qquad E\,w^0 = \frac{E\,a^2}{1000\,K\,\pi}(-\,1{,}1) = -\,4{,}8$$

Die Verformungen der unteren durch n_φ^0 und m_φ^0 belasteten Schalenhälfte werden mit

The deformations of the lower shell half, subjected to n_φ^0 and m_φ^0, become with

$$n_\varphi^0 = \frac{1}{\pi\,a}\cdot 0{,}58 = \frac{1}{\pi\,7{,}11}\cdot 0{,}58 = +\,0{,}0260 \qquad \text{und/and} \qquad m_\varphi^0 = \frac{1}{100\,\pi}\cdot 3{,}9 = +\,0{,}0124$$

bei Benutzung der vierten und letzten Spalte der Tabelle 8 sowie

and by using the fourth and the last column of Table 8, as well as,

$$\psi_{Nyy} = 1{,}0 \qquad \text{und/and} \qquad \psi_{Myx} = 0{,}12\,\frac{7{,}11}{1{,}41^2} = 0{,}429:$$

$$E\,u^u = E\,v_{xx}\,(N_{yy} = 1)\,\frac{n_\varphi^0}{\psi_{Nyy}} + E\,v_{xx}\,(M_{yx} = 1)\,\frac{m_\varphi^0}{\psi_{Myx}} = +\,200{,}1\,\frac{0{,}0260}{1{,}0} - 22{,}8\,\frac{0{,}0124}{0{,}429} = +\,4{,}54$$

$$E\,v^u = -\,857{,}0\,\frac{0{,}0260}{1{,}0} + 161{,}1\,\frac{0{,}0124}{0{,}429} = -\,17{,}6; \qquad E\,w^u = +\,2286\,\frac{0{,}0260}{1{,}0} - 824\,\frac{0{,}0124}{0{,}429} = +\,35{,}6; \qquad E\,\chi_\varphi = -\,369{,}2\,\frac{0{,}0260}{1{,}0} + 364{,}9\,\frac{0{,}0124}{0{,}429} = +\,0{,}95$$

Die Klaffungen sind bei den symmetrischen Verschiebungen u_{part} und w_{part}, wobei $w = -v_{zz}$ zu berücksichtigen ist, die Werte des oberen abzüglich des unteren Schalenrandes und bei den antimetrischen Verformungen die Summe aus v_{part} und $\chi_{\varphi part}$; bei $\bar{\varphi} = 0$ wird jedoch $v^0 = \chi_\varphi^0 = 0$. Somit folgt schließlich:

For the symmetric displacements u_{part} and w_{part}, the gaps are, considering herein $w = -v_{zz}$, the values of the upper minus those of the lower shell boundary, and for the anti-symmetric deformations, the sum of v_{part} and $\chi_{\varphi\, part}$; however, it becomes $v^0 = \chi_\varphi^0 = 0$ for $\bar{\varphi} = 0$. Thus, it is finally:

$$E\,u_{part} = +0{,}66 - 4{,}54 = -3{,}88 \qquad E\,v_{part} = -17{,}6$$

$$E\,w_{part} = +4{,}8 - 35{,}6 = -30{,}8 \qquad E\,\chi_{\varphi part} = +0{,}95$$

Die Addition der Verformungen des oberen und unteren Schalenrandes infolge der Unbekannten X_1 bis X_4 liefert wegen der Symmetrie die verdoppelten Werte der Tabelle 8 (bzw. Null) und schließlich das Gleichungssystem:

The summation of the deformations of the upper and lower shell boundaries, due to the unknowns X_1 to X_4, produces, because of the symmetry, the values of Table 8 doubled (or zero, respectively) and finally the set of equations:

$$\begin{aligned}
-3{,}88 + 0 - 144{,}6\,X_2 - 317{,}6\,X_3 + 0 &= 0\\
-17{,}6 - 1714\,X_1 + 0 + 0 + 322{,}2\,X_4 &= 0\\
-30{,}8 + 0 - 2742\,X_2 - 1752\,X_3 + 0 &= 0\\
+0{,}95 - 738{,}4\,X_1 + 0 + 0 + 729{,}8\,X_4 &= 0
\end{aligned}$$

Mit den Unbekannten

With the unknowns

$$X_1 = -0{,}0130 \qquad X_2 = -0{,}0048 \qquad X_3 = -0{,}0100 \qquad X_4 = -0{,}0144$$

lassen sich die vier Anteile des Längsrandeinflusses mit den Tafeln 3 aus [22] – bei $\varepsilon = 0{,}24$ – zumindest bis $\bar{\varphi} = 7{,}5$ schnell ermitteln. Für die untere Rohrhälfte liefern die Unbekannten X_2 und X_3, die antimetrischen Schnittgrößen entsprechen, bei symmetrischen Zustandsgrößen negative Werte. Zu diesen Ordinaten werden noch die Anteile aus n_φ^0 und m_φ^0 addiert, die sich mit den gleichen Tafeln 3 ermitteln lassen.

Hier soll nur für n_φ und m_φ der Schnitt $x = 0$ berechnet werden (vgl. Tabelle 9). Wegen der Zustandsflächen im Anhang werden die Werte $\pi a\, n_\varphi$ bzw. $100\pi\, m_\varphi$ er-

the four contributions due to the influence of the longitudinal boundary can be determined quickly with Table 3 of [22]—for $\varepsilon = 0{,}24$—to, at least, $\bar{\varphi} = 7{,}5$. The unknowns X_2 and X_3, which correspond to anti-symmetric stress resultants, produce, for the lower pipe half, negative values for symmetrical condition quantities. To these ordinates, the contributions from n_φ^0 and m_φ^0 are added, which can be obtained with the same Tables 3.

Only section $x = 0$ shall be analyzed, herein, for n_φ and m_φ (see Table 9). Because of the condition surfaces in the Appendix, the values $\pi\, a\, n_\varphi$ and $100\,\pi\, m_\varphi$

Tabelle 9. Längsrandeinfluß für n_φ und m_φ.

Table 9. Influence of the longitudinal boundary on n_φ and m_φ

		für $N_{yy} \triangleq n_\varphi$ ($\psi_{Nyy} = 1$)									
$\pm\bar{\varphi}$	$\triangleq \pm\bar{\varphi}^0$	X_1	X_2	X_3	X_4	Σ	$\pi a n_\varphi^{oben}$	$\pi a n_\varphi^{unten}$	n_φ^0	m_φ^0	$\pi a \Delta n_\varphi^{unten}$
0	0°	−0,290	0	0	0	−0,290	−0,288	−0,282	+0,580	0	+0,573
1,0	20°	−0,056	−0,025	+0,040	−0,016	−0,057	−0,056	−0,080	+0,113	+0,032	+0,139
2,0	40°	+0,121	−0,029	−0,018	−0,007	+0,067	+0,062	+0,150	−0,242	+0,014	−0,213
3,0	60°	+0,094	−0,009	−0,031	+0,011	+0,065	+0,050	+0,108	−0,187	−0,022	−0,156
4,0	80°	+0,008	+0,006	−0,012	+0,015	+0,017	+0,004	−0,010	−0,016	−0,029	+0,006
4,5	90°	−0,020	+0,008	−0,002	+0,012	−0,002	−0,004	−0,028	+0,040	−0,023	+0,034
5,0	100°	−0,033	+0,008	+0,005	+0,007	−0,013	+0,004	−0,010	+0,066	−0,015	+0,006
6,0	120°	−0,026	+0,003	+0,008	−0	−0,015	+0,050	+0,108	+0,052	+0,001	−0,156
7,0	140°	−0,005	−0,001	+0,004	−0,003	−0,005	+0,062	+0,150	+0,010	+0,005	−0,213
8,0	160°	+0,005	−0,002	−0,001	−0,001	+0,001	−0,056	−0,080	−0,009	+0,003	+0,139
9,0	180°	+0,004	−0,001	−0,002	+0,001	+0,002	−0,288	−0,282	−0,005	−0,002	+0,573
1	2	3	4	5	6	7	8	9	10	11	12

		für $M_{yx} \triangleq m_\varphi$ ($\psi_{Myx} = 0{,}429$)									
$\pm\bar{\varphi}$	$\triangleq \pm\bar{\varphi}^0$	X_1	X_2	X_3	X_4	Σ	$100\pi m_\varphi^{oben}$	$100\pi m_\varphi^{unten}$	n_φ^0	m_φ^0	$100\pi \Delta m_\varphi^{unten}$
0	0°	0	0	0	−1,94	−1,94	−1,93	−1,91	0	+3,90	+3,87
1,0	20°	+0,16	−0,25	+0,08	−1,06	−1,07	−1,08	−0,76	−0,32	+2,13	+1,84
2,0	40°	+1,02	−0,31	−0,17	−0,51	+0,03	−0,03	+0,83	−2,05	+1,02	−0,81
3,0	60°	+1,03	−0,18	−0,29	−0,14	+0,42	+0,33	+1,13	−2,07	+0,27	−1,48
4,0	80°	+0,45	−0,03	−0,19	+0,04	+0,27	+0,28	+0,72	−0,90	−0,08	−1,01
4,5	90°	+0,16	+0,01	−0,11	+0,06	+0,12	+0,24	+0,64	−0,32	−0,13	−0,90
5,0	100°	−0,05	+0,04	−0,04	+0,06	+0,01	+0,28	+0,72	+0,09	−0,12	−1,01
6,0	120°	−0,18	+0,03	+0,04	+0,02	−0,09	+0,33	+1,13	+0,37	−0,05	−1,48
7,0	140°	−0,10	+0,01	+0,04	−0,01	−0,06	−0,03	+0,83	+0,20	+0,02	−0,81
8,0	160°	−0,01	−0	+0,01	−0,01	−0,01	−1,08	−0,76	+0,02	+0,01	+1,84
9,0	180°	+0,01	+0	−0,01	+0,01	+0,01	−1,93	−1,91	−0,02	−0,01	+3,87
1	2	13	14	15	16	17	18	19	20	21	22

mittelt und deshalb bereits X_1 bis X_4 bzw. n_φ^0 und m_φ^0 mit den Multiplikatoren – einschließlich ψ_w bzw. $1/\psi_w$ bei n_φ^0 und m_φ^0 – vervielfacht. Für die obere Schalenhälfte finden sich die Einzelanteile aus X_1 bis X_4 und deren Summation in den Spalten 3 bis 7 bzw. 13 bis 17. Der Einfluß des gegenüberliegenden Schalenrandes ist in Spalte 8 bzw. 18 berücksichtigt.

Die Addition der mit umgekehrten Vorzeichen bei X_2 und X_3 berechneten Ordinaten liefert die entsprechenden Werte der unteren Rohrhälfte. In den Spalten 9 und 19 ist der gegenüberliegende Randeinfluß bereits enthalten. Schließlich finden sich in den Spalten 12 und 22 die in ähnlicher Weise ermittelten Werte infolge der freigewordenen Schnittgrößen n_φ^0 und m_φ^0.

7.3.4 Endgültige Schnittgrößen

Im Bild 18 sind die Einzelanteile von Tabelle 9 zur Ermittlung von n_φ und m_φ aufgetragen. Zusätzlich benötigt man die Ordinaten aus den Zustandsflächen. Diese können durch graphische Extrapolation entsprechend Abschnitt 7.3.2 (vgl. Bild 17) aus den Flächen im Anhang gewonnen werden. Mit dem Lastumordnungsverfahren lassen sich noch die Randbedingung des Querrandes erfüllen. Die Summation der drei Anteile für $\bar{\varphi} > 0$ ($\varphi < 90°$) und der zwei Werte für $\bar{\varphi} < 0$ ($\varphi > 90°$) liefert die endgültigen Schnittgrößen, die zur Kontrolle bei $\bar{\varphi} = 0$ ($\varphi = \pm 90°$) eigentlich stetig verlaufen müssen.

are being determined, and therefore X_1 to X_4 or n_φ^0 and m_φ^0, respectively, are already multiplied by the multipliers – including ψ_w or $1/\psi_w$, respectively, for n_φ^0 and m_φ^0 –. For the upper shell half, the contributions from X_1 to X_4 and their summation are found in columns 3 to 7 and 13 to 17, respectively. The influence of the opposite shell boundary is considered in columns 8 and 18, respectively.

The summation of the ordinates, computed with the opposite sign for X_2 and X_3, produces the corresponding values of the lower shell half. The influence of the opposite boundary is already considered in columns 9 and 19. Finally, the values due to the freed stress resultants n_φ^0 and m_φ^0, determined similarily, are found in columns 12 and 22.

7.3.4 Final Section Stress Resultants

The individual contributions from Table 9 for the determination of n_φ and m_φ are plotted in Fig. 18. The ordinates from the condition surfaces are required, in addition. They can be obtained from the surfaces of the Appendix by graphical extrapolation according to Section 7.3.2 (see Fig. 17). The boundary conditions at the transverse boundary can be satisfied with the load rearrangement method. The summation of the three parts for $\bar{\varphi} > 0$ ($\varphi < 90°$) and the two values for $\bar{\varphi} < 0$ ($\varphi > 90°$) produce the final stress resultants which, as a check, should be continuous at $\bar{\varphi} = 0$ ($\varphi = \pm 90°$).

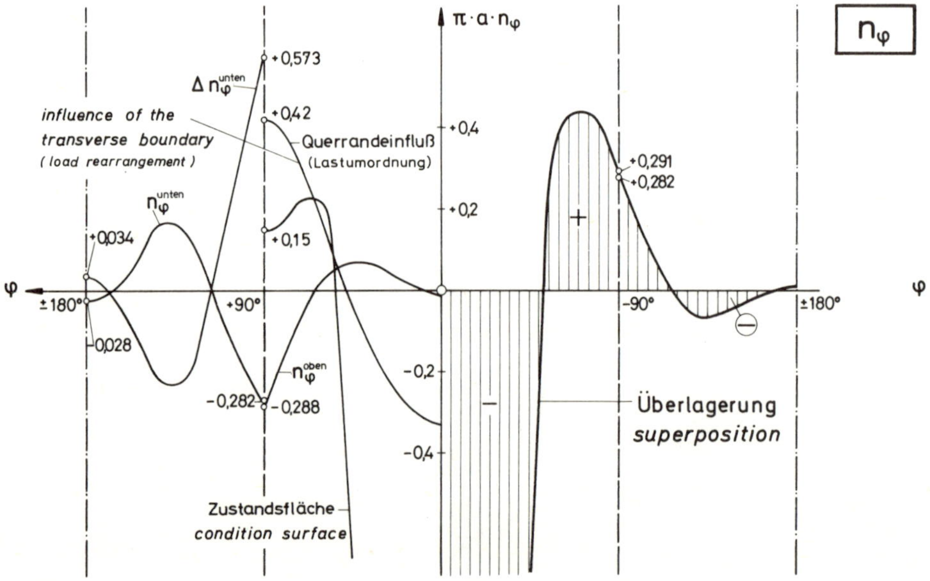

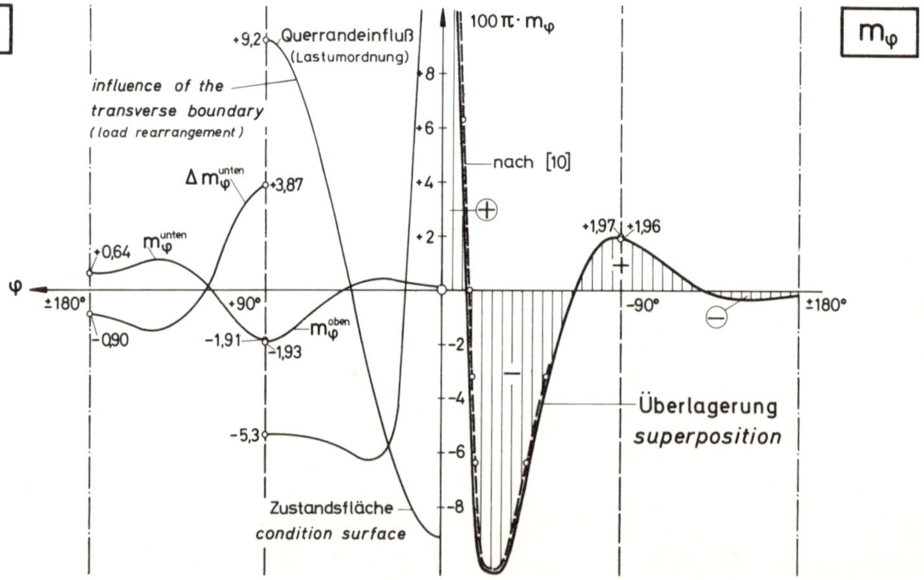

Bild 18. Endgültige Schnittgrößen n_φ und m_φ für $x = 0$.

Figure 18. Final stress resultants n_φ and m_φ at $x = 0$.

Man erkennt aus Bild 18, daß bei den Schnittgrößen der Längsrandeinfluß klein gegenüber dem Querrandeinfluß (Lastumordnung) und den Ordinaten der Zustandsflächen ist.

Für die Momente sind die Einflußflächen auch bei Holand [11] zu finden. Für das hier vorliegende Rohr wird der Parameter $l/\sqrt{r\,h}$ ungefähr 10. Die bis $\varphi \approx 60°$ aufgezeichneten Werte für m_φ sind zusätzlich gestrichelt in Bild 18 eingetragen.

One recognizes from Fig. 18 that, for the stress resultants, the influence of the longitudinal boundary is small when compared with that of the transverse boundary (load rearrangement) and with the ordinates of the condition surfaces.

The influence surfaces for the moments can be found, also, in Holand [11]. For the pipe, considered herein, the parameter $l/\sqrt{r \cdot h}$ becomes approximately 10. The values for m_φ, up to $\varphi \approx 60°$, are shown dashed in Fig. 18 in addition.

8. Literatur

8. References

[1] Aas-Jakobsen, A.: Einzellasten auf Kreiszylinderschalen. Bauingenieur 22 (1941) 343–346

[2] Lundgren, H.: Einzellasten auf Zylinderschalen mit Ringversteifungen. Bygningsstatiske Meddelelser 12 (1942) 1–24

[3] Kotulla, B., May, B., Nowak, B.: Traglastuntersuchungen zylindrischer Stahlbetonschalen unter Einzellasten. Konstruktiver Ingenieurbau, Ber. (Untersuchungen an ausgew. Tragwerken) 9 (1971) 5–30

[4] Peters, H. L.: Bemessung und Konstruktion der Einzelbauteile. Konstruktiver Ingenieurbau, Ber. (Naturzug-Kühltürme) 1 (1970) 76–83

[5] Wlassow, W. S.: Allgemeine Schalentheorie und ihre Anwendung in der Technik. Berlin: Akademie-Verlag 1958

[6] Bieger, K.-W.: Einflußflächen der Kreiszylinderschalen. Diss. T. U. Berlin 1959

[7] Koepcke, W.: Die Berechnung von Kreiszylinderschalen unter Flächen-, Linien- und Einzellasten. Habil.-Schrift T. U. Berlin 1949

[8] Flügge, W.: Statik und Dynamik der Schalen. 3. Aufl. Berlin: Springer 1962

[9] Yuan, S.W.: Thin cylindrical shells subjected to concentrated loads. Quart. Appl. Math. 4 (1946) 13–26

[10] Kemper, J., Sheng, J., v. Pohle, F. V.: Tables and curves for deformations and stresses in circular cylindrical shells under localized loadings. J. Aeronaut. Sciences 24 (1957) 119–129

[11] Holand, I.: Influence surfaces for bending moments in circular cylindrical shells or curved plates. Intern. Verein. f. Brücken- und Hochbau, Abh. 21 (1961) 111–128

[12] Holand, I.: Bending moments in circular cylindrical shells subjected to local radial loads. World Conf. on Shell Structures, San Francisco 1962, Nat. Academy of Science, Washington, D.C. 1964, 553–562

[13] Moe, J.: Über die Theorie zylindrischer Schalen. Intern. Verein. f. Brücken- und Hochbau, Abh. 13 (1953) 283–296

[14] Bijlaard, P. P.: Stresses from local loadings in cylindrical pressure vessels. Trans. ASME 77 (1955) 805–814.

[15] Hoff, N. J.: The accuracy of Donnell's equations. J. Appl. Mech. 22 (1955) 329–334

[16] Bijlaard, P. P.: Stresses from radial loads in cylindrical pressure vessels. Welding J. 33 (1954) 615s–623s

[17] Yuan, S.W., Ting, L.: On radial deflections of a cylinder subjected to equal and opposite concentrated radial loads. J. Appl. Mech. 24 (1957) 278–282

[18] Ting, L., Yuan, S.W.: On radial deflection of a cylinder of finite length with various end conditions. J. Aeronaut. Sciences 25 (1958) 230–234

[19] Morley, L. S. D.: An improvement on Donnell's approximation for thin-walled circular cylinders. Quart. J. Mech. Appl. Math. 12 (1959) 89–99

[20] Morley, L. S. D.: The thin-walled circular cylinder subjected to concentrated radial loads. Quart. J. Mech. Appl. Math. 13 (1960) 24–37

[21] Bieger, K.-W.: Die Kreiszylinderschale unter konzentrierten Belastungen. Ing.-Arch. 30 (1961) 57–62

[22] Rüdiger, D., Urban, J.: Kreiszylinderschalen – Circular Cylindrical Shells. Leipzig: Teubner 1955

[23] Dischinger, F.: Die strenge Theorie der Kreiszylinderschale in ihrer Anwendung auf die Zeiss-Dywidag-Schalen. Beton u. Eisen 34 (1935) 257–264, 283–294

[24] Aas-Jakobsen, A.: Die Berechnung der Zylinderschalen. Berlin: Springer 1958

[25] Hansen, H.: Vorgespannte kreiszylindrische Tonnendächer. Düsseldorf: Werner-Verlag 1974

[26] Holand, J.: Design of Circular Cylindrical Shells. Oslo: University Press 1957

[27] Dischinger, F.: Das durchlaufende ausgesteifte zylindrische Rohr oder Zeiss-Dywidag-Dach. Zürich: Intern. Verein. f. Brücken- und Hochbau, Abh. 4 (1936) 227–248

[28] ASCE-Manual No. 31: Design of Cylindrical Concrete Shell Roofs. New York: ASCE 1952

[29] Olsen, O.: Kontinnerlige Skall. Diss. N. T. H. Trondheim 1951

[30] Morice, P. B.: An approximate solution to the problem of longitudinally continuous shells. Mag. Concrete Res. 9 (1957) 95–104

[31] Huber, M.T.: Einige Anwendungen der Biegungstheorie orthotroper Platten. Z. f. angew. Math. Mech. 6 (1926) 228–231

[32] Soare, M.: Application of Finite Difference Equations to Shell Analysis. Oxford: Pergamon 1967

[33] Peters, H. L.: Zur Berechnung allgemeiner Flächentragwerke mit Hilfe der dynamischen Relaxation. Ing.-Arch. 42 (1972) 42–57
[34] Fette, H., May, B.: Finite Elemente für gekrümmte Flächentragwerke. Konstruktiver Ingenieurbau, Ber. 11 (1971) 1–80
[35] Beles, A. A., Soare, M. V.: Berechnung von Schalentragwerken. Wiesbaden: Bauverlag 1972
[36] Aas-Jakobsen, A.: Über das Randstörungsproblem an Kreiszylinderschalen. Bauingenieur 20 (1939) 394–405
[37] Lundgren, H.: Cylindrical Shells. 3. Aufl. Copenhagen: Danish Technical Press 1960
[38] Rabich, R.: Die Statik der Schalenträger. Bauplanung und Bautechnik 9 (1955) 115–125, 162–167
[39] Rabich, R.: Randwerttabellen für Zylinderschalen. Berlin: Verlag für Bauwesen 1960
[40] Tottenham, H.: A simplified method of design for cylindrical shell roofs. Struct. Eng. 32 (1954) 161–180
[41] Mehmel, A., Kruse, W., Samaan, S., Schwarz, H.: Schnittkrafttafeln für den Entwurf kreiszylindrischer Tonnendächer. Heft 216, Deutscher Ausschuß für Stahlbeton. Berlin: Ernst & Sohn 1971
[42] Girkmann, K.: Flächentragwerke. 5. Aufl. Wien: Springer 1959

9. Tafelübersicht

9. Chart Classification

Verformung bzw. Schnittgröße Deformation or section quantity	Verschiebung Displacement			Verdrehung Rotation	Normalkraft Normal force		Randschubkraft Boundary shearing force	Biegemoment Bending moment		Drillmoment Twisting moment	Randquerkraft Transverse boundary force		Wurzeln für $n = 2$ Roots for $n = 2$		Parameter $k = t^2/12\,a^2$
	u	v	w	χ_φ	n_x	n_φ	$\bar{n}_{x\varphi}$	m_x	m_φ	$m_{x\varphi}$	$\bar{q}_x$	$\bar{q}_\varphi$	$\varkappa_1$	μ_1	
Faktor – Factor	$a^2/1000 \cdot \pi \cdot K$			$a/100\,\pi\,K$	$1/\pi\,a$			$1/100\,\pi$			$1/10\,\pi\,a$				
$k = 1{,}0 \cdot 10^{-6}$	1	2	3	4	5	6	7	8	9	10	11	12	0,077761	0,078249	$1{,}0 \cdot 10^{-6}$
$k = 2{,}5 \cdot 10^{-6}$	13	14	15	16	17	18	19	20	21	22	23	24	0,097598	0,098568	$2{,}5 \cdot 10^{-6}$
$k = 5{,}0 \cdot 10^{-6}$	25	26	27	28	29	30	31	32	33	34	35	36	0,115819	0,117450	$5{,}0 \cdot 10^{-6}$
$k = 7{,}5 \cdot 10^{-6}$	37	38	39	40	41	42	43	44	45	46	47	48	0,127964	0,130174	$7{,}5 \cdot 10^{-6}$
$k = 1{,}0 \cdot 10^{-5}$	49	50	51	52	53	54	55	56	57	58	59	60	0,137315	0,140057	$1{,}0 \cdot 10^{-5}$
$k = 2{,}5 \cdot 10^{-5}$	61	62	63	64	65	66	67	68	69	70	71	72	0,171605	0,177050	$2{,}5 \cdot 10^{-5}$
$k = 5{,}0 \cdot 10^{-5}$	73	74	75	76	77	78	79	80	81	82	83	84	0,202619	0,211762	$5{,}0 \cdot 10^{-5}$
$k = 7{,}5 \cdot 10^{-5}$	85	86	87	88	89	90	91	92	93	94	95	96	0,222975	0,235348	$7{,}5 \cdot 10^{-5}$
$k = 1{,}0 \cdot 10^{-4}$	97	98	99	100	101	102	103	104	105	106	107	108	0,238445	0,253771	$1{,}0 \cdot 10^{-4}$

$1000\,\pi\,K/a^2 \cdot u$	**1**
$k = 1{,}0 \cdot 10^{-6}$	

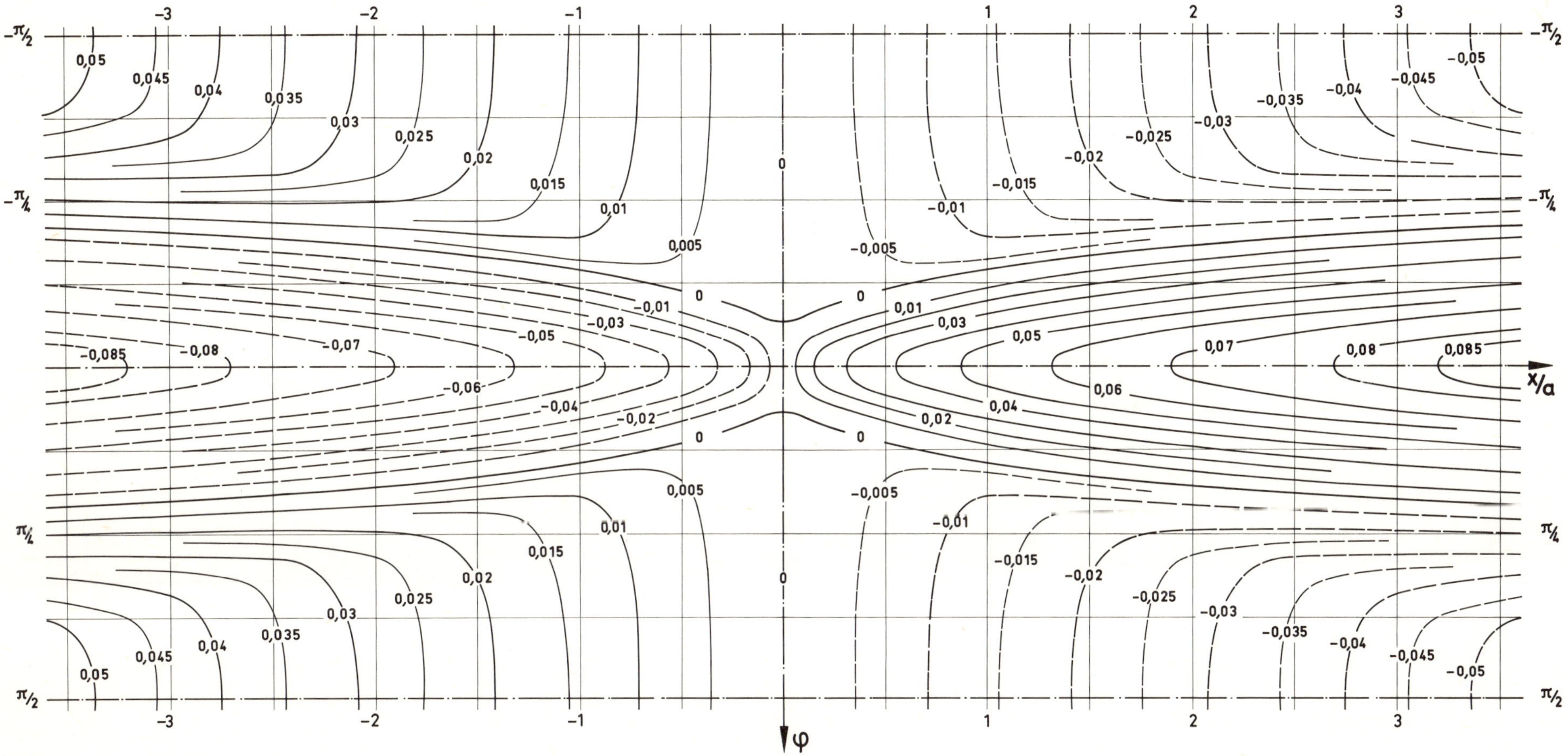

Einflußfläche (negative Zustandsfläche) der Verschiebung u

Influence surface (negative condition surface) for the displacement u

2 | $1000\,\pi\,K/a^2 \cdot v$ — $k = 1{,}0 \cdot 10^{-6}$

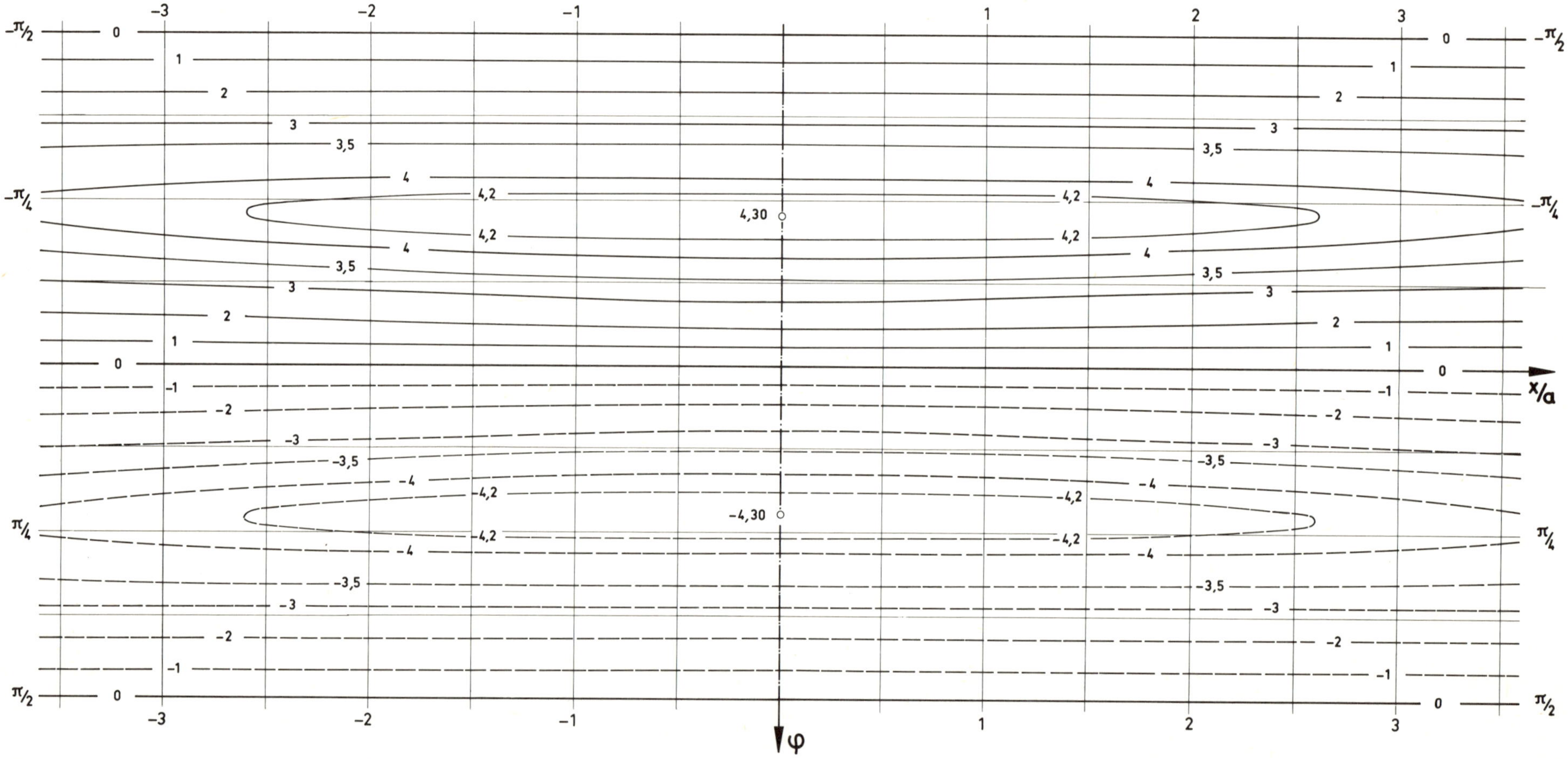

Einflußfläche (negative Zustandsfläche) der Verschiebung v

Influence surface (negative condition surface) for the displacement v

$1000\,\pi\,K/a^2 \cdot w$	**3**
$k = 1{,}0 \cdot 10^{-6}$	

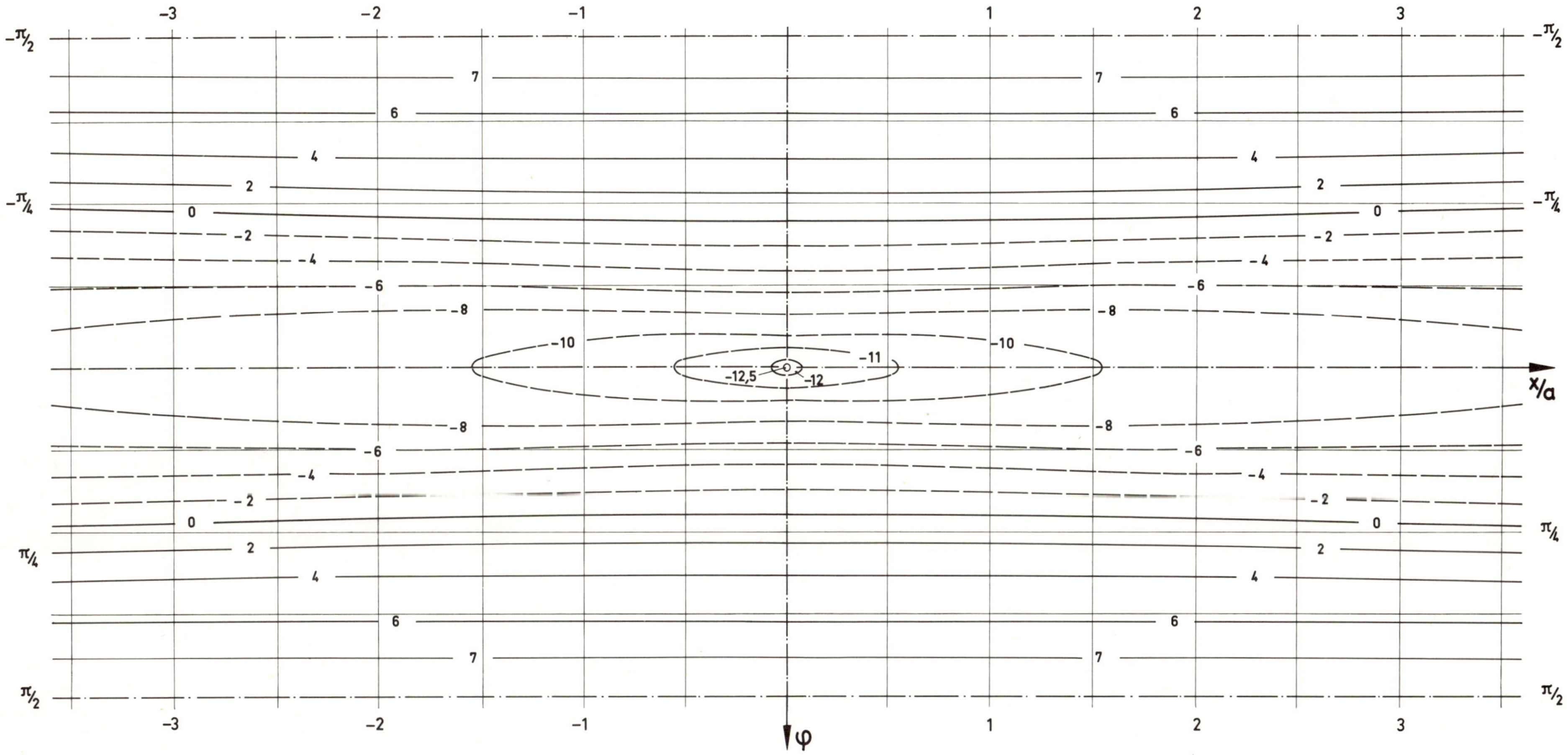

Einflußfläche (Zustandsfläche) der Verschiebung *w*

Influence surface (condition surface) for the displacement *w*

4	$100\,\pi\,K/a \cdot \chi_{\varphi}$
	$k = 1{,}0 \cdot 10^{-6}$

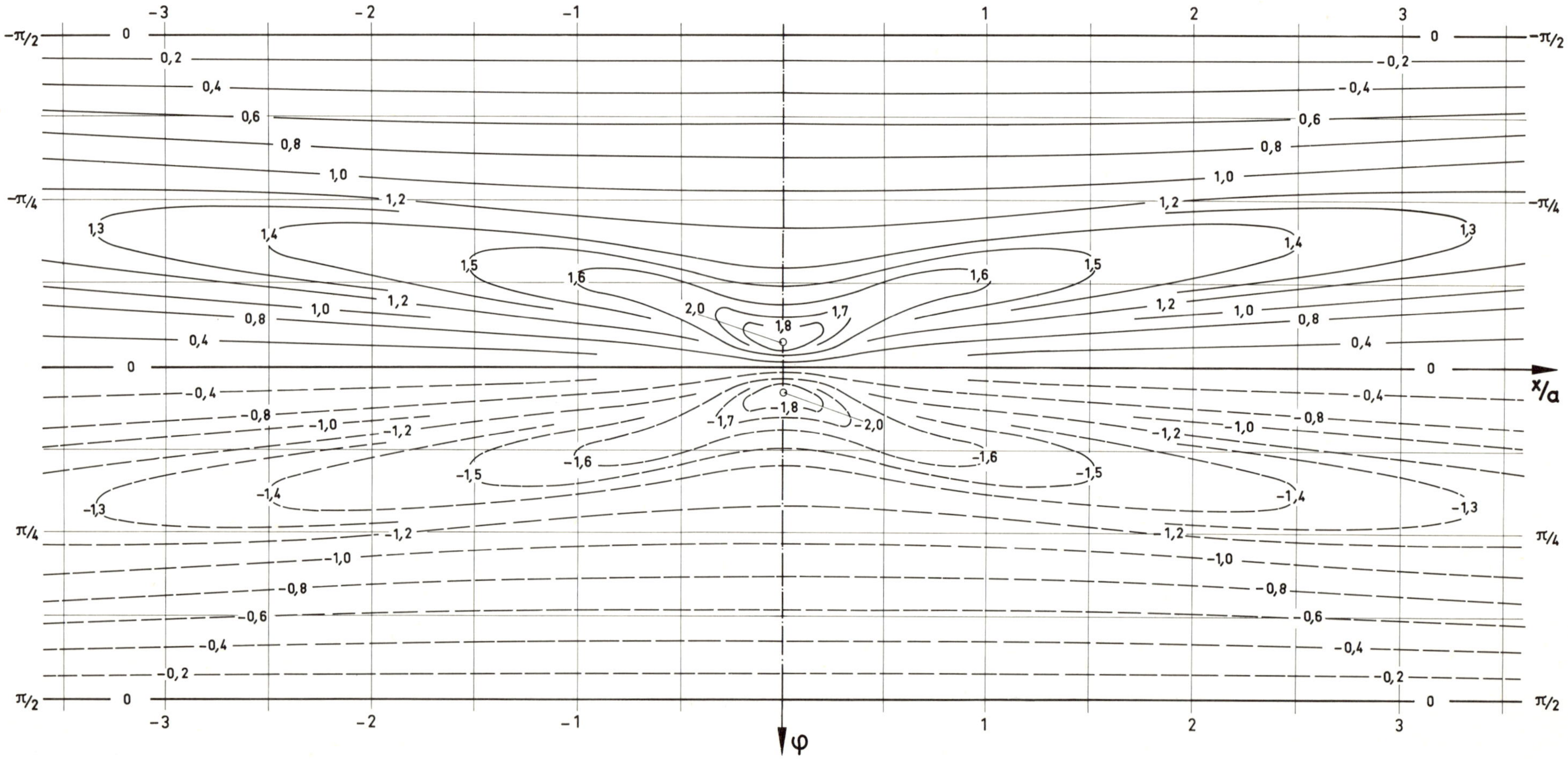

Einflußfläche (negative Zustandsfläche) der Verdrehung χ_{φ}

Influence surface (negative condition surface) for the rotation χ_{φ}

$\pi a \cdot n_x$	**5**
$k = 1{,}0 \cdot 10^{-6}$	

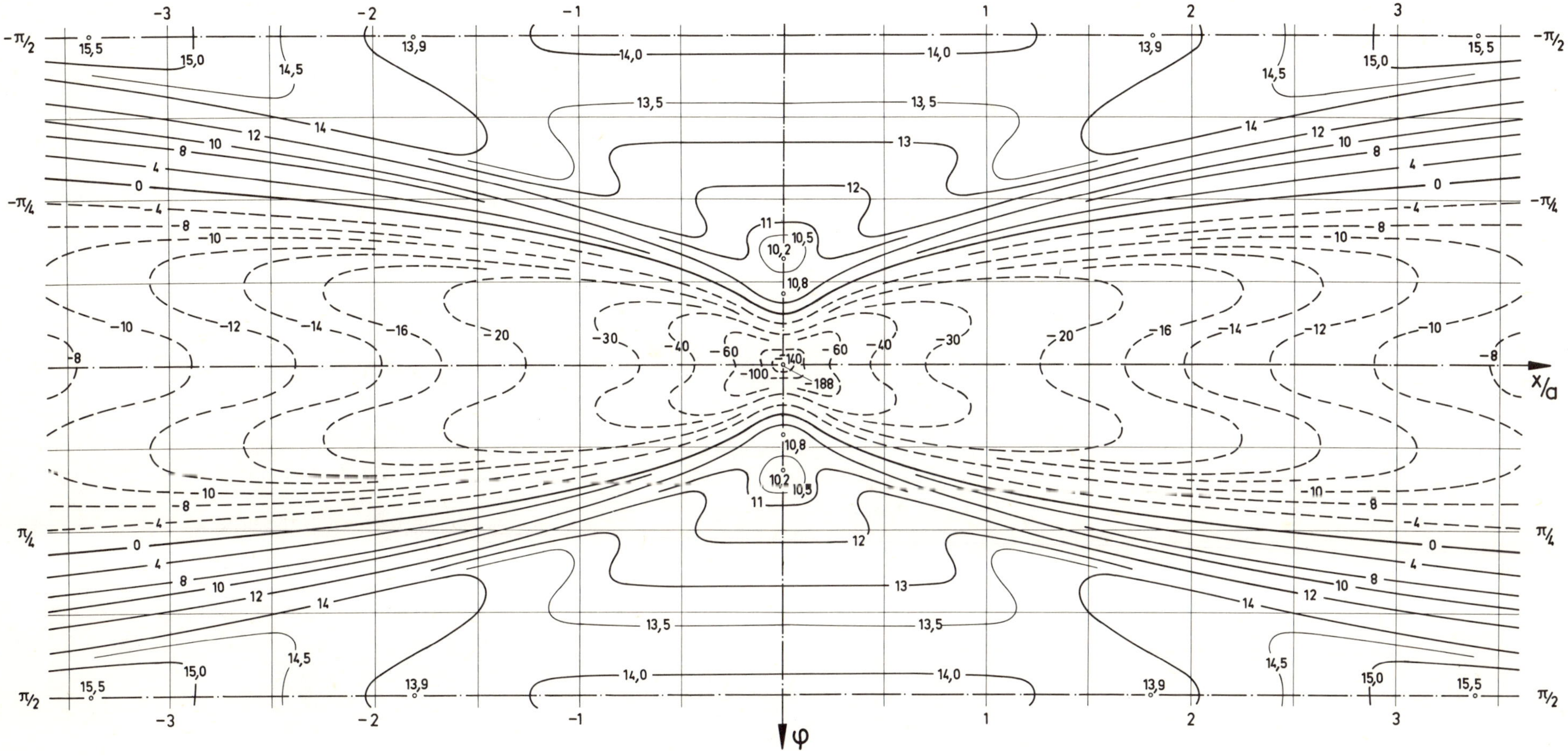

Einflußfläche (Zustandsfläche) der Normalkraft n_x

Influence surface (condition surface) for the normal force n_x

6	$\pi a \cdot n_\varphi$
	$k = 1{,}0 \cdot 10^{-6}$

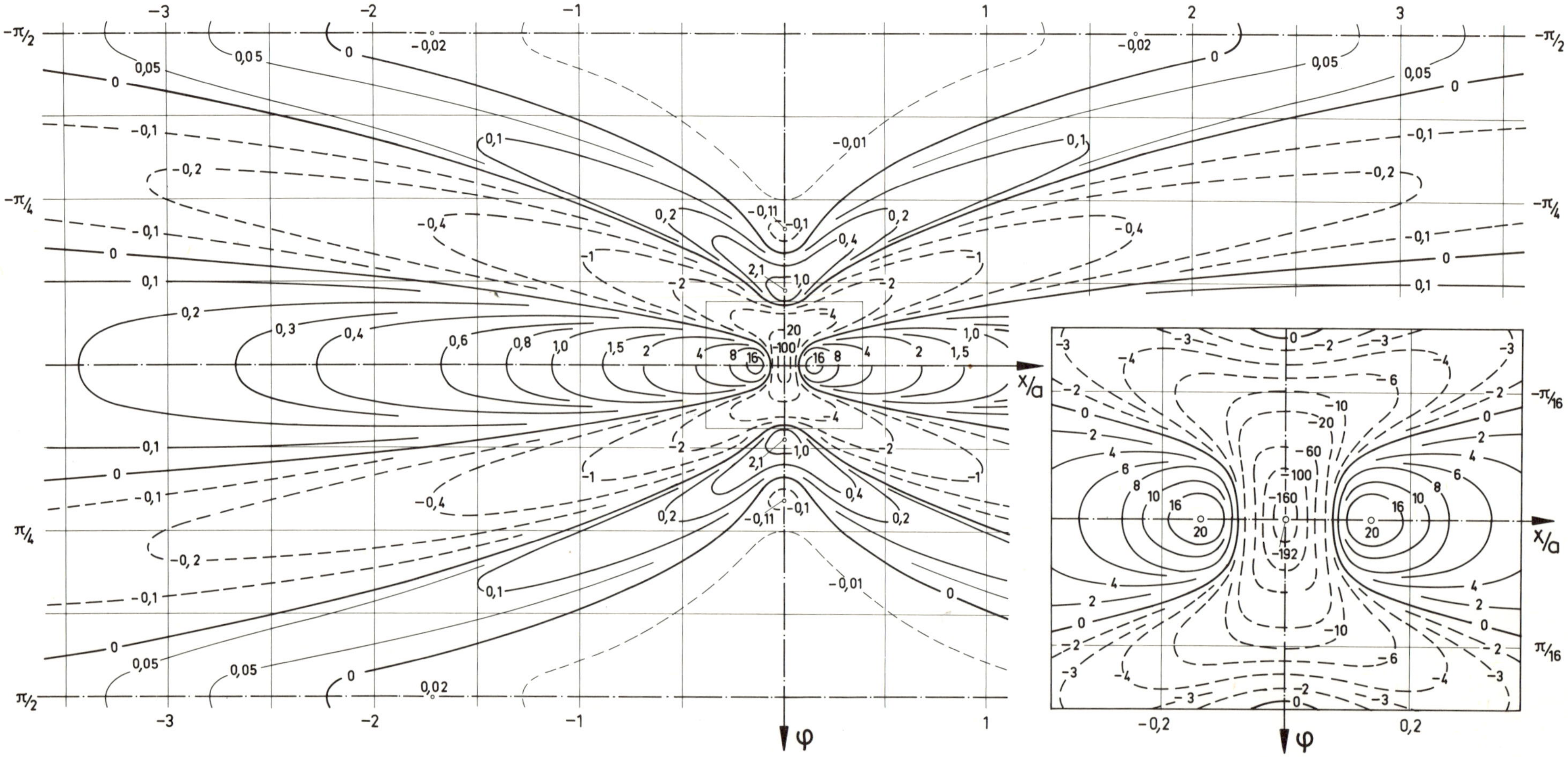

Einflußfläche (Zustandsfläche) der Normalkraft n_φ

Influence surface (condition surface) for the normal force n_φ

$\pi a \cdot \bar{n}_{x\varphi}$	**7**
$k = 1{,}0 \cdot 10^{-6}$	

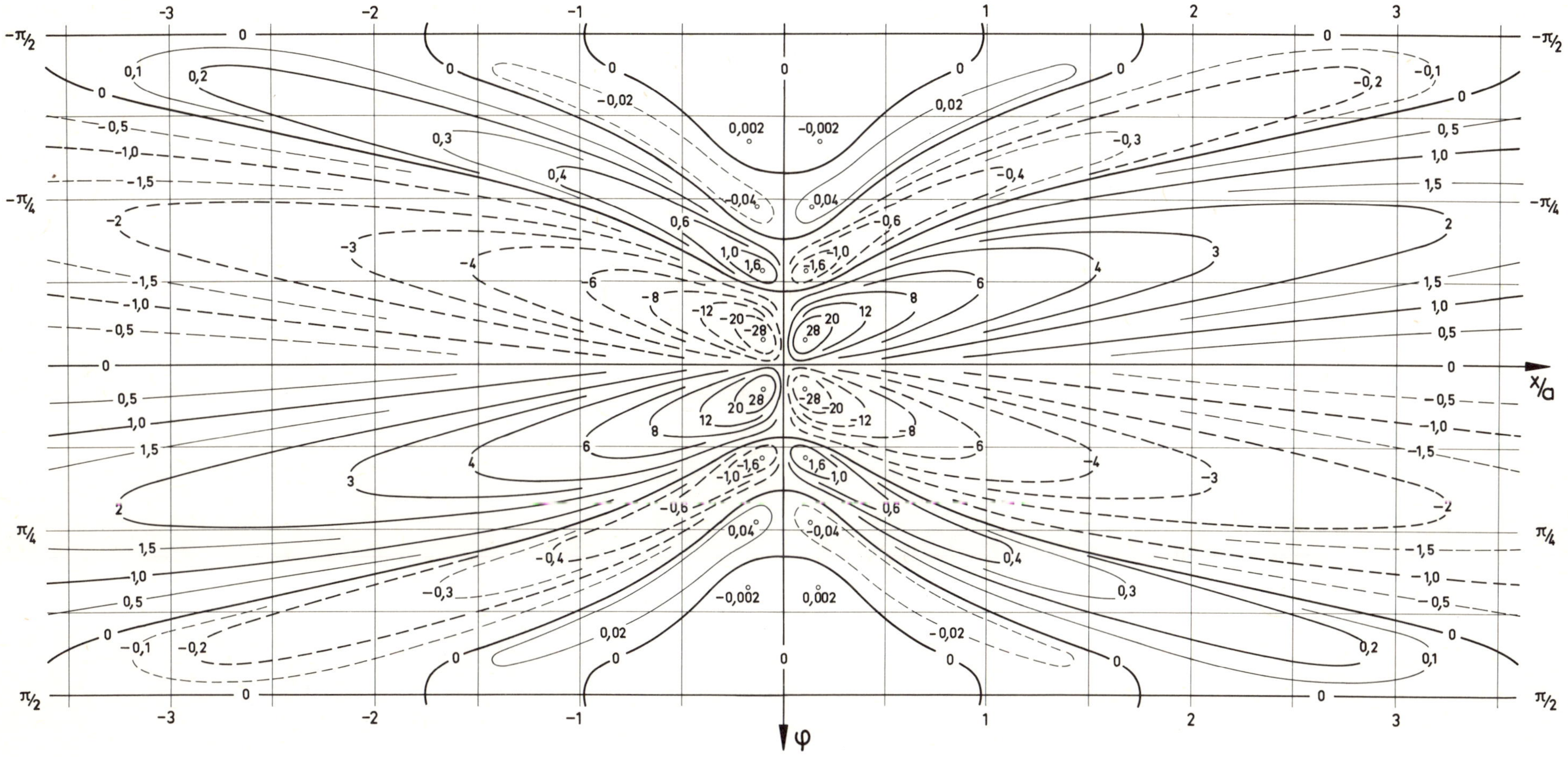

Einflußfläche (Zustandsfläche) der Randschubkraft $\bar{n}_{x\varphi}$

Influence surface (condition surface) for the boundary shearing force $\bar{n}_{x\varphi}$

8	$100\,\pi \cdot m_x$
	$k = 1{,}0 \cdot 10^{-6}$

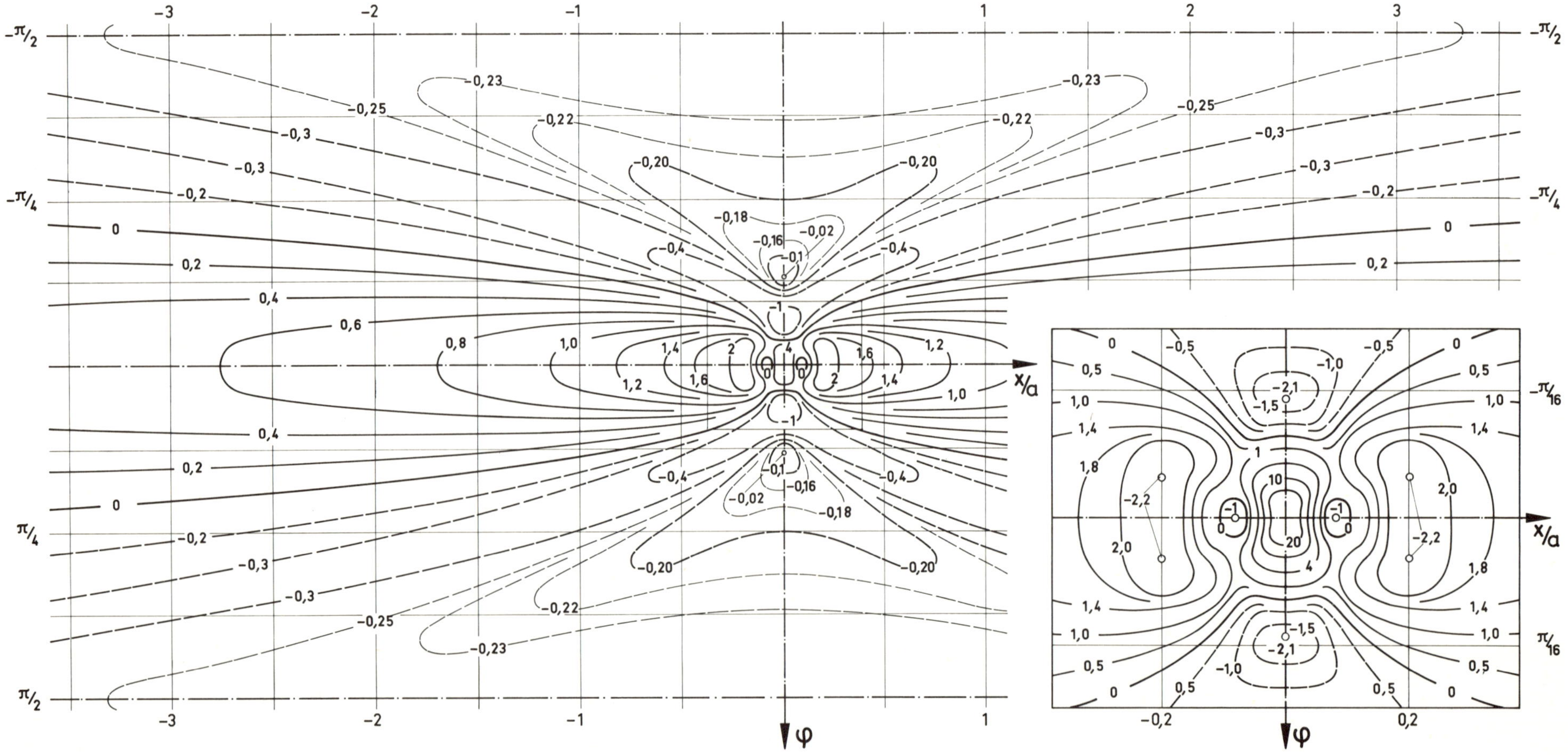

Einflußfläche (Zustandsfläche) des Biegemomentes m_x

Influence surface (condition surface) for the bending moment m_x

$100\,\pi \cdot m_\varphi$	**9**
$k = 1{,}0 \cdot 10^{-6}$	

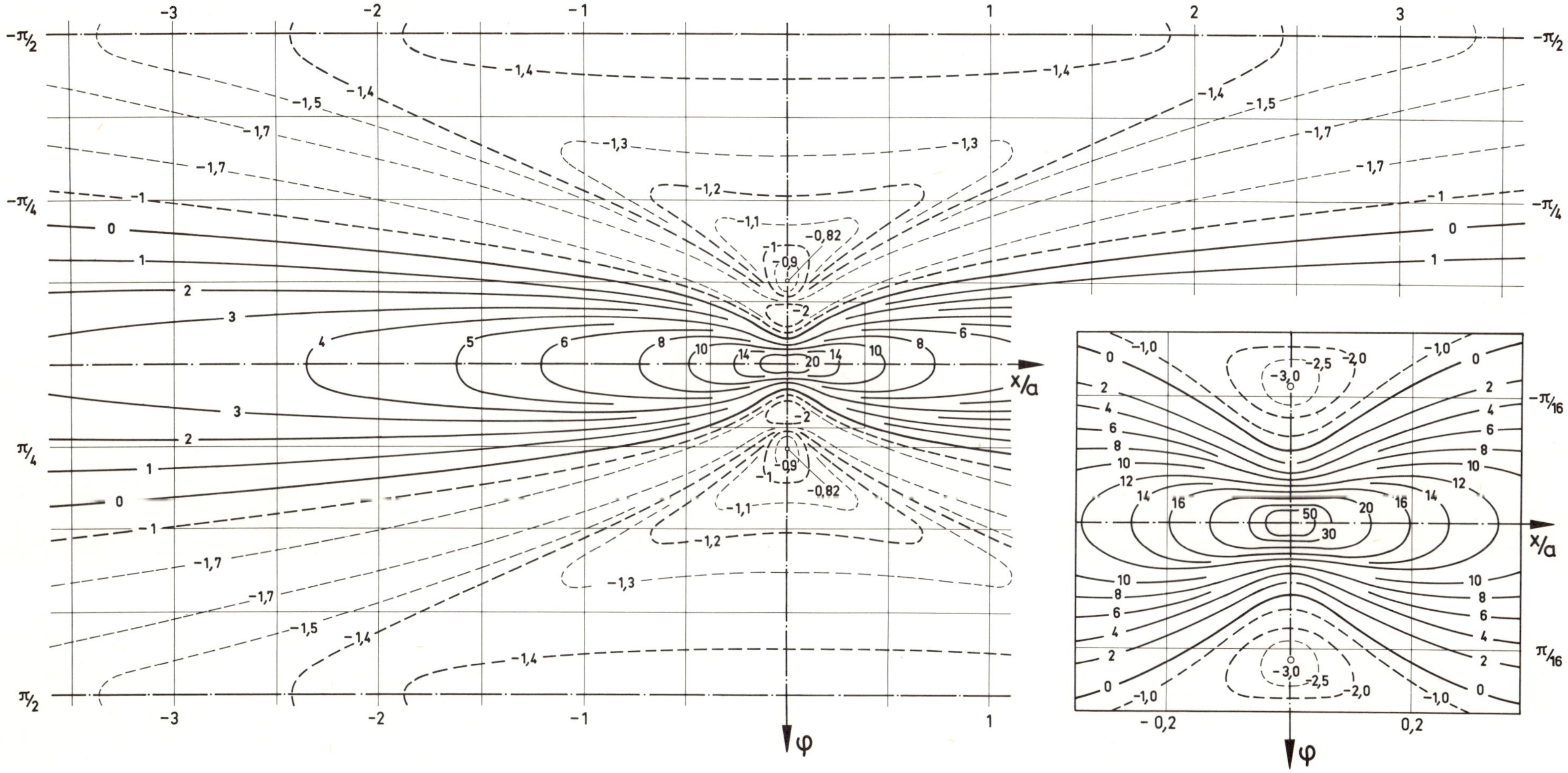

Einflußfläche (Zustandsfläche) des Biegemomentes m_φ

Influence surface (condition surface) for the bending moment m_φ

10	$100\,\pi \cdot m_{x\varphi}$
	$k = 1{,}0 \cdot 10^{-6}$

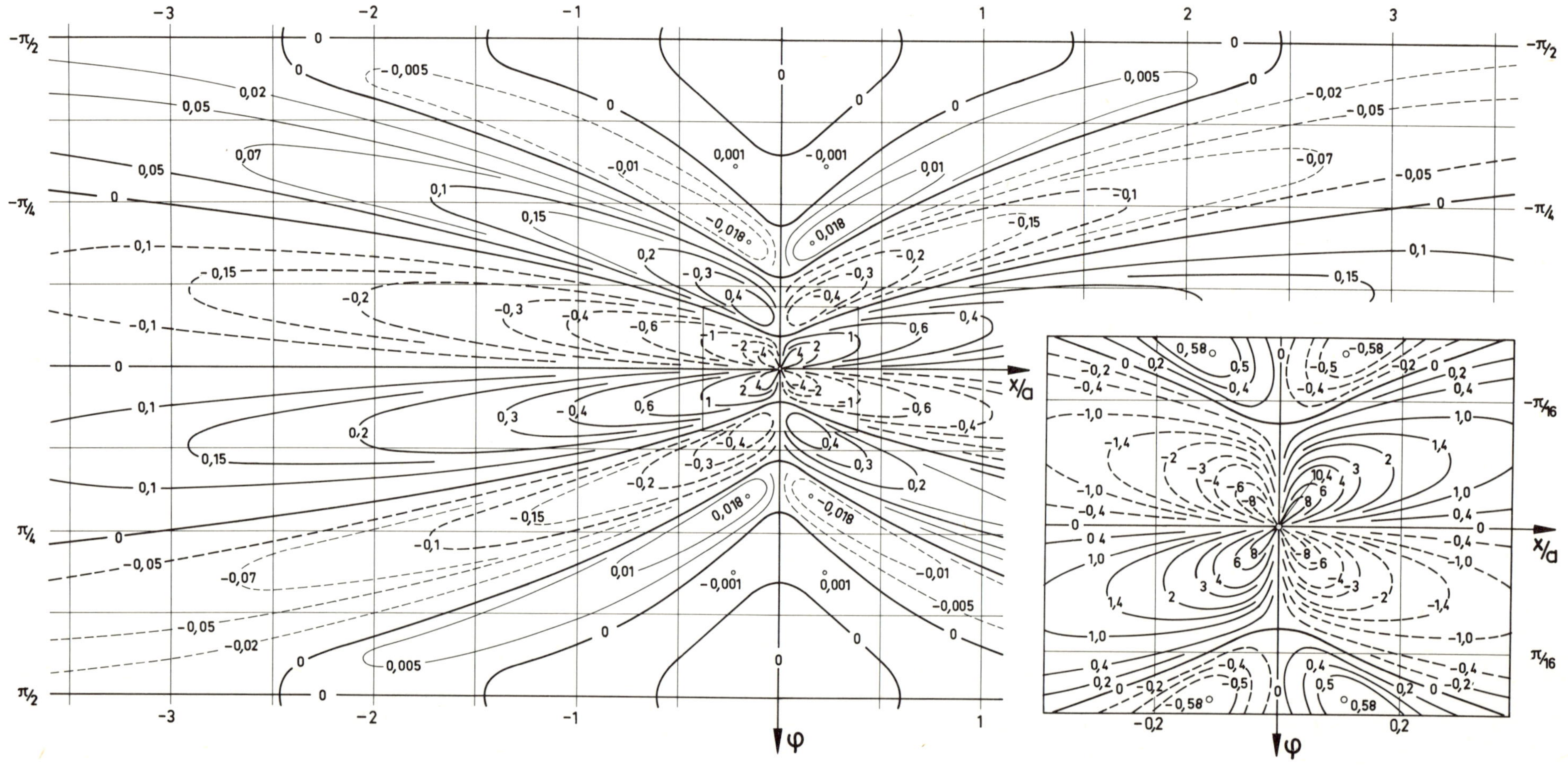

Einflußfläche (Zustandsfläche) des Drillmomentes $m_{x\varphi}$

Influence surface (condition surface) for the twisting moment $m_{x\varphi}$

$10\,\pi\,a \cdot \bar{q}_x$	**11**
$k = 1{,}0 \cdot 10^{-6}$	

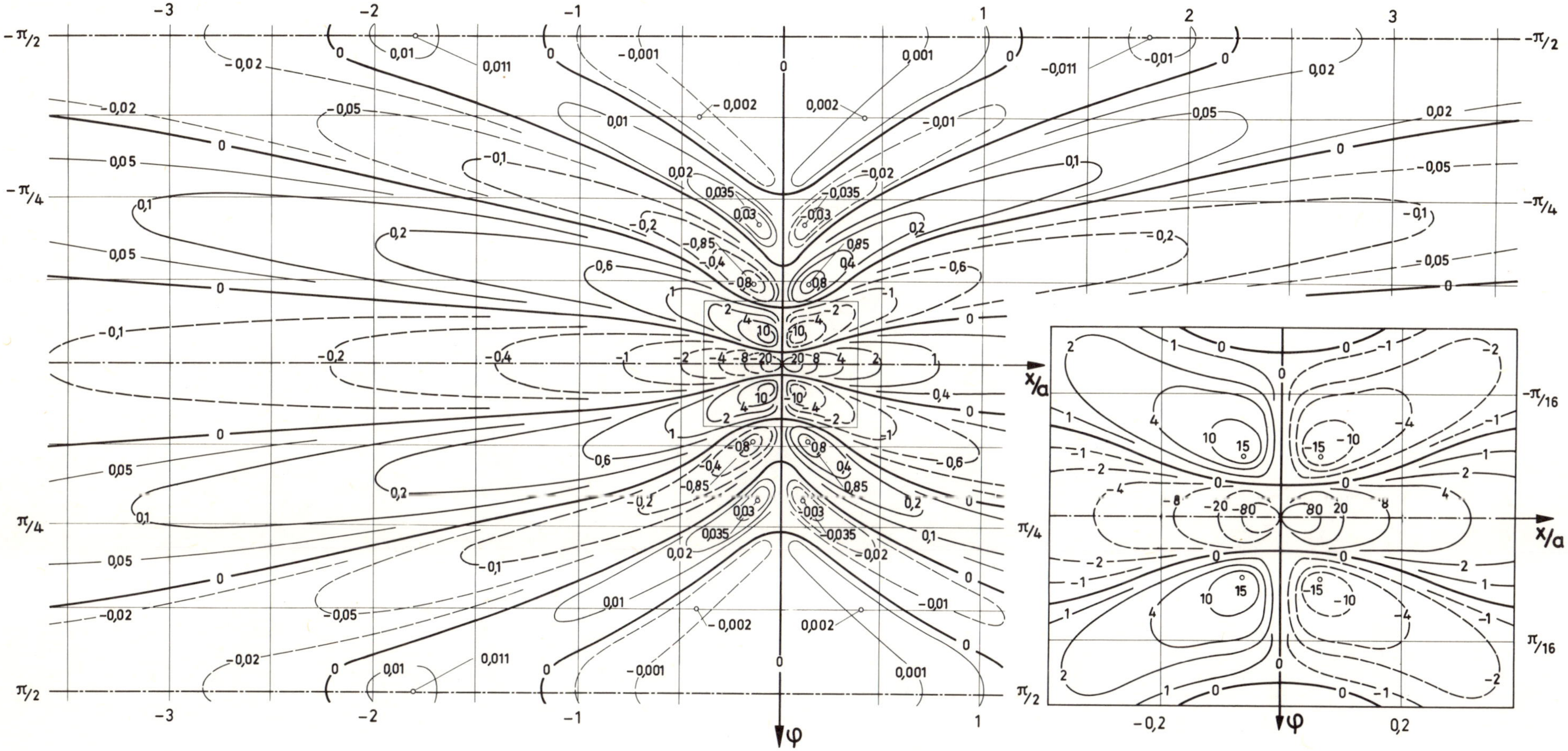

Einflußfläche (negative Zustandsfläche) der Randquerkraft $\bar{q}_x$

Influence surface (negative condition surface) for the transverse boundary force $\bar{q}_x$

12	$10\,\pi\, a \cdot \bar{q}_{\varphi}$
	$k = 1{,}0 \cdot 10^{-6}$

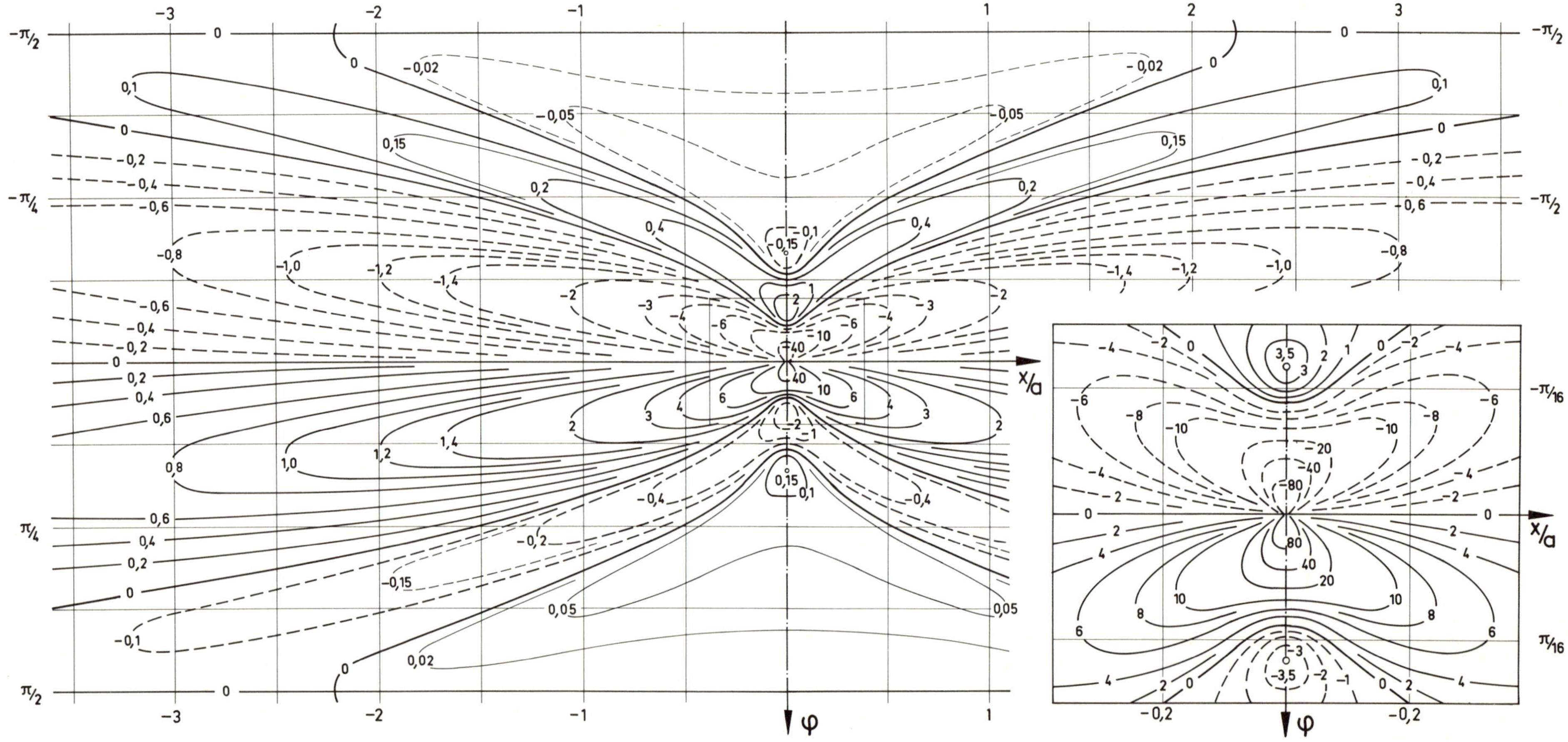

Einflußfläche (negative Zustandsfläche) der Randquerkraft $\bar{q}_{\varphi}$

Influence surface (negative condition surface) for the transverse boundary force $\bar{q}_{\varphi}$

$1000\,\pi\,K/a^2 \cdot u$	**13**
$k = 2{,}5 \cdot 10^{-6}$	

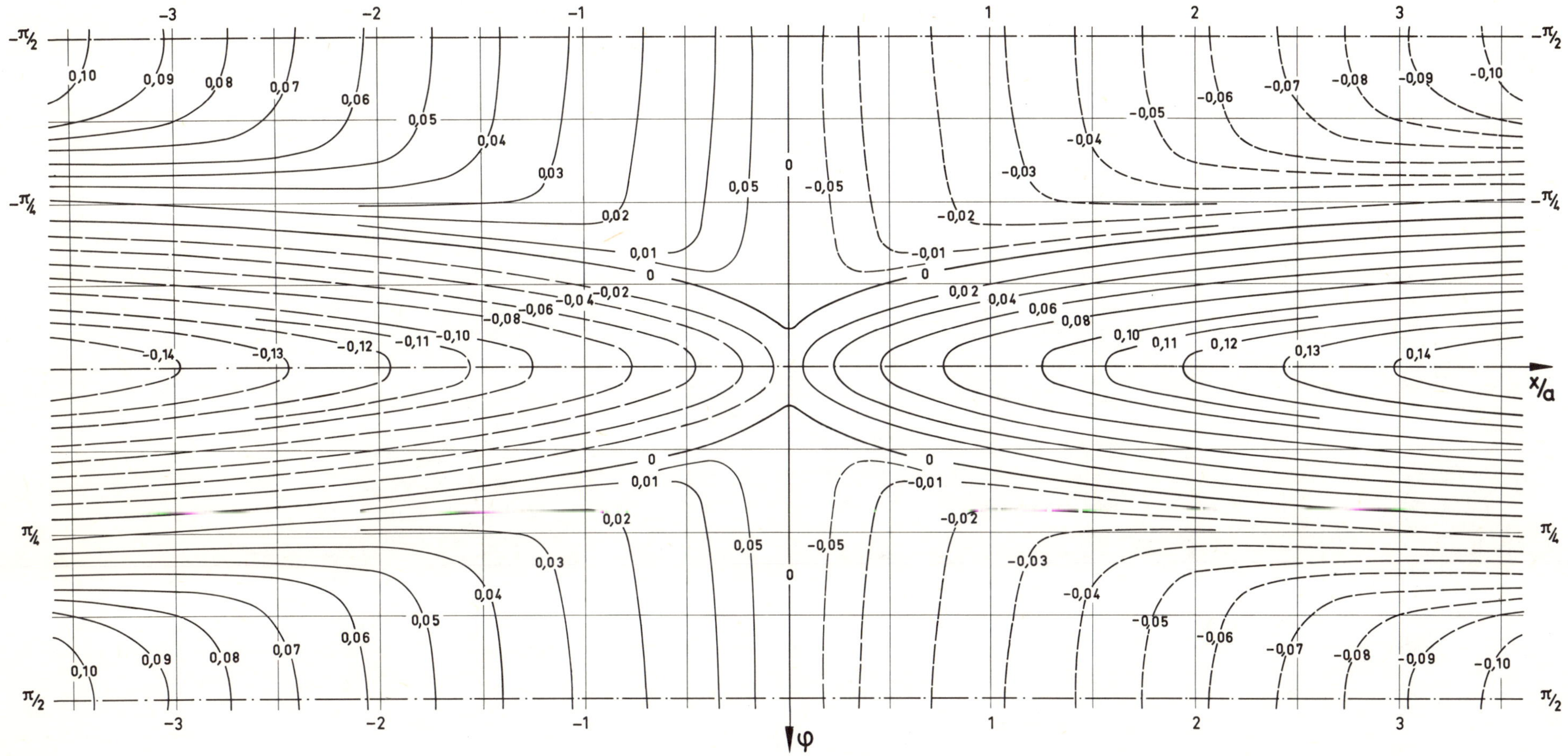

Einflußfläche (negative Zustandsfläche) der Verschiebung u

Influence surface (negative condition surface) for the displacement u

14	$1000\,\pi\, K/a^2 \cdot v$
	$k = 2{,}5 \cdot 10^{-6}$

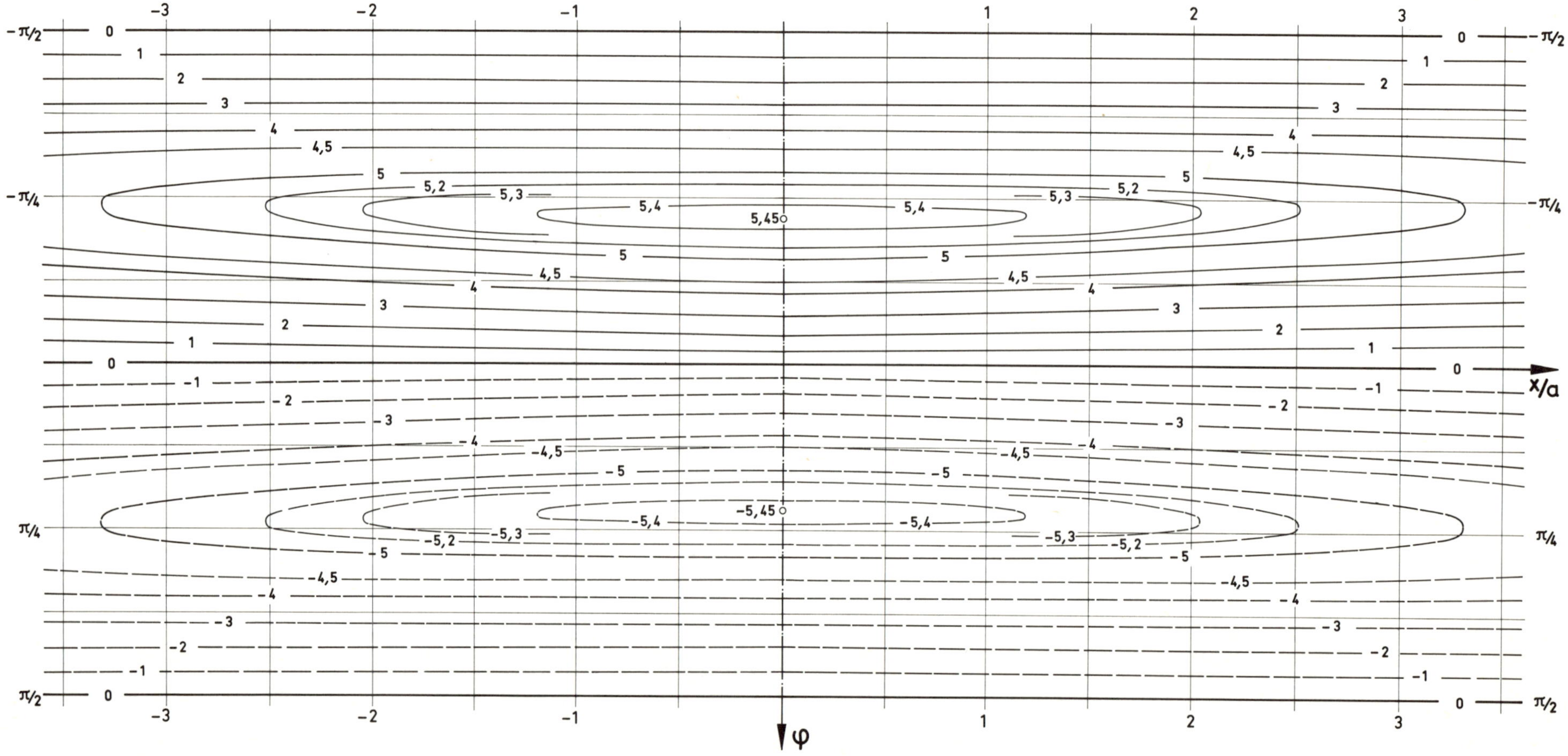

Einflußfläche (negative Zustandsfläche) der Verschiebung v

Influence surface (negative condition surface) for the displacement v

$1000\,\pi\,K/a^2 \cdot w$	**15**
$k = 2{,}5 \cdot 10^{-6}$	

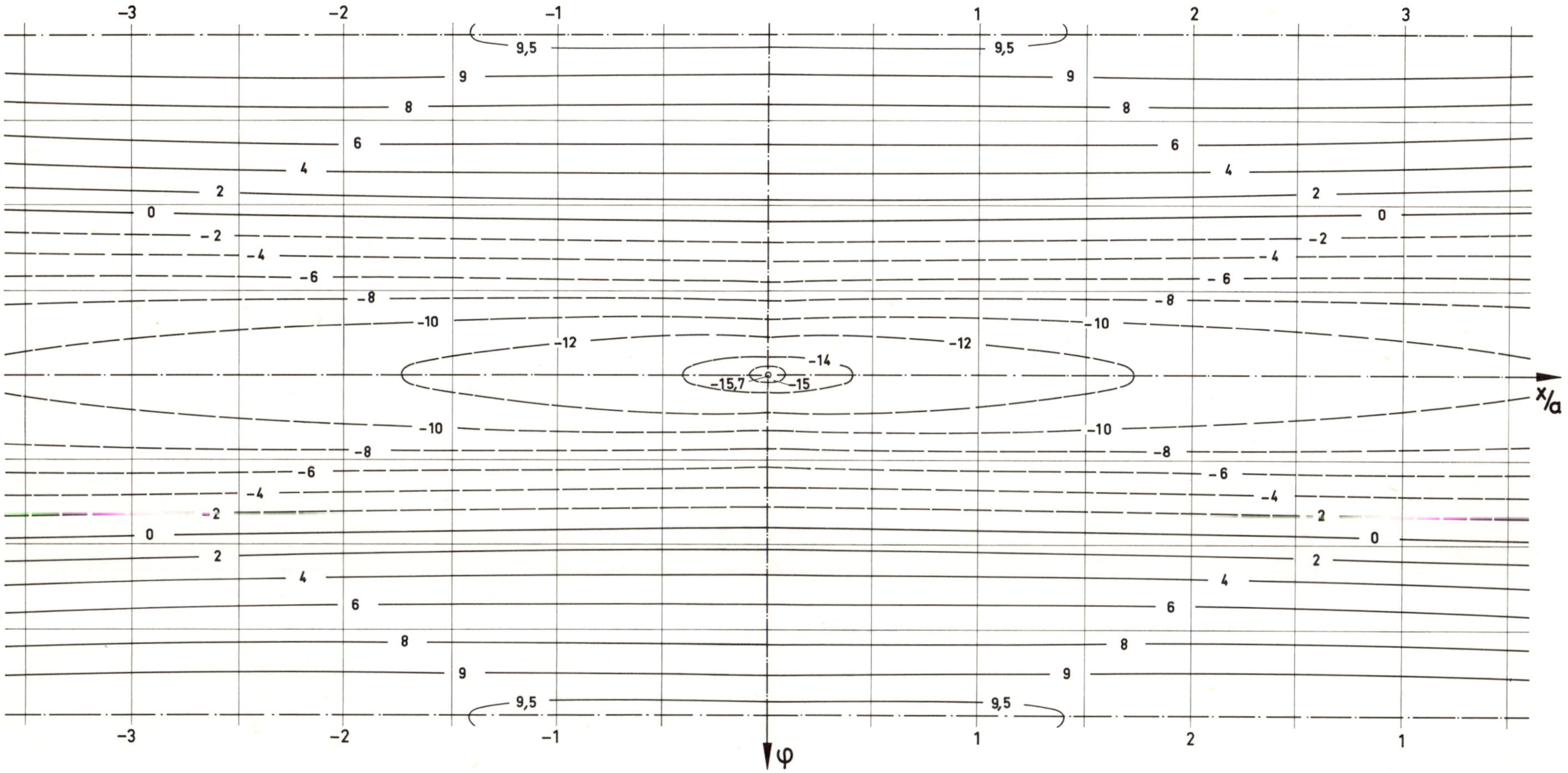

Einflußfläche (Zustandsfläche) der Verschiebung w

Influence surface (condition surface) for the displacement w

16	$100\,\pi\,K/a \cdot \chi_\varphi$
	$k = 2{,}5 \cdot 10^{-6}$

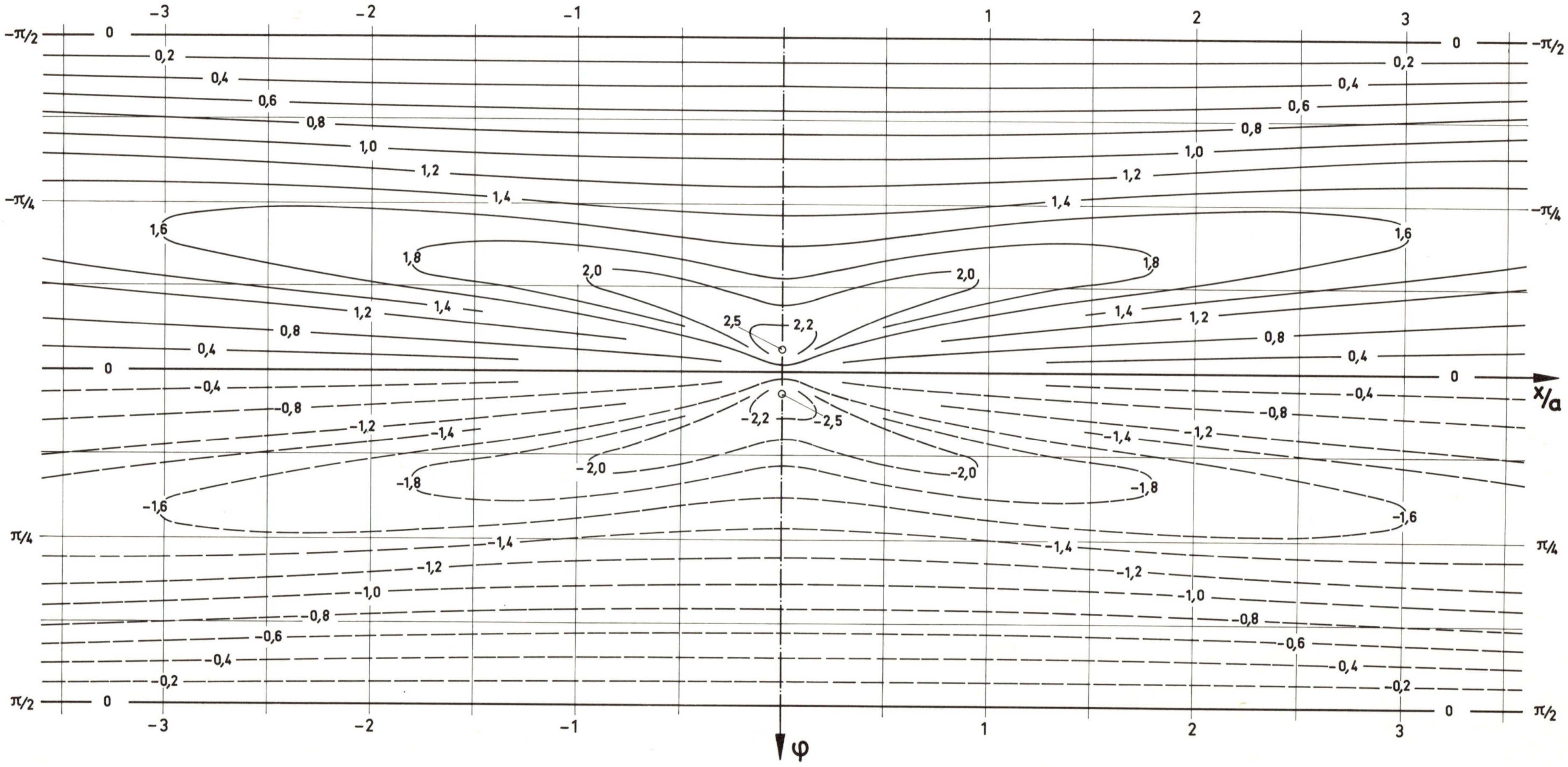

Einflußfläche (negative Zustandsfläche) der Verdrehung χ_φ

Influence surface (negative condition surface) for the rotation χ_φ

$\pi a \cdot n_x$	**17**
$k = 2{,}5 \cdot 10^{-6}$	

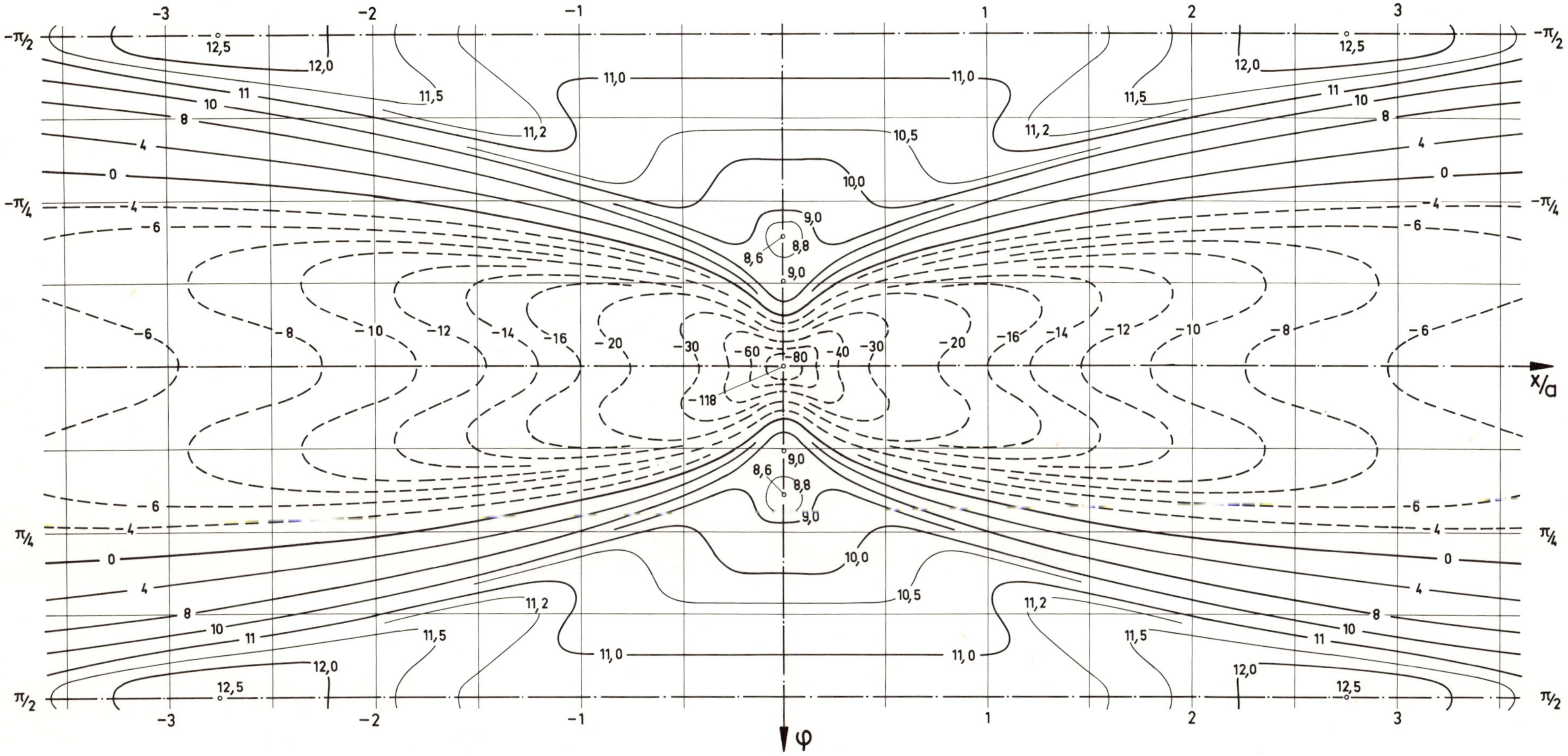

Einflußfläche (Zustandsfläche) der Normalkraft n_x

Influence surface (condition surface) for the normal force n_x

18	$\pi a \cdot n_{\varphi}$
	$k = 2{,}5 \cdot 10^{-6}$

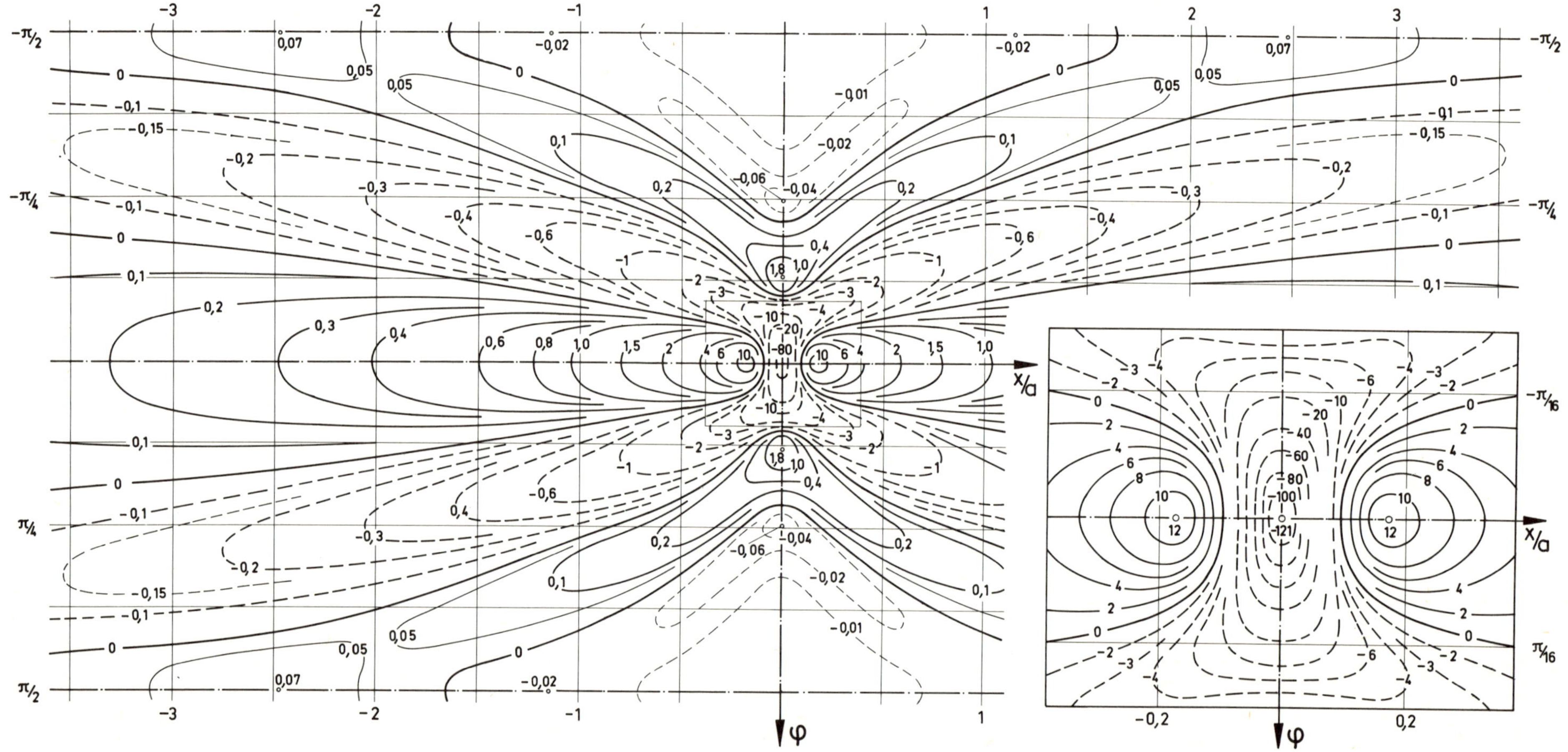

Einflußfläche (Zustandsfläche) der Normalkraft n_{φ}

Influence surface (condition surface) for the normal force n_{φ}

$\pi a \cdot \bar{n}_{x\varphi}$	**19**
$k = 2{,}5 \cdot 10^{-6}$	

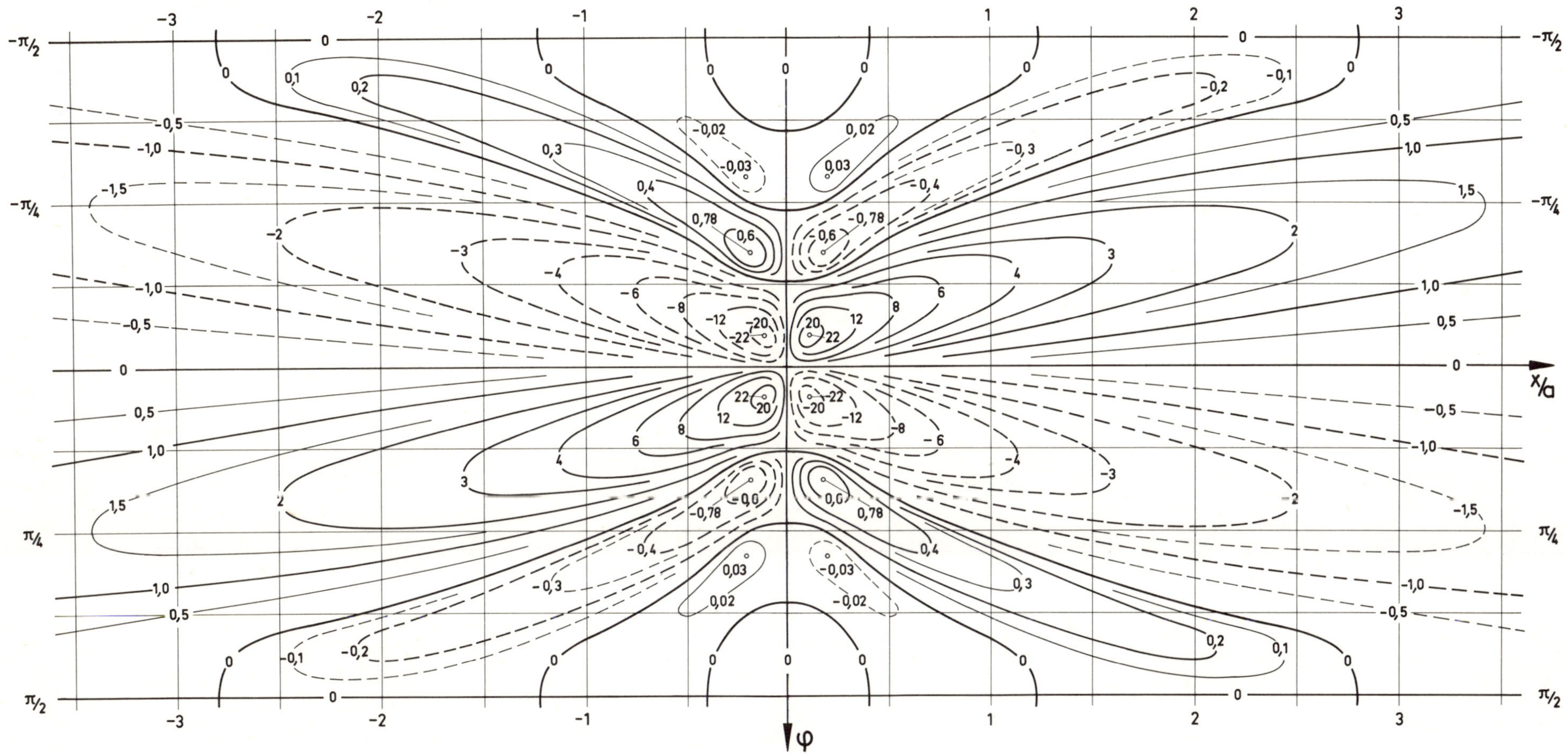

Einflußfläche (Zustandsfläche) der Randschubkraft $\bar{n}_{x\varphi}$

Influence surface (condition surface) for the boundary shearing force $\bar{\boldsymbol{n}}_{\boldsymbol{x\varphi}}$

20	$100\,\pi \cdot m_x$
	$k = 2{,}5 \cdot 10^{-6}$

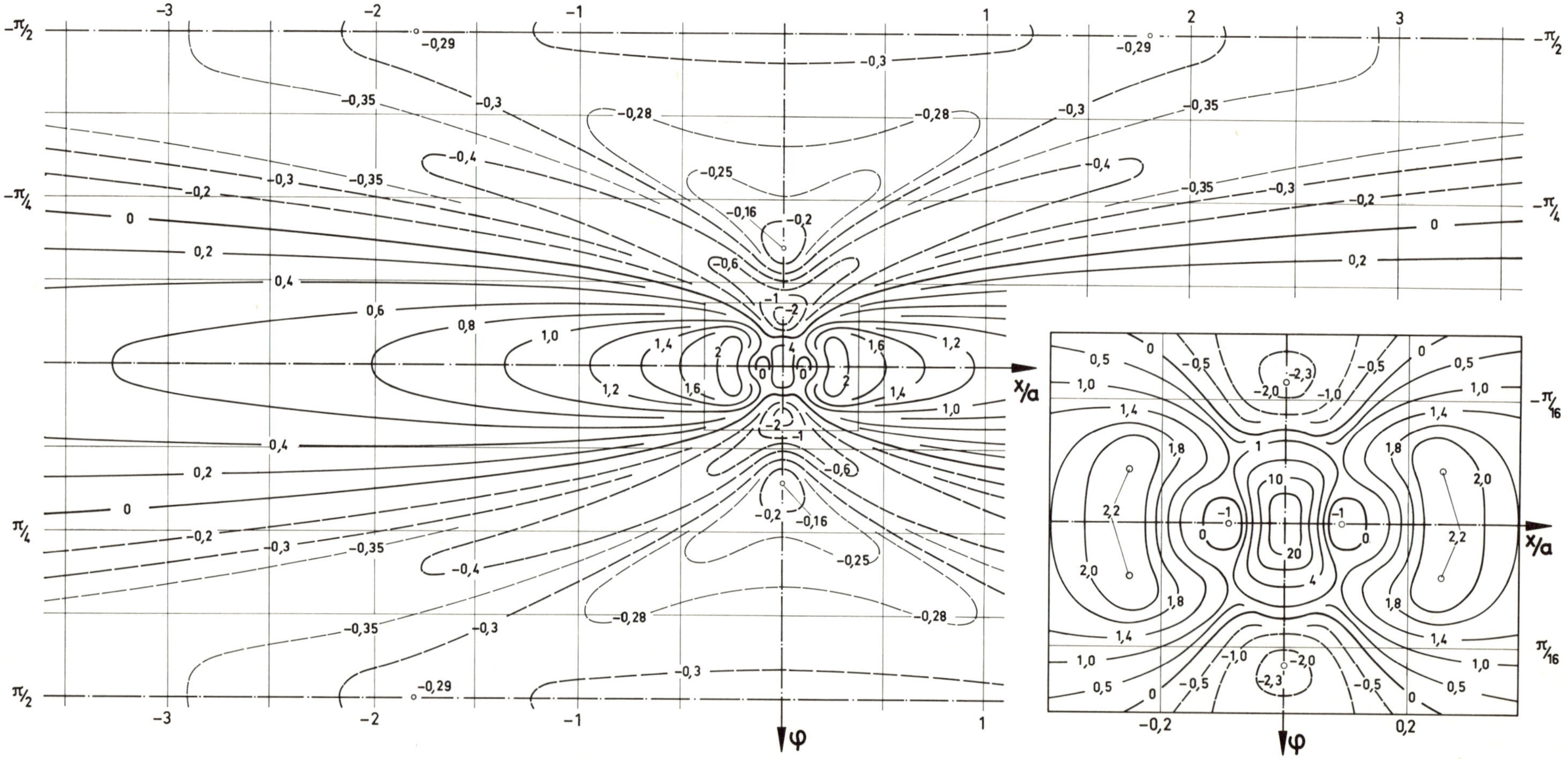

Einflußfläche (Zustandsfläche) des Biegemomentes m_x

Influence surface (condition surface) for the bending moment m_x

$100\,\pi \cdot m_{\varphi}$	**21**
$k = 2{,}5 \cdot 10^{-6}$	

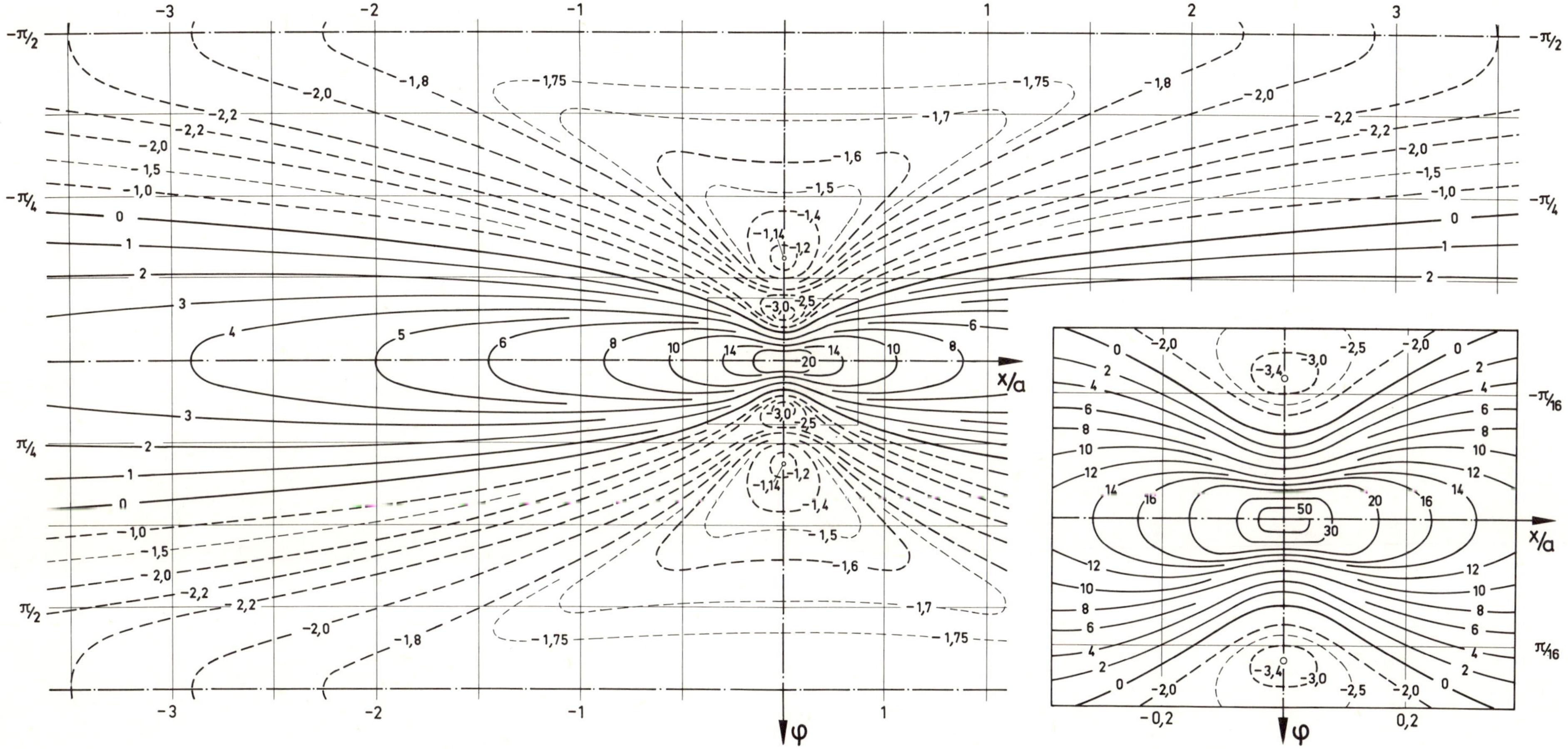

Einflußfläche (Zustandsfläche) des Biegemomentes m_{φ}

Influence surface (condition surface) for the bending moment m_{φ}

22	$100\,\pi \cdot m_{x\varphi}$
	$k = 2{,}5 \cdot 10^{-6}$

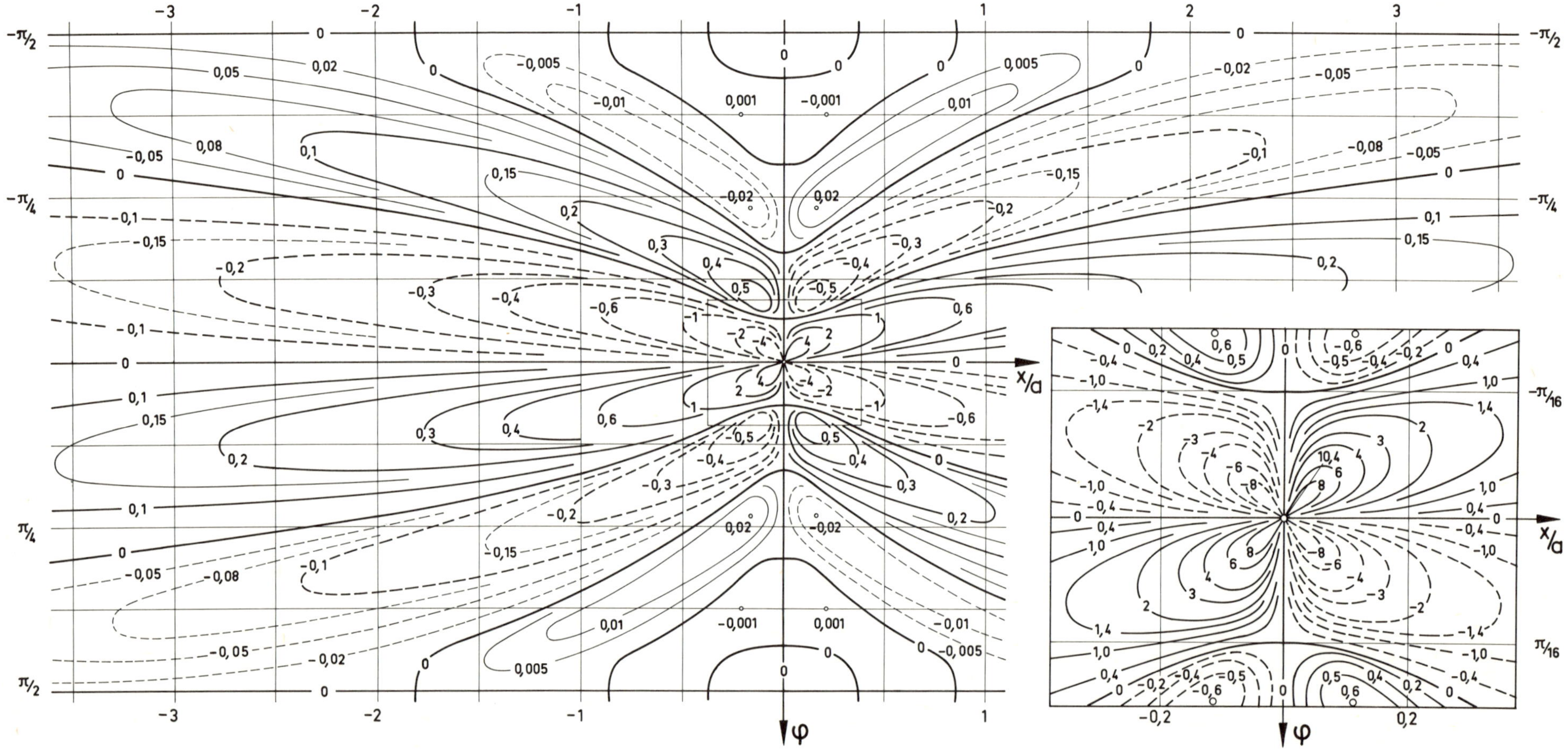

Einflußfläche (Zustandsfläche) des Drillmomentes $m_{x\varphi}$

Influence surface (condition surface) for the twisting moment $m_{x\varphi}$

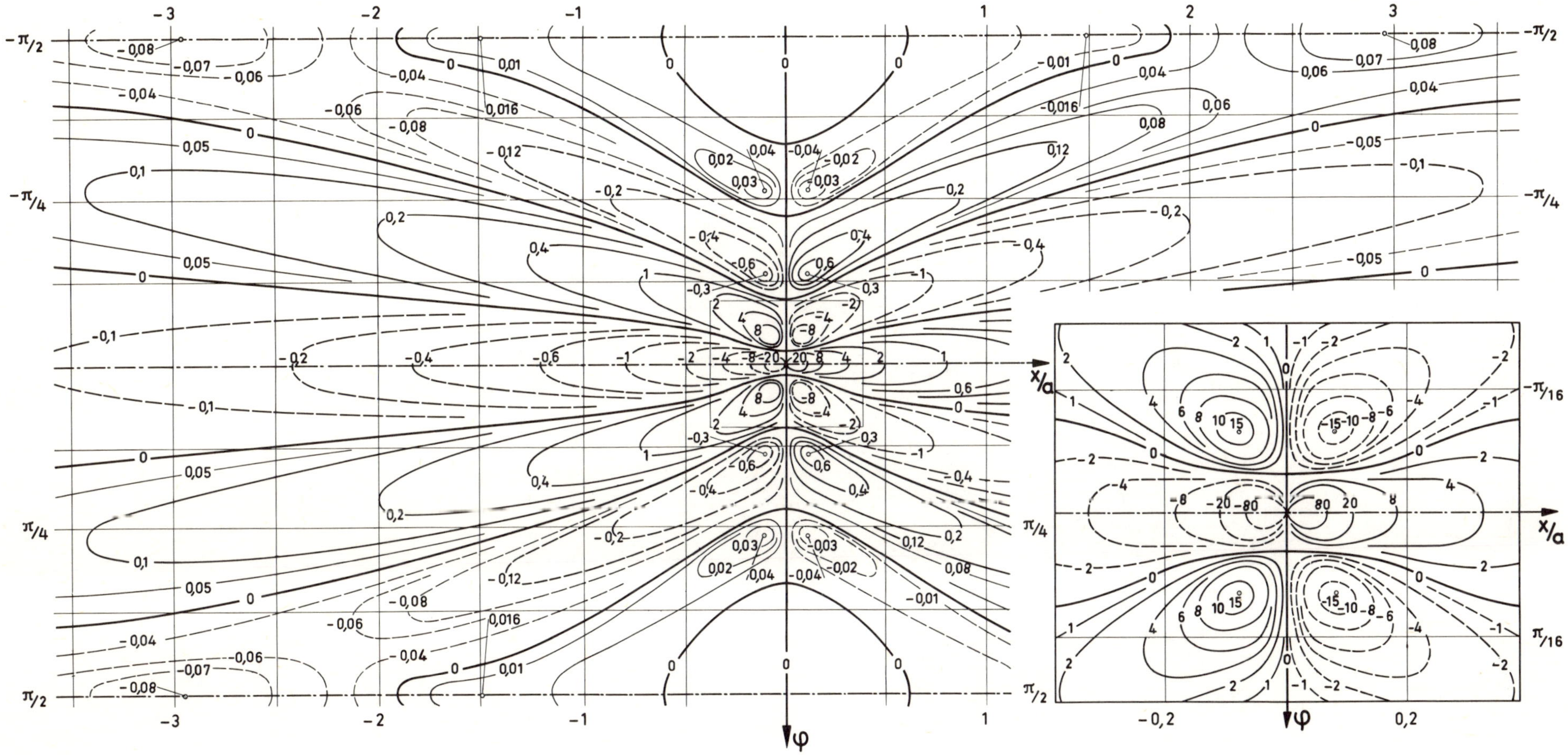

Einflußfläche (negative Zustandsfläche) der Randquerkraft $\bar{q}_x$

Influence surface (negative condition surface) for the transverse boundary force $\bar{q}_x$

24 | $10\,\pi\, a \cdot \bar{q}_\varphi$ / $k = 2{,}5 \cdot 10^{-6}$

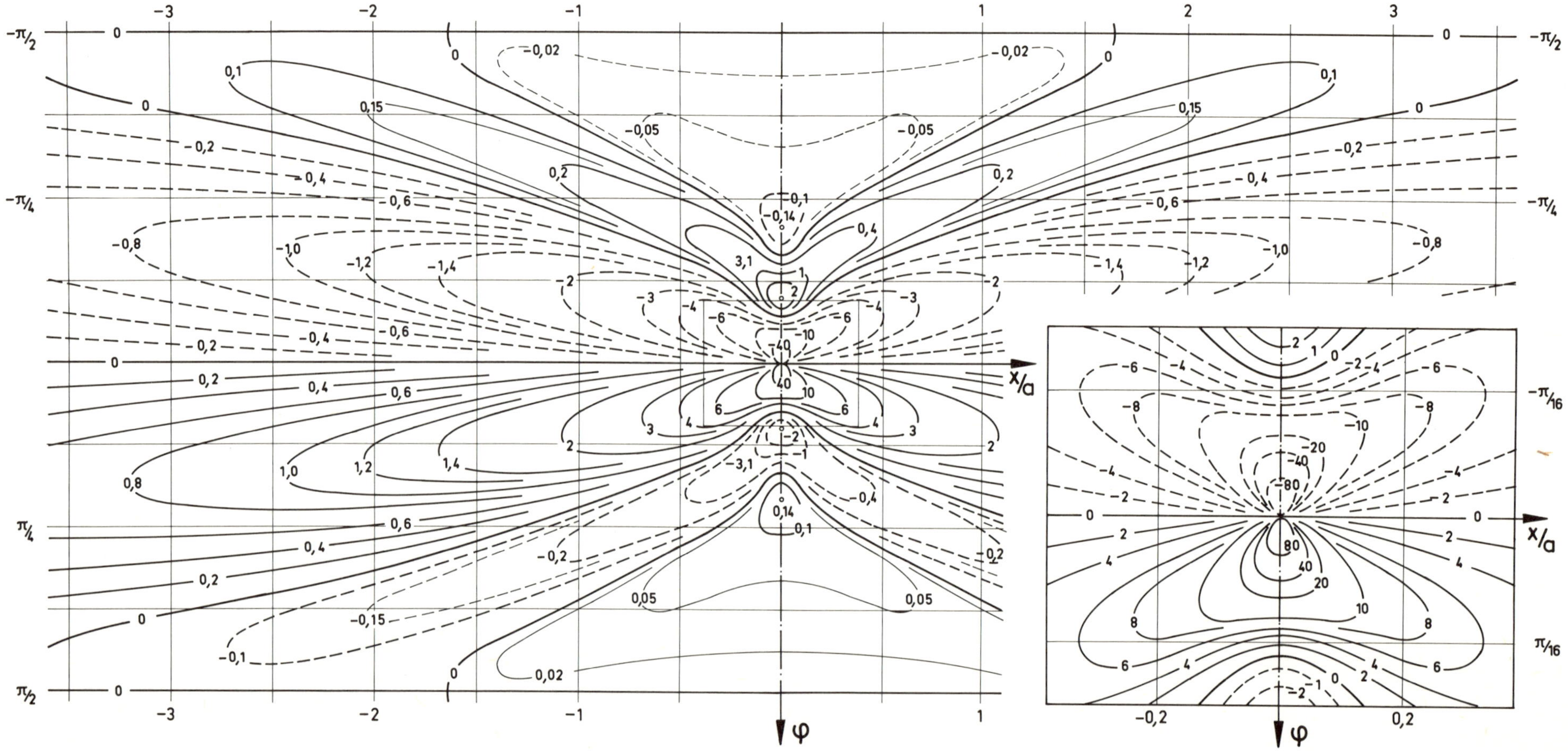

Einflußfläche (negative Zustandsfläche) der Randquerkraft $\bar{q}_\varphi$

Influence surface (negative condition surface) for the transverse boundary force $\bar{q}_\varphi$

$1000\,\pi\,K/a^2 \cdot u$	**25**
$k = 5{,}0 \cdot 10^{-6}$	

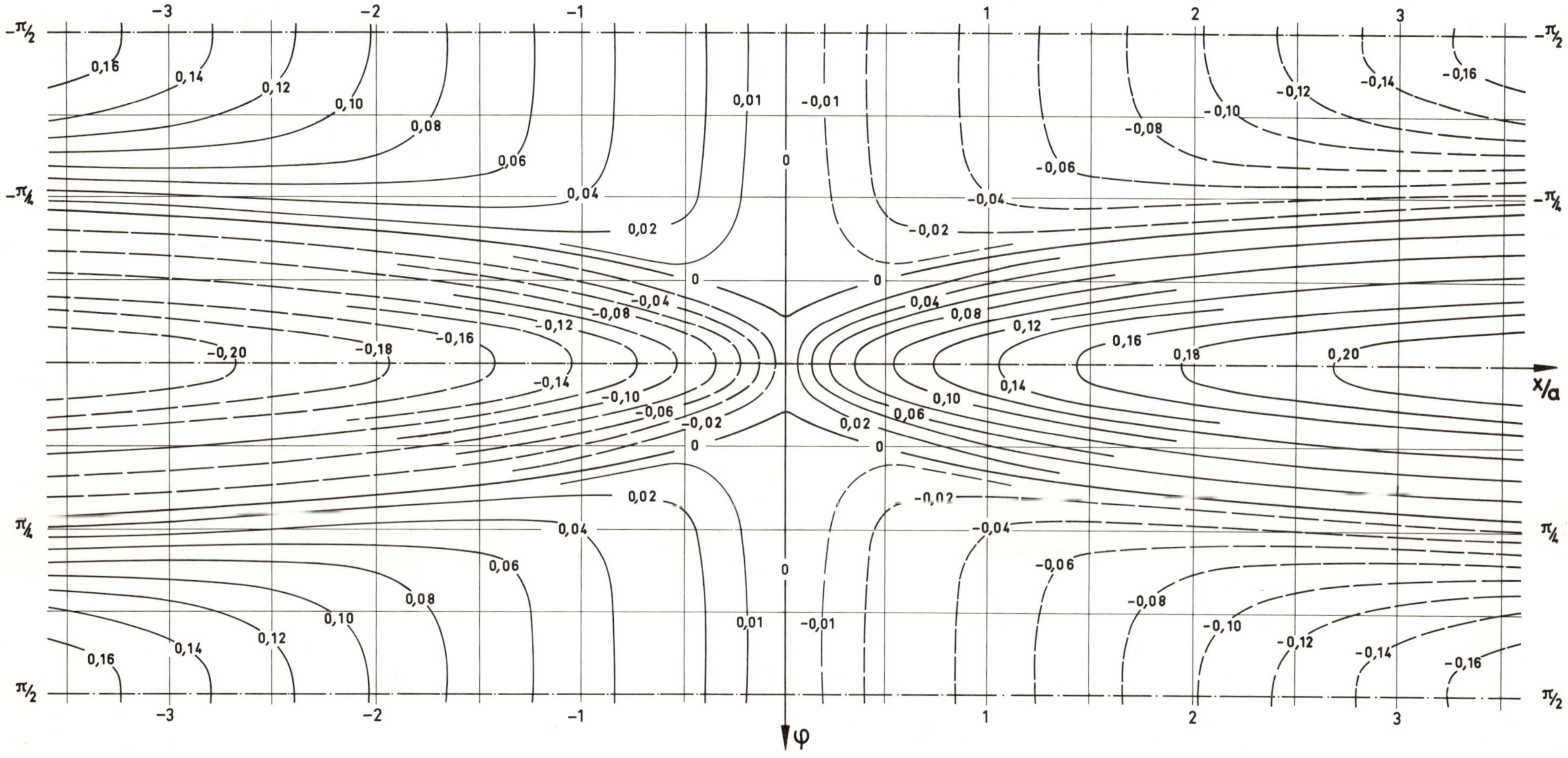

Einflußfläche (negative Zustandsfläche) der Verschiebung *u*

Influence surface (negative condition surface) for the displacement *u*

26 | $1000\,\pi\,K/a^2 \cdot v$ — $k = 5{,}0 \cdot 10^{-6}$

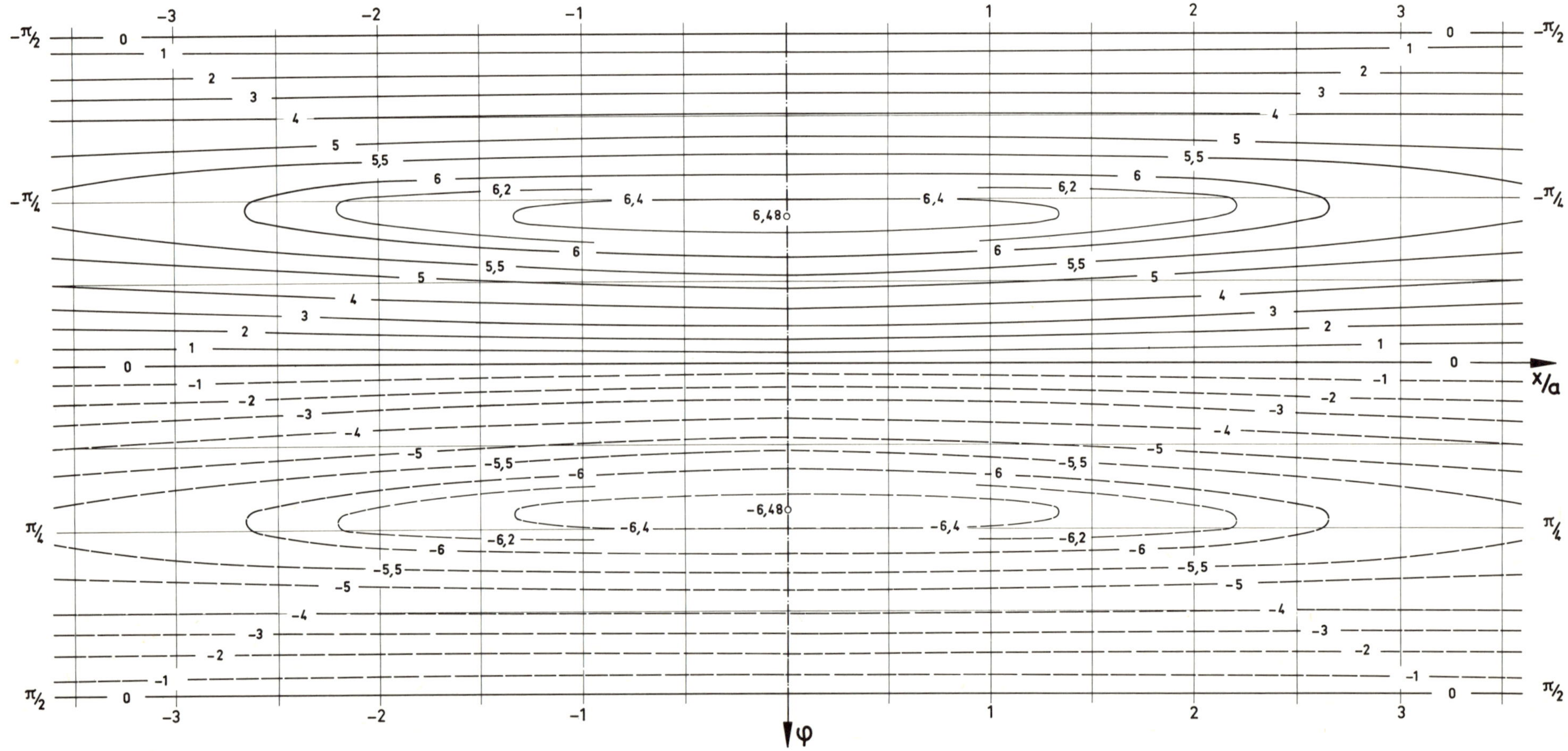

Einflußfläche (negative Zustandsfläche) der Verschiebung v

Influence surface (negative condition surface) for the displacement v

$1000\,\pi\,K/a^2 \cdot w$	**27**
$k = 5{,}0 \cdot 10^{-6}$	

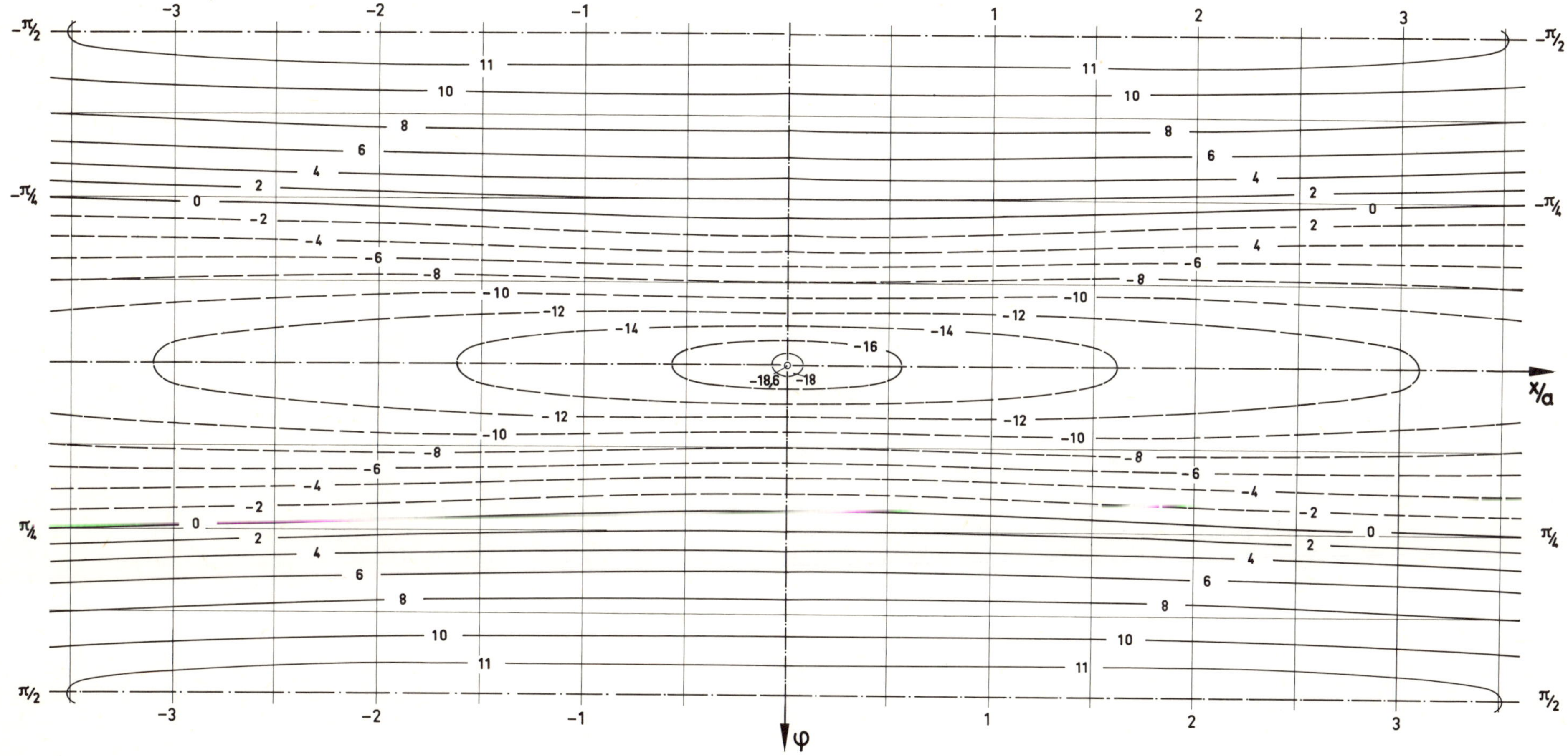

Einflußfläche (Zustandsfläche) der Verschiebung w

Influence surface (condition surface) for the displacement w

28	$100\,\pi\,K/a \cdot \chi_\varphi$
	$k = 5{,}0 \cdot 10^{-6}$

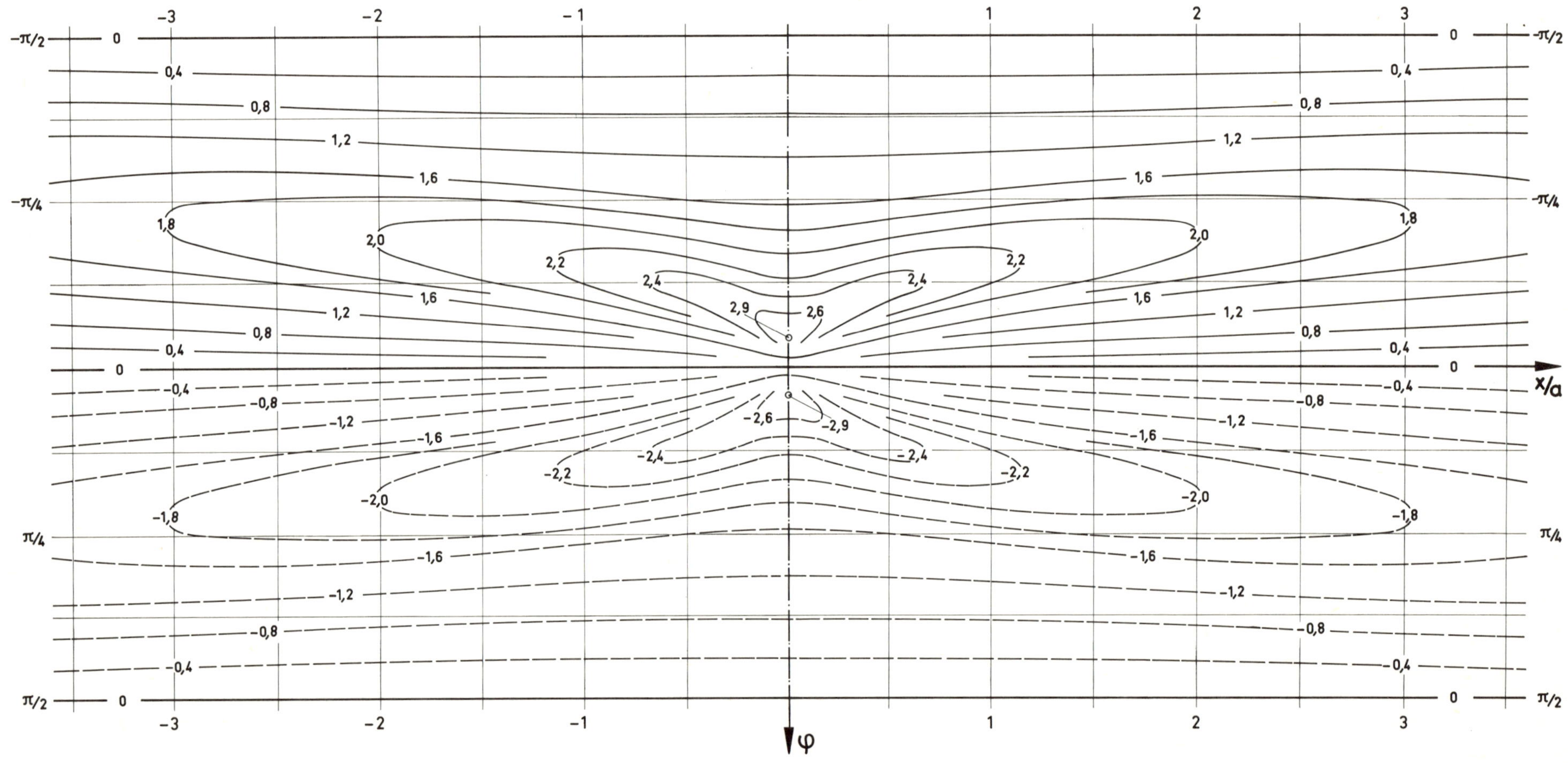

Einflußfläche (negative Zustandsfläche) der Verdrehung χ_φ

Influence surface (negative condition surface) for the rotation χ_φ

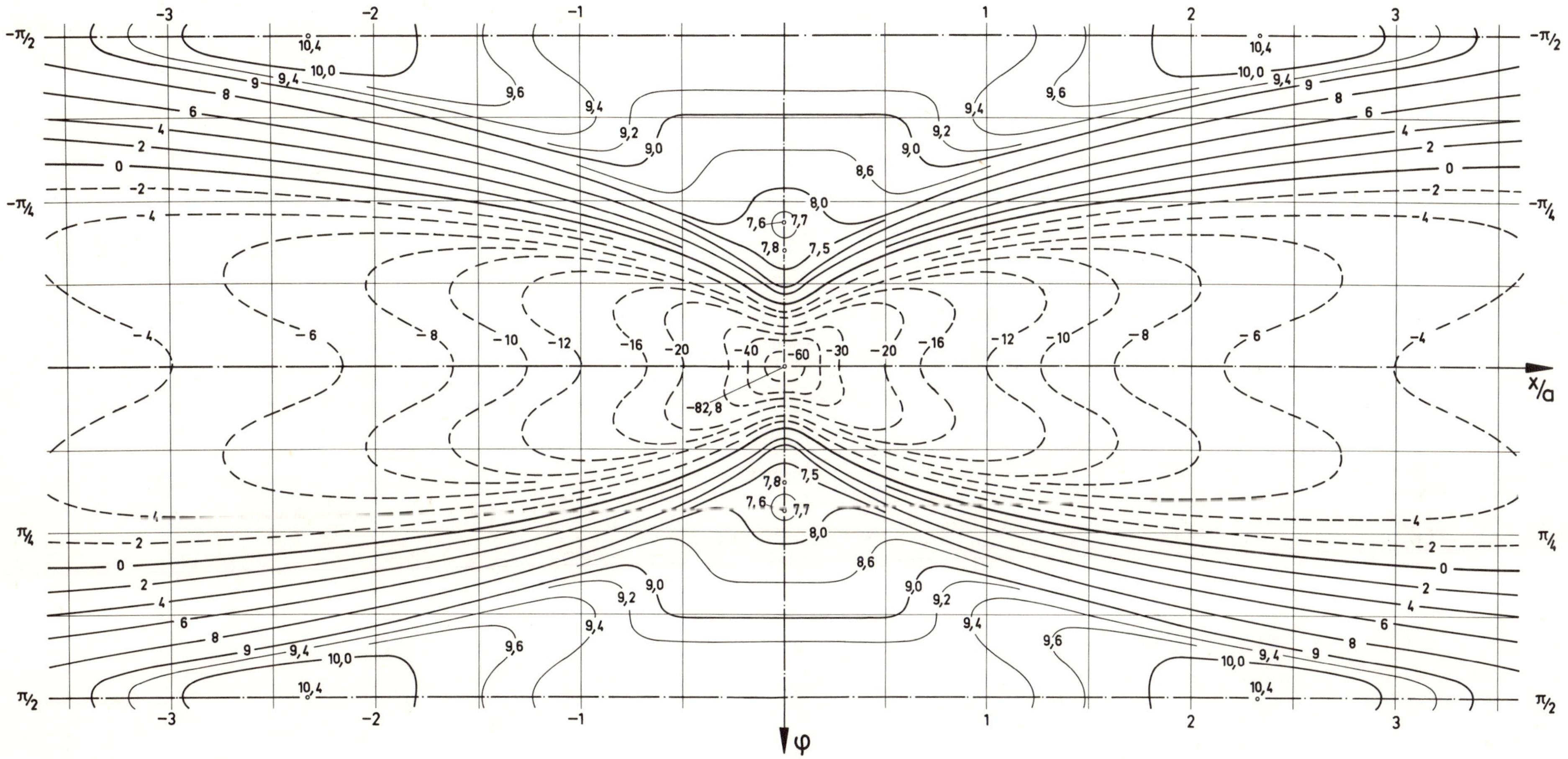

Einflußfläche (Zustandsfläche) der Normalkraft n_x

Influence surface (condition surface) for the normal force n_x

30	$\pi a \cdot n_\varphi$
	$k = 5{,}0 \cdot 10^{-6}$

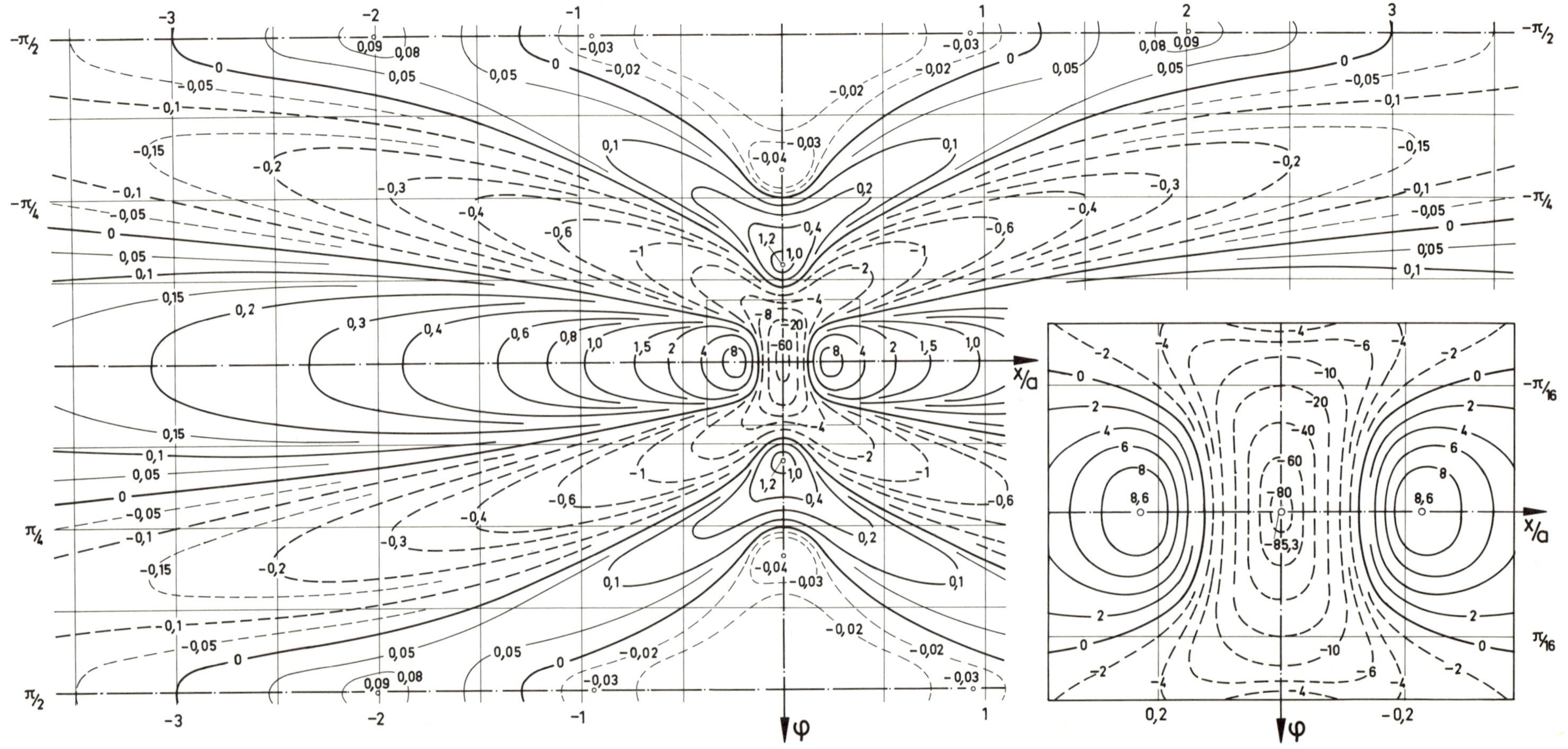

Einflußfläche (Zustandsfläche) der Normalkraft n_φ

Influence surface (condition surface) for the normal force n_φ

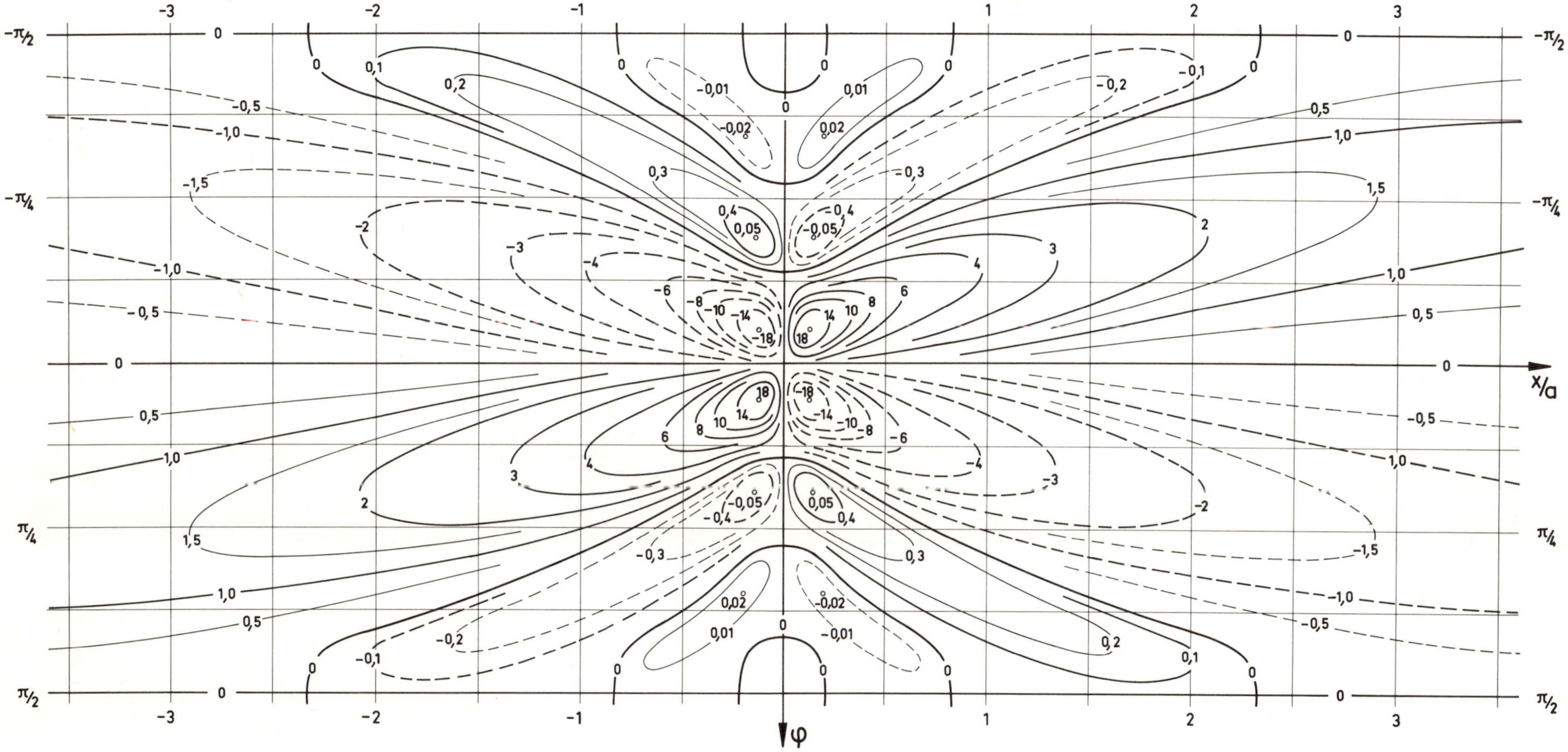

Einflußfläche (Zustandsfläche) der Randschubkraft $\bar{n}_{x\varphi}$

Influence surface (condition surface) for the boundary shearing force $\bar{n}_{x\varphi}$

32	$100\,\pi \cdot m_x$
	$k = 5{,}0 \cdot 10^{-6}$

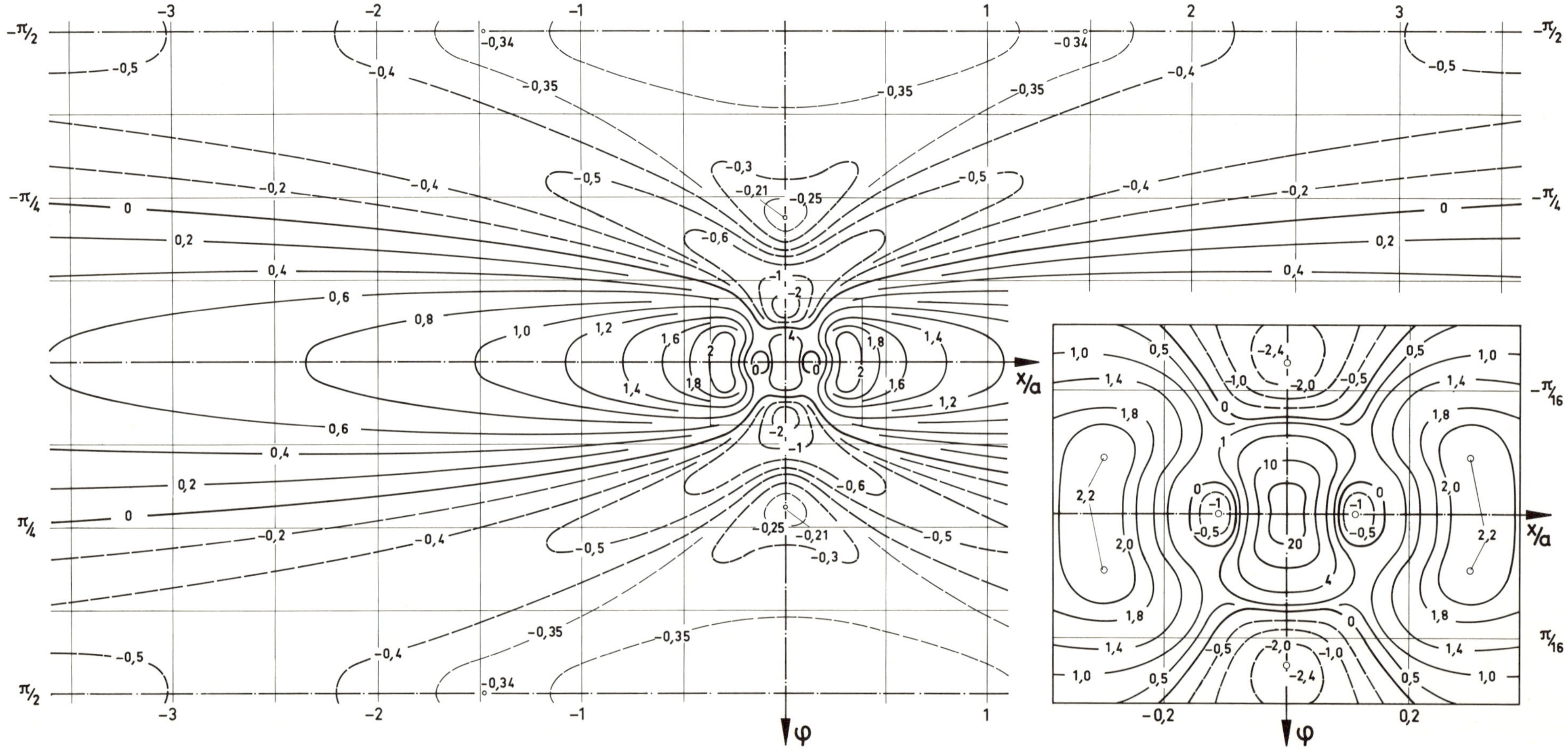

Einflußfläche (Zustandsfläche) des Biegemomentes m_x

Influence surface (condition surface) for the bending moment m_x

$100\,\pi \cdot m_\varphi$	**33**
$k = 5{,}0 \cdot 10^{-6}$	

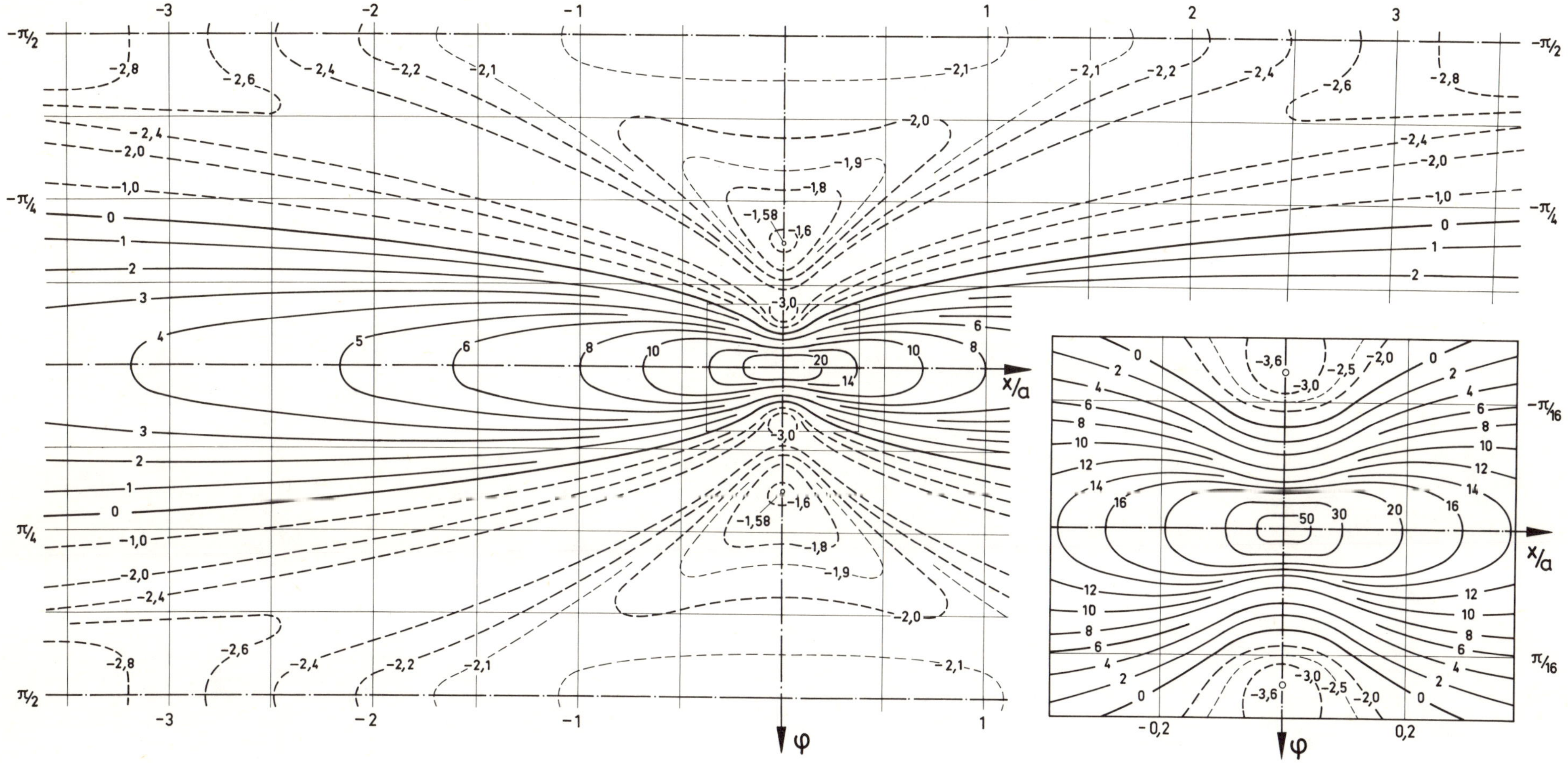

Einflußfläche (Zustandsfläche) des Biegemomentes m_φ

Influence surface (condition surface) for the bending moment m_φ

34	$100\,\pi \cdot m_{x\varphi}$
	$k = 5{,}0 \cdot 10^{-6}$

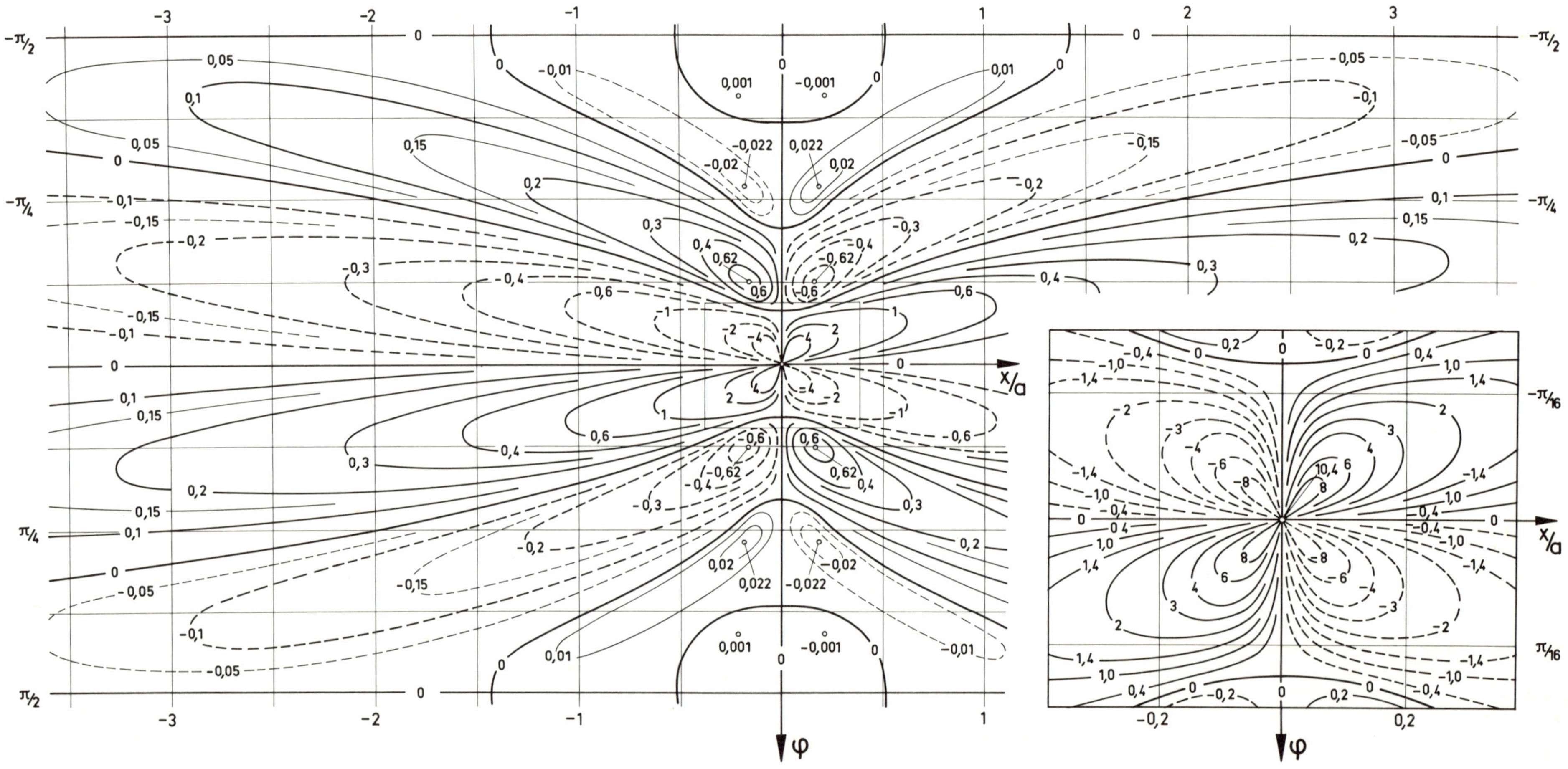

Einflußfläche (Zustandsfläche) des Drillmomentes $m_{x\varphi}$

Influence surface (condition surface) for the twisting moment $m_{x\varphi}$

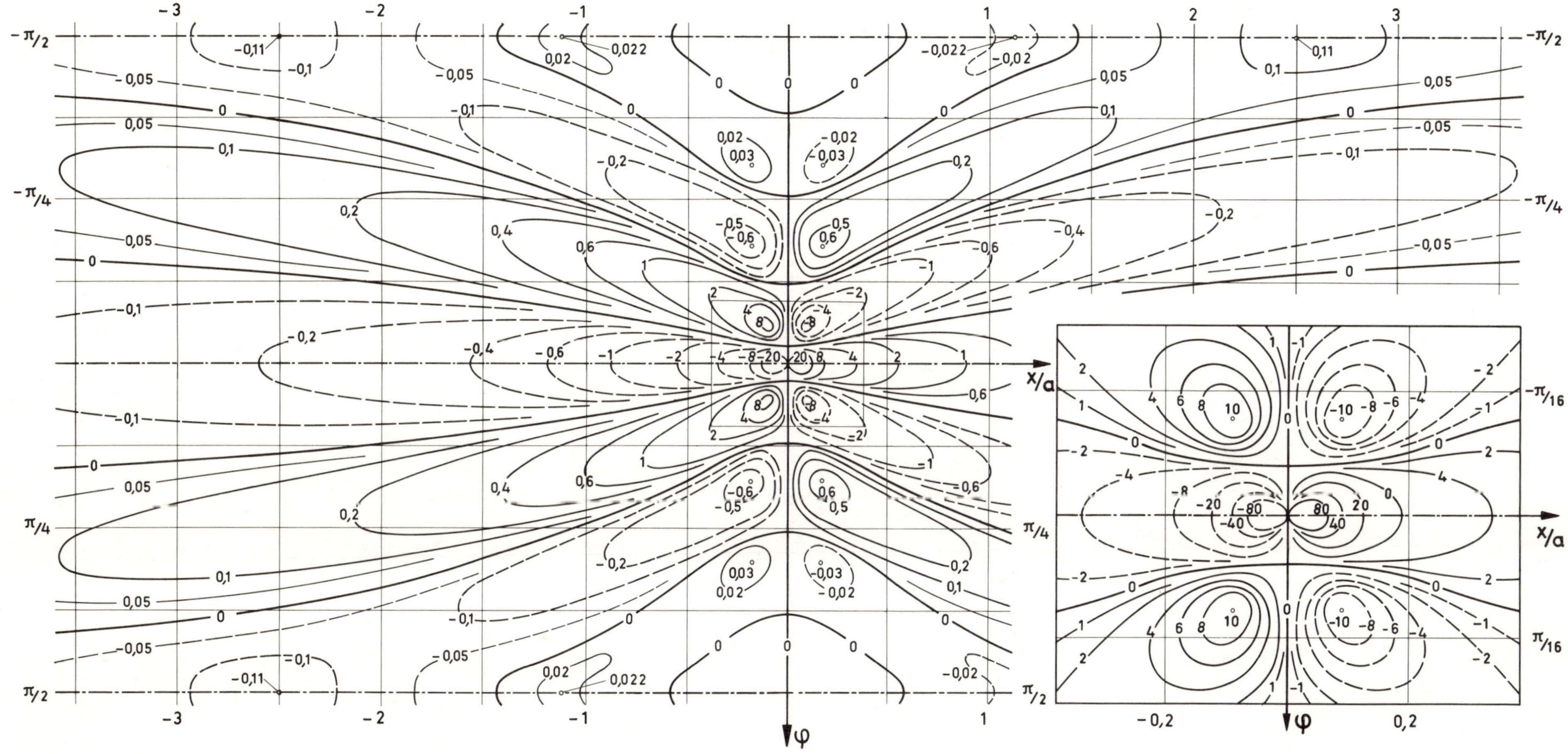

Einflußfläche (negative Zustandsfläche) der Randquerkraft $\bar{q}_x$

Influence surface (negative condition surface) for the transverse boundary force $\bar{q}_x$

36	$10\,\pi\, a \cdot \bar{q}_{\varphi}$
	$k = 5{,}0 \cdot 10^{-6}$

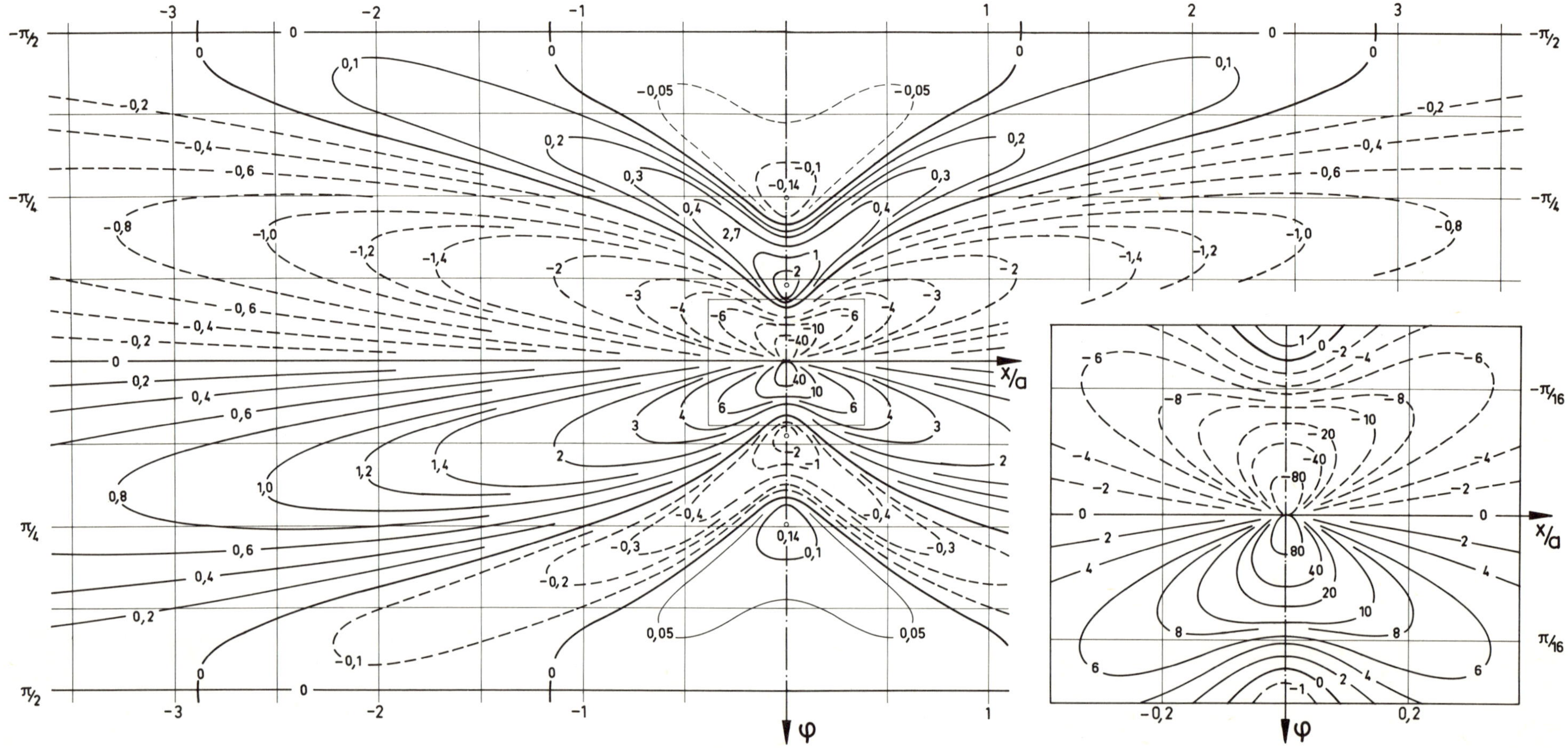

Einflußfläche (negative Zustandsfläche) der Randquerkraft $\bar{q}_{\varphi}$

Influence surface (negative condition surface) for the transverse boundary force $\bar{q}_{\varphi}$

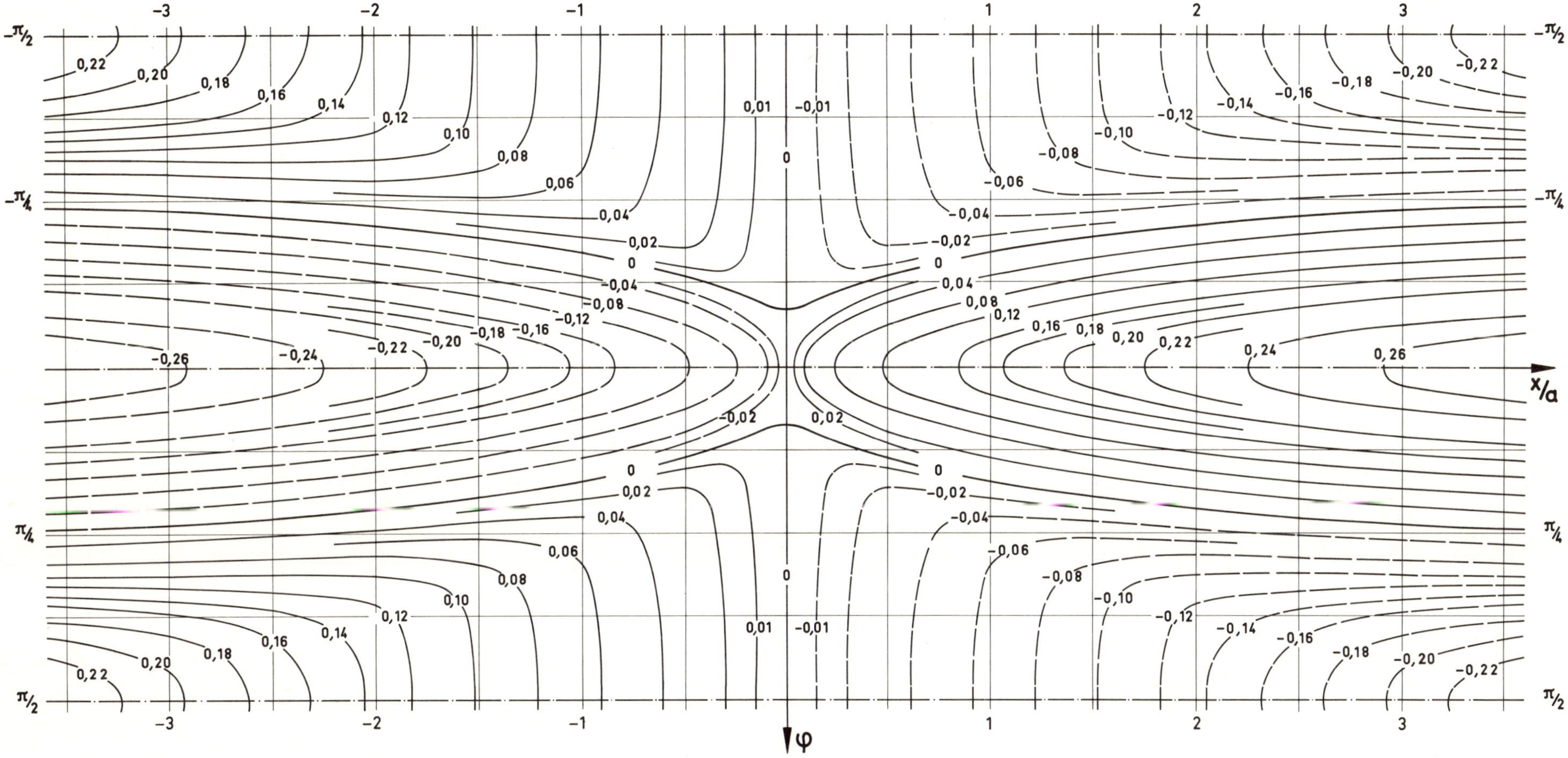

Einflußfläche (negative Zustandsfläche) der Verschiebung u

Influence surface (negative condition surface) for the displacement u

38	$1000\,\pi\,K/a^2 \cdot v$
	$k = 7{,}5 \cdot 10^{-6}$

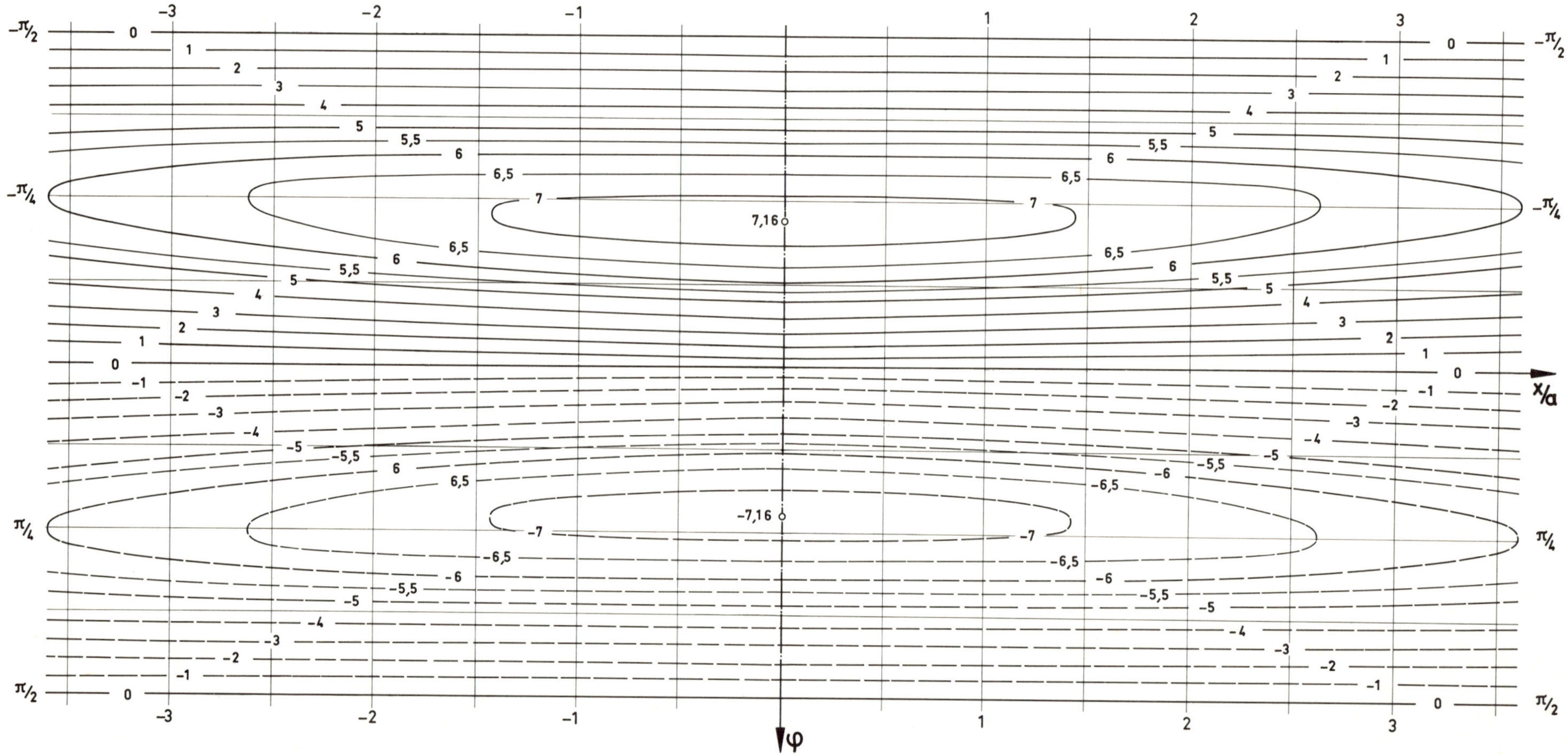

Einflußfläche (negative Zustandsfläche) der Verschiebung v

Influence surface (negative condition surface) for the displacement v

$1000\,\pi\,K/a^2 \cdot w$	**39**
$k = 7{,}5 \cdot 10^{-6}$	

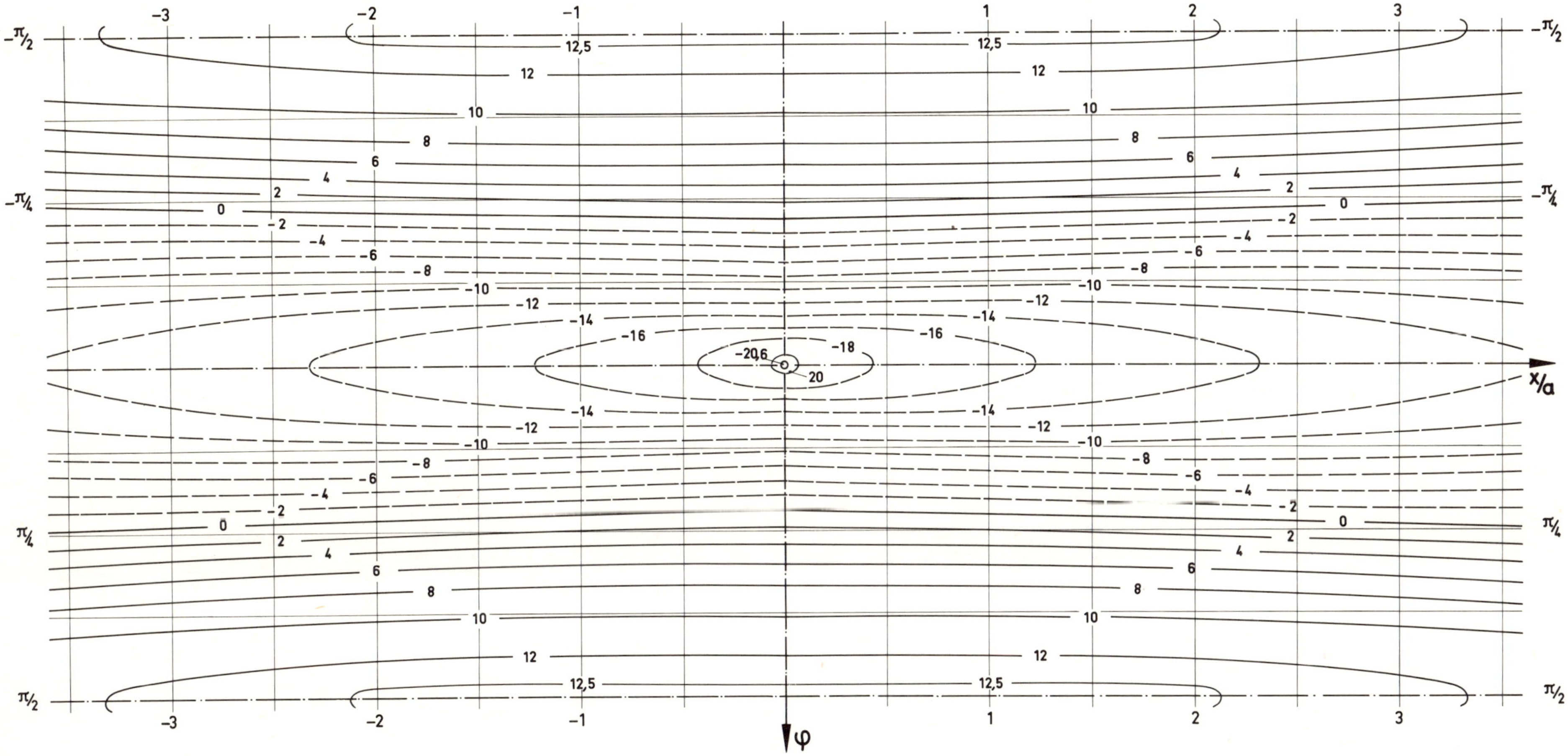

Einflußfläche (Zustandsfläche) der Verschiebung w

Influence surface (condition surface) for the displacement w

40	$100\,\pi\, K/a \cdot \chi_\varphi$
	$k = 7{,}5 \cdot 10^{-6}$

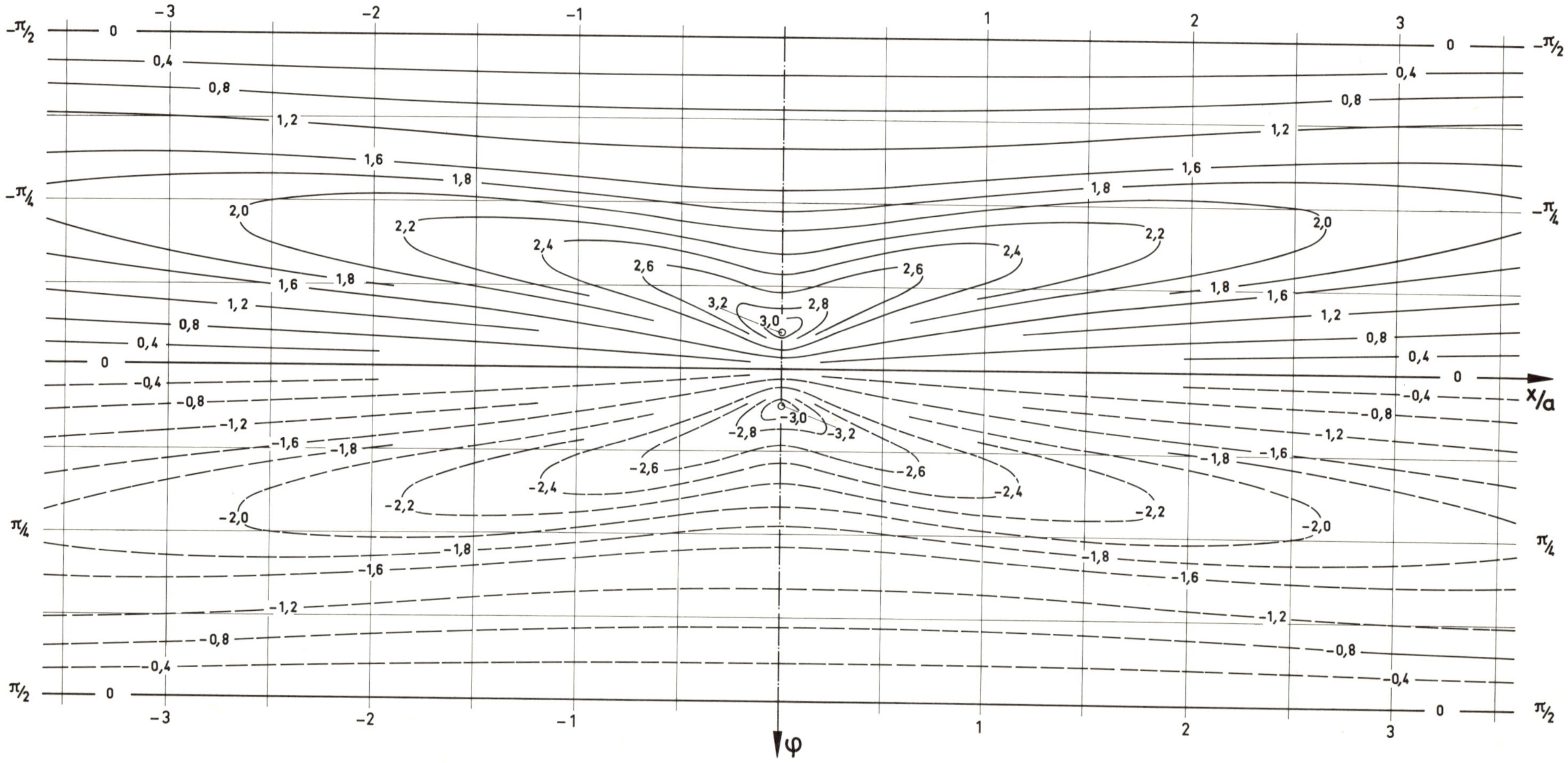

Einflußfläche (negative Zustandsfläche) der Verdrehung χ_φ

Influence surface (negative condition surface) for the rotation χ_φ

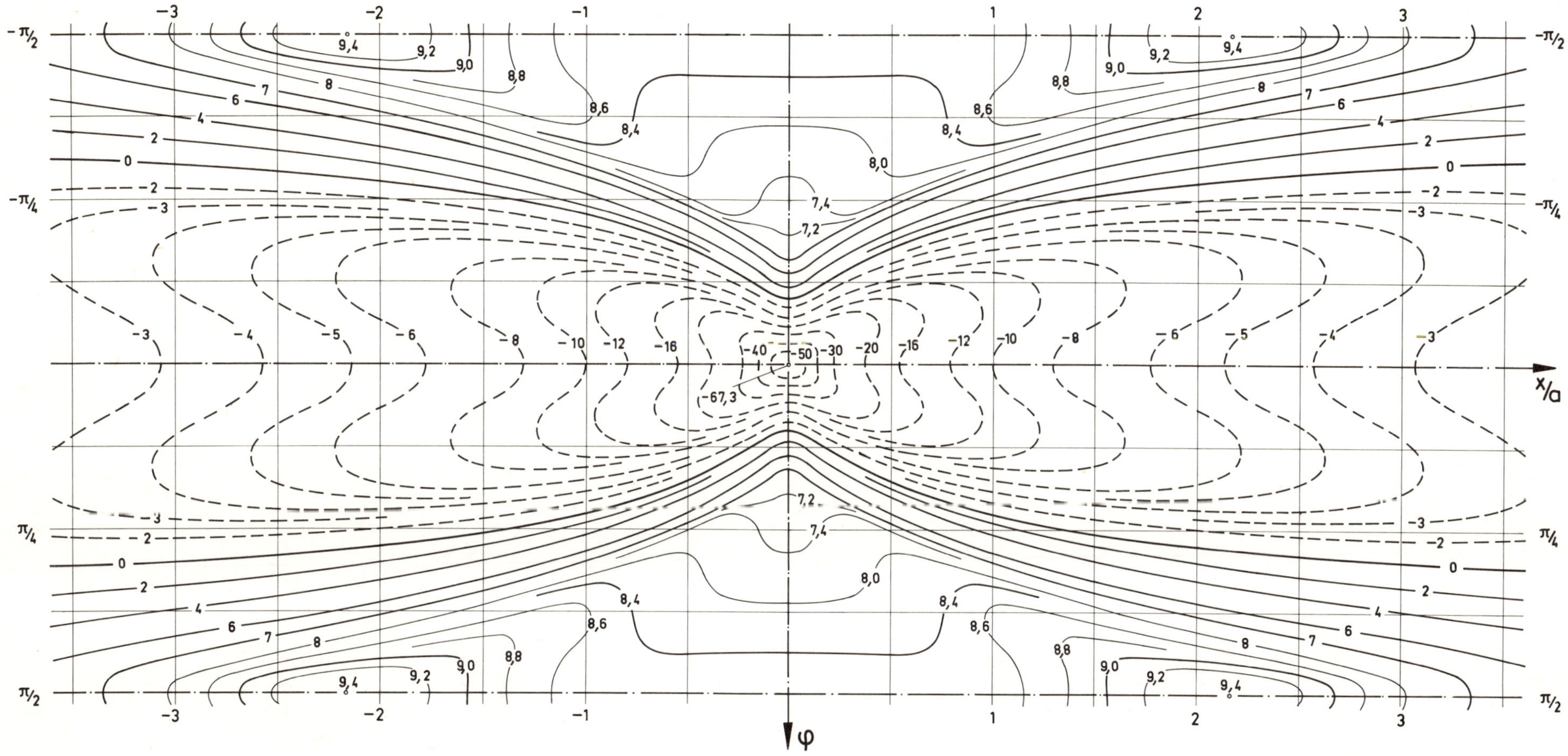

Einflußfläche (Zustandsfläche) der Normalkraft n_x

Influence surface (condition surface) for the normal force n_x

42	$\pi a \cdot n_\varphi$
	$k = 7{,}5 \cdot 10^{-6}$

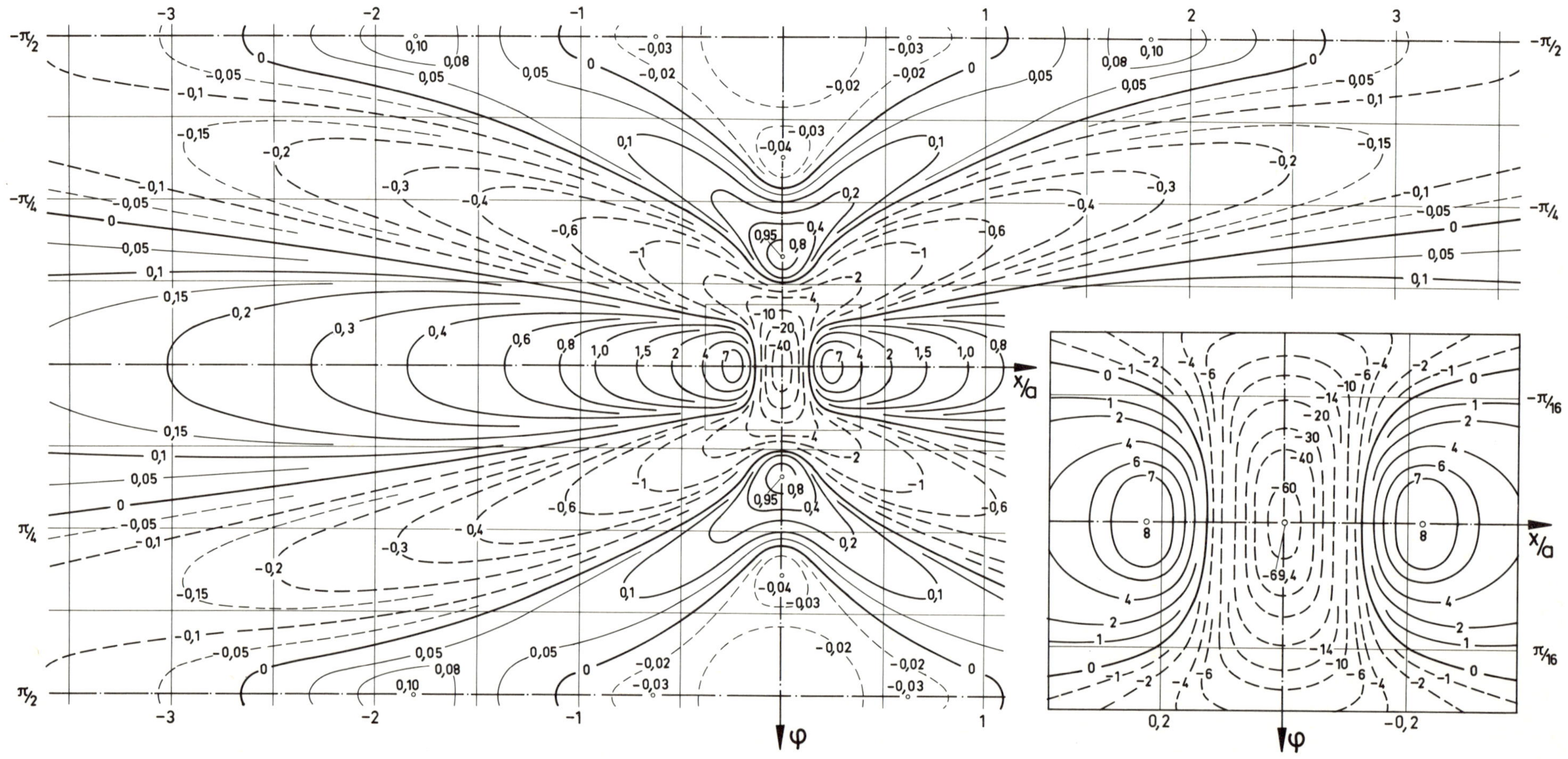

Einflußfläche (Zustandsfläche) der Normalkraft n_φ

Influence surface (condition surface) for the normal force n_φ

$\pi a \cdot \bar{n}_{x\varphi}$	**43**
$k = 7{,}5 \cdot 10^{-6}$	

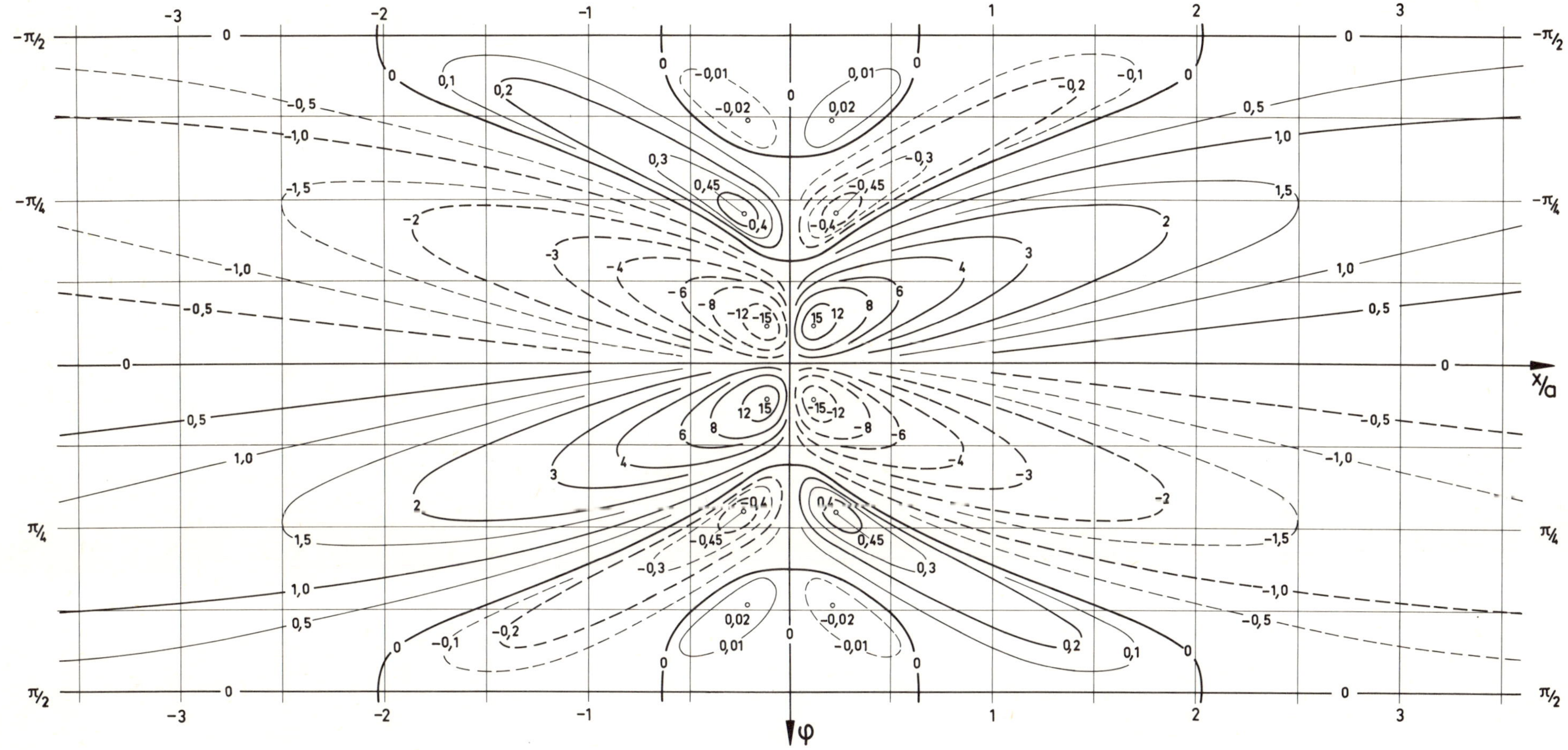

Einflußfläche (Zustandsfläche) der Randschubkraft $\bar{n}_{x\varphi}$

Influence surface (condition surface) for the boundary shearing force $\bar{n}_{x\varphi}$

44	$100\,\pi \cdot m_x$
	$k = 7{,}5 \cdot 10^{-6}$

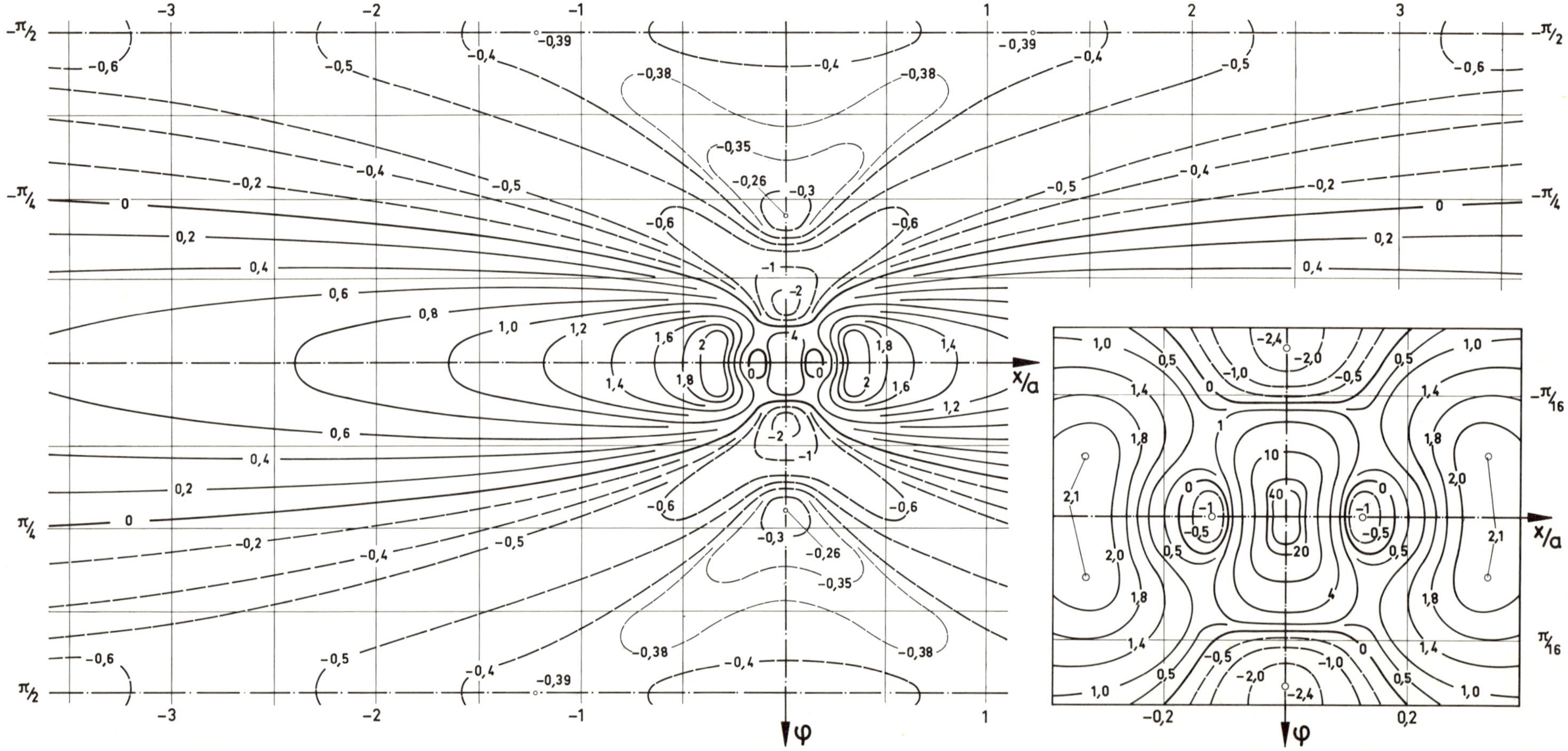

Einflußfläche (Zustandsfläche) des Biegemomentes m_x

Influence surface (condition surface) for the bending moment m_x

$100\,\pi \cdot m_{\varphi}$	**45**
$k = 7{,}5 \cdot 10^{-6}$	

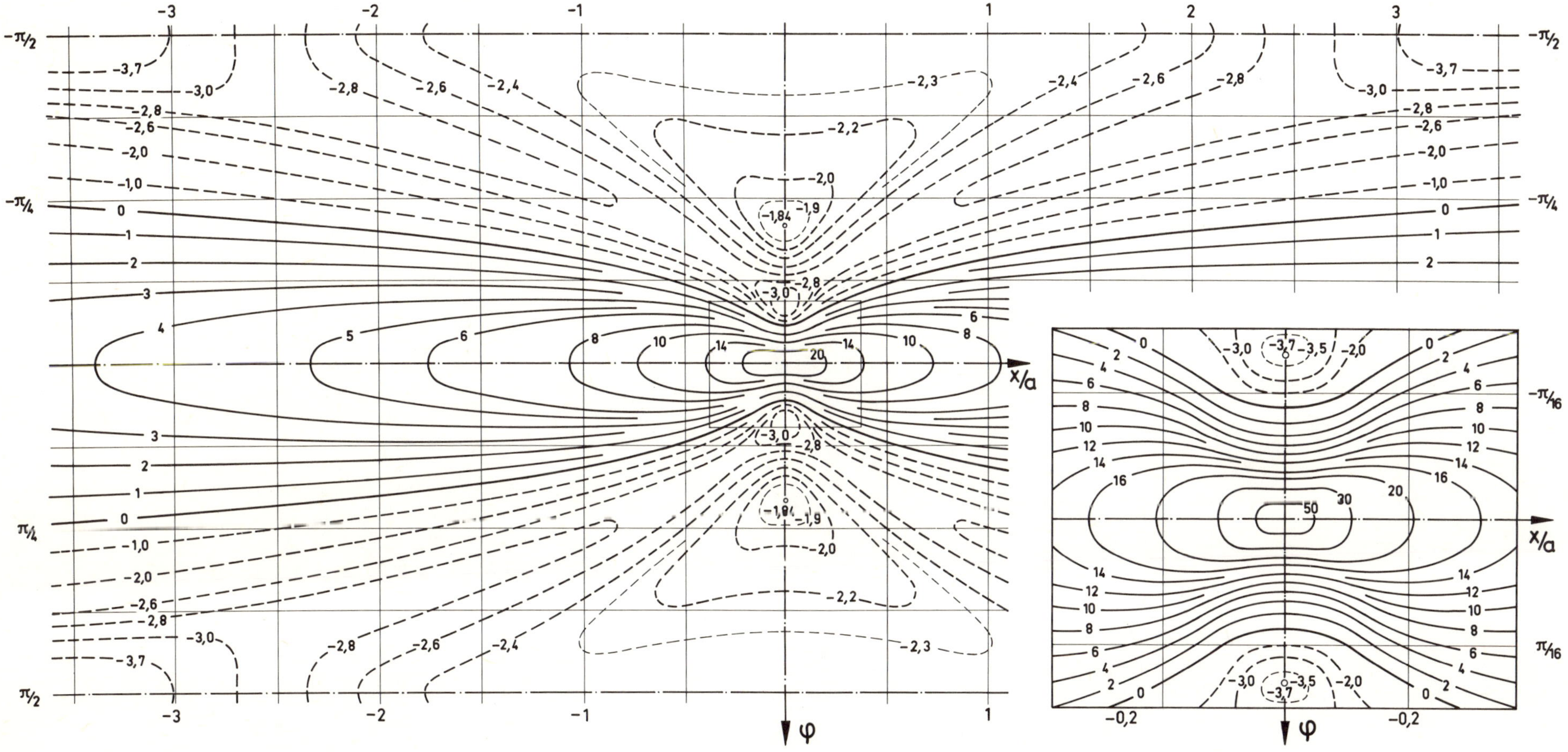

Einflußfläche (Zustandsfläche) des Biegemomentes m_{φ}

Influence surface (condition surface) for the bending moment m_{φ}

46	$100\,\pi \cdot m_{x\varphi}$
	$k = 7{,}5 \cdot 10^{-6}$

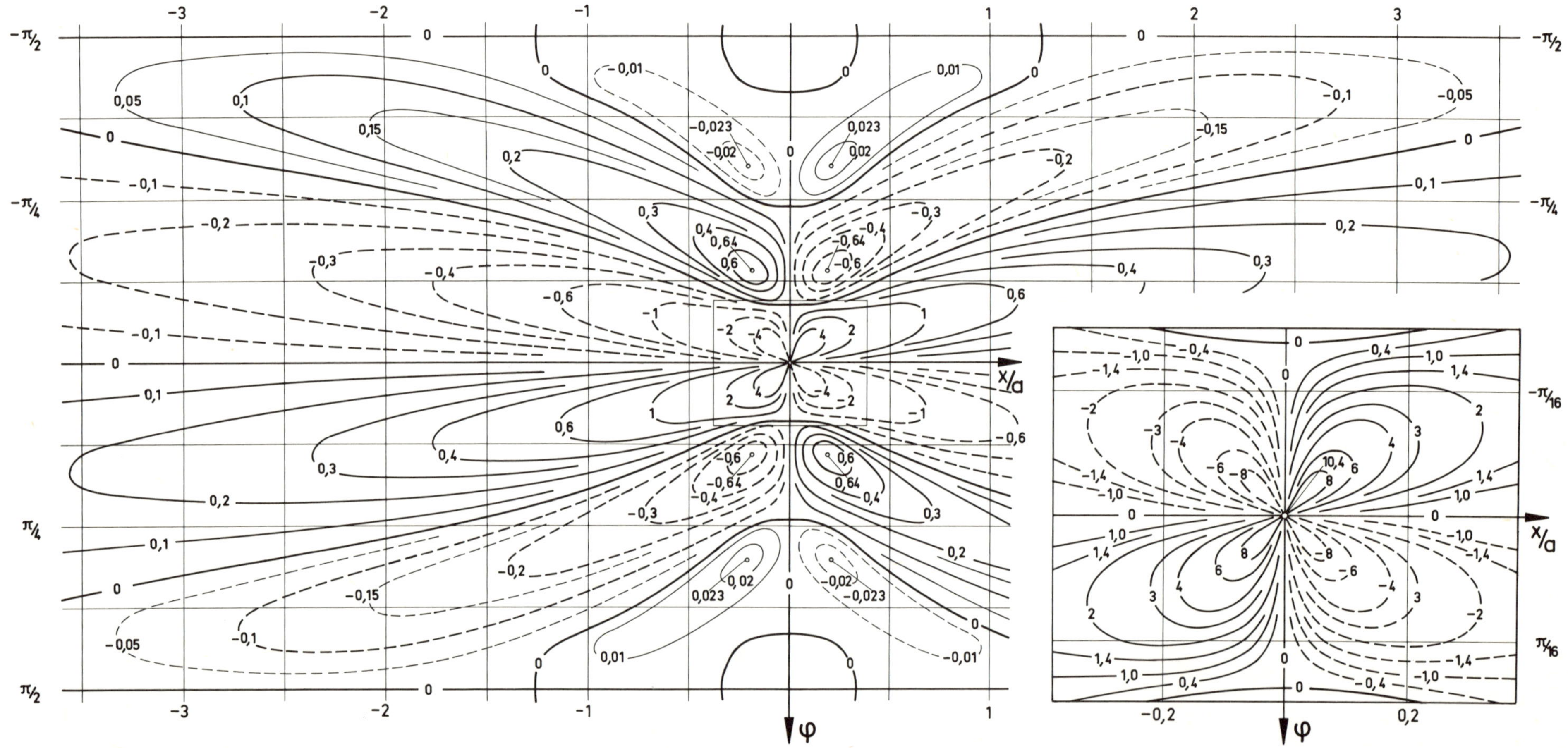

Einflußfläche (Zustandsfläche) des Drillmomentes $m_{x\varphi}$

Influence surface (condition surface) for the twisting moment $m_{x\varphi}$

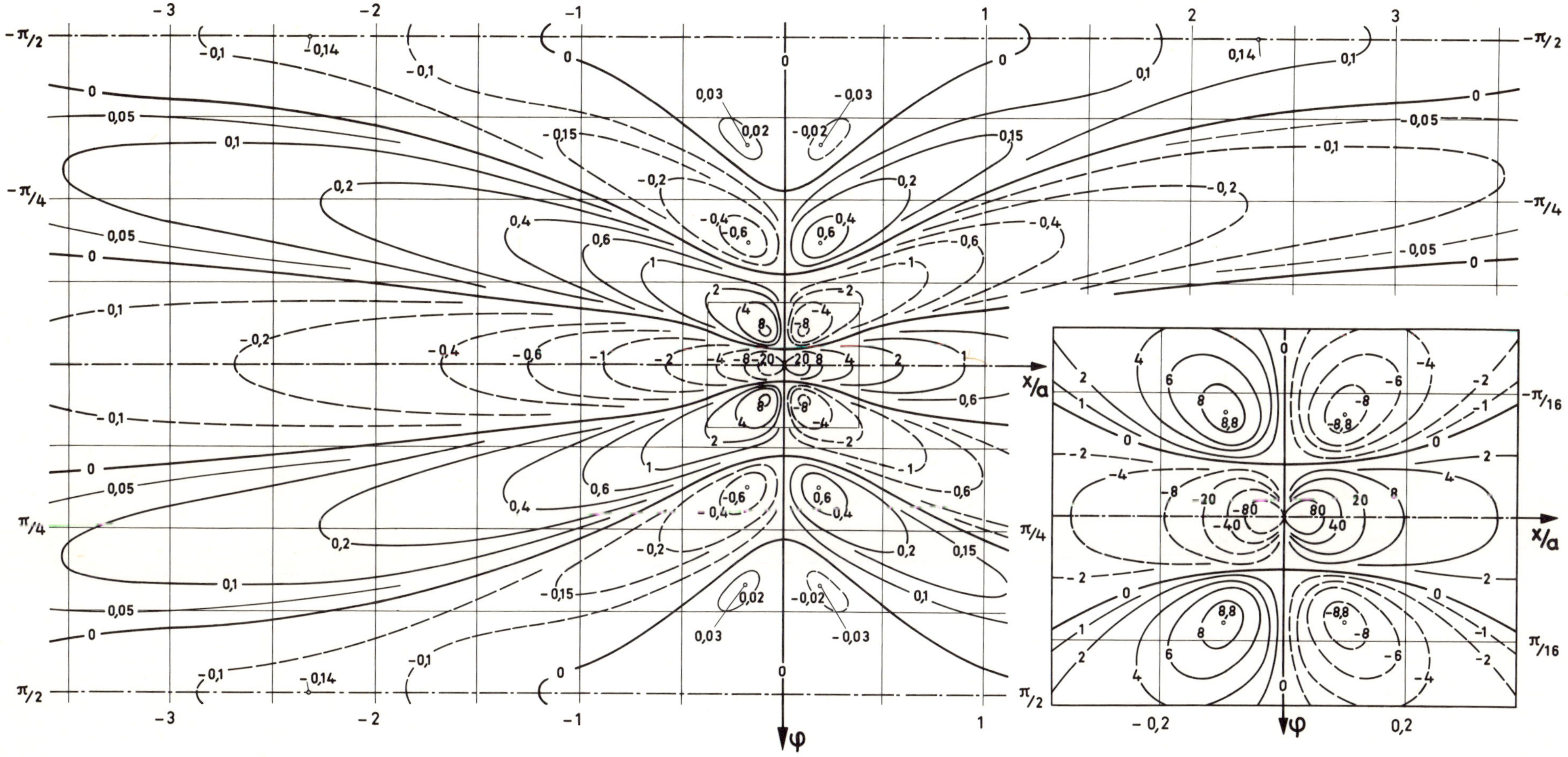

Einflußfläche (negative Zustandsfläche) der Randquerkraft $\bar{q}_x$

Influence surface (negative condition surface) for the transverse boundary force $\bar{q}_x$

48	$10\,\pi\, a \cdot \bar{q}_{\varphi}$
	$k = 7{,}5 \cdot 10^{-6}$

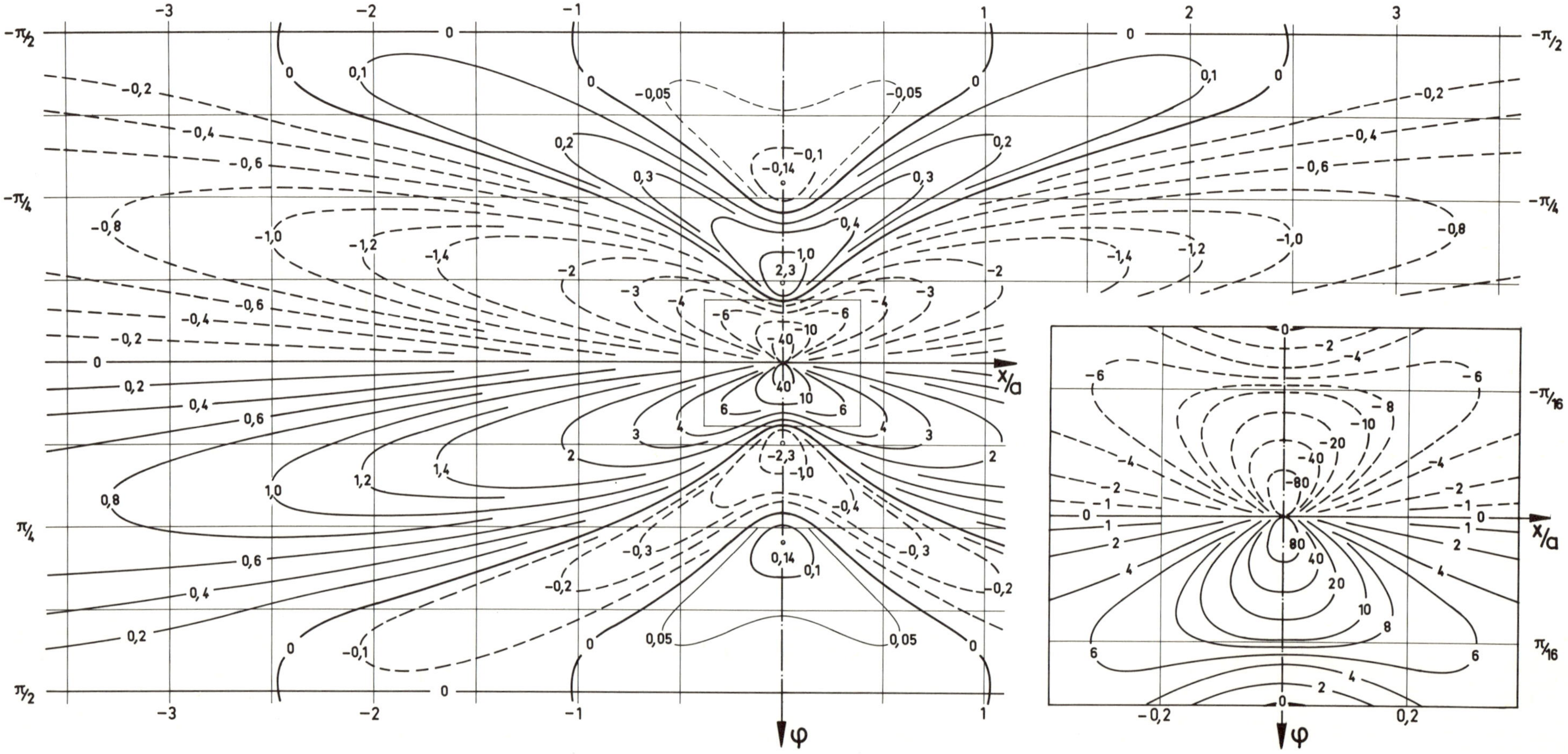

Einflußfläche (negative Zustandsfläche) der Randquerkraft $\bar{q}_{\varphi}$ Influence surface (negative condition surface) for the transverse boundary force $\bar{q}_{\varphi}$

$1000\,\pi\,K/a^2 \cdot u$	**49**
$k = 1{,}0 \cdot 10^{-5}$	

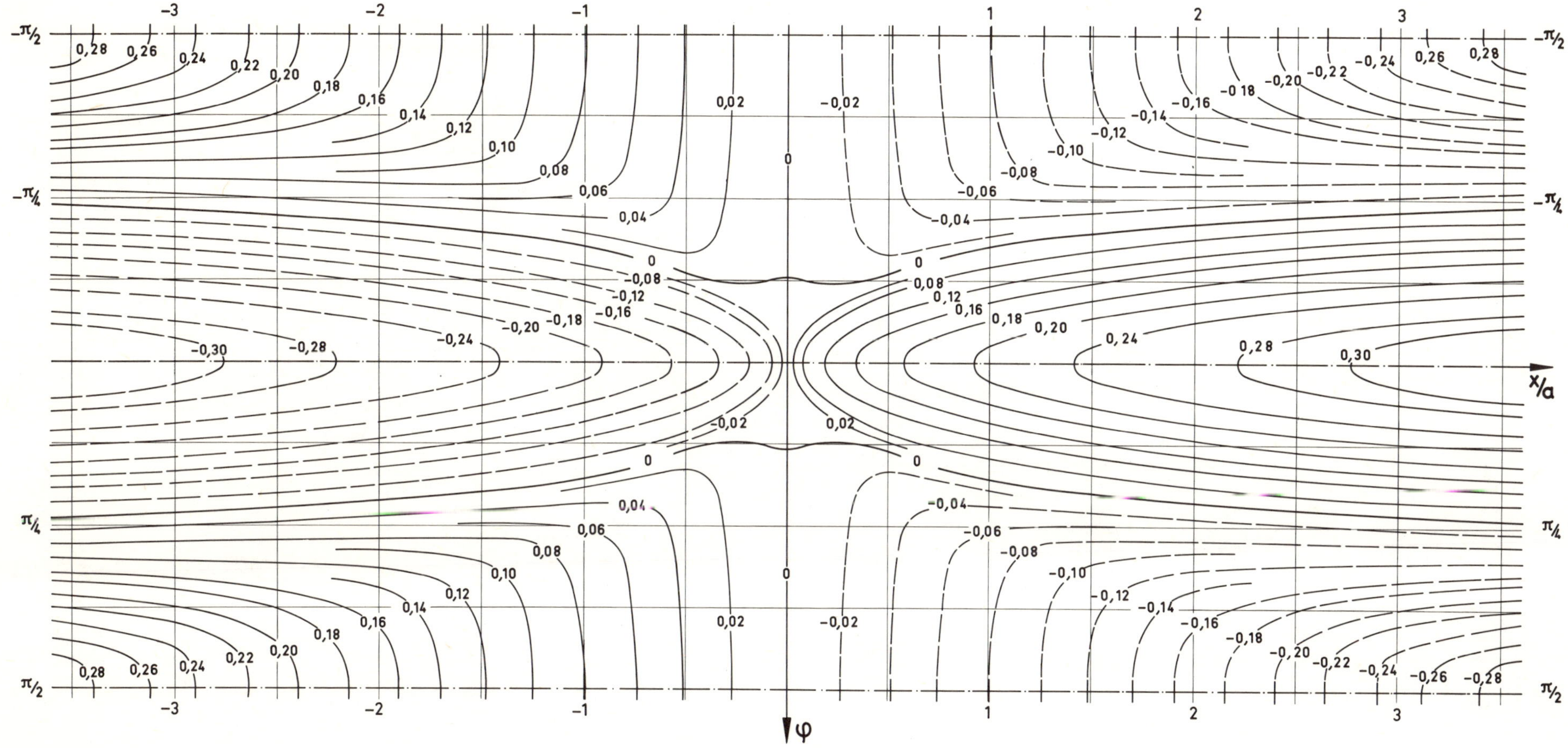

Einflußfläche (negative Zustandsfläche) der Verschiebung u

Influence surface (negative condition surface) for the displacement u

50 | $1000\,\pi\,K/a^2 \cdot v$ — $k = 1{,}0 \cdot 10^{-5}$

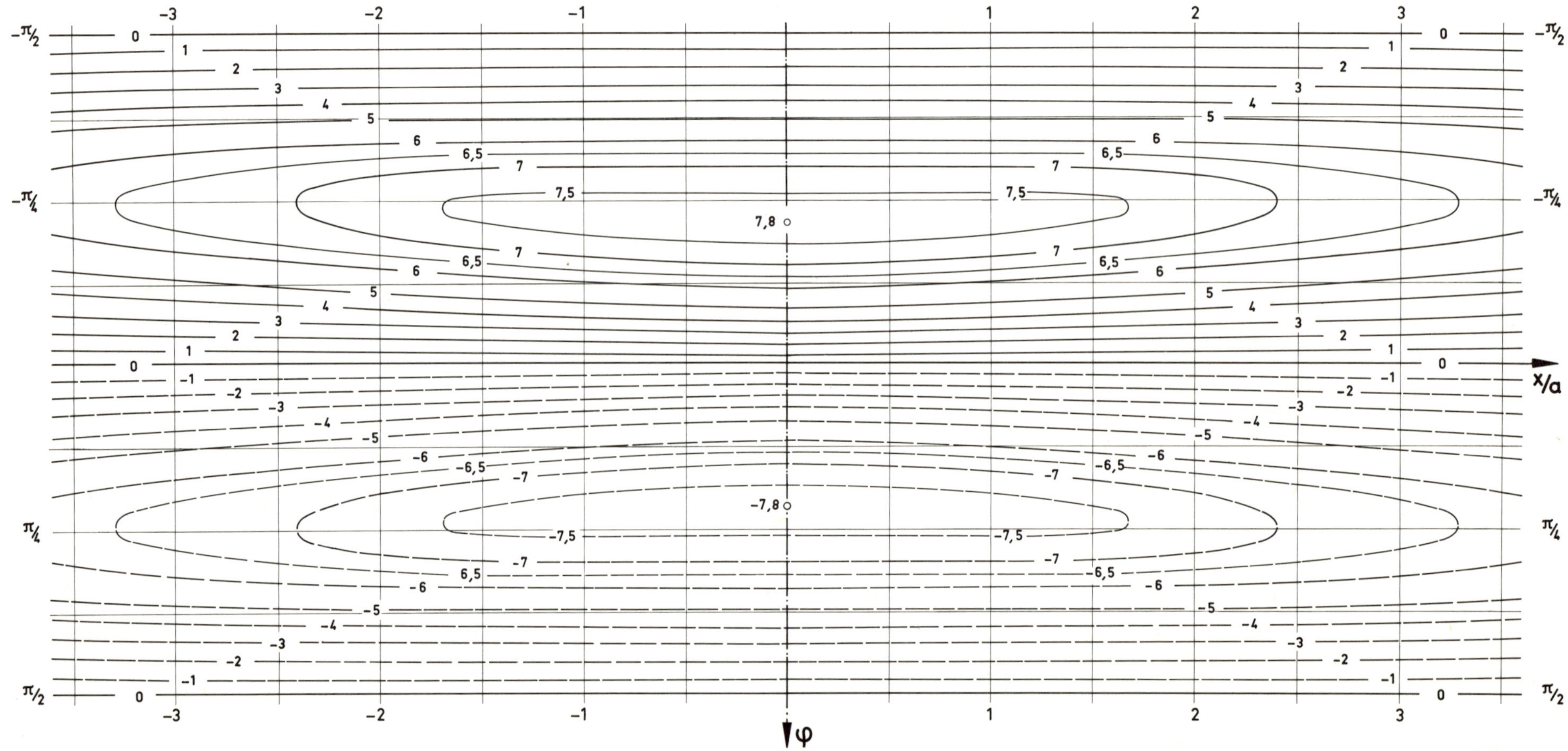

Einflußfläche (negative Zustandsfläche) der Verschiebung v

Influence surface (negative condition surface) for the displacement v

$1000\,\pi\,K/a^2 \cdot w$	**51**
$k = 1{,}0 \cdot 10^{-5}$	

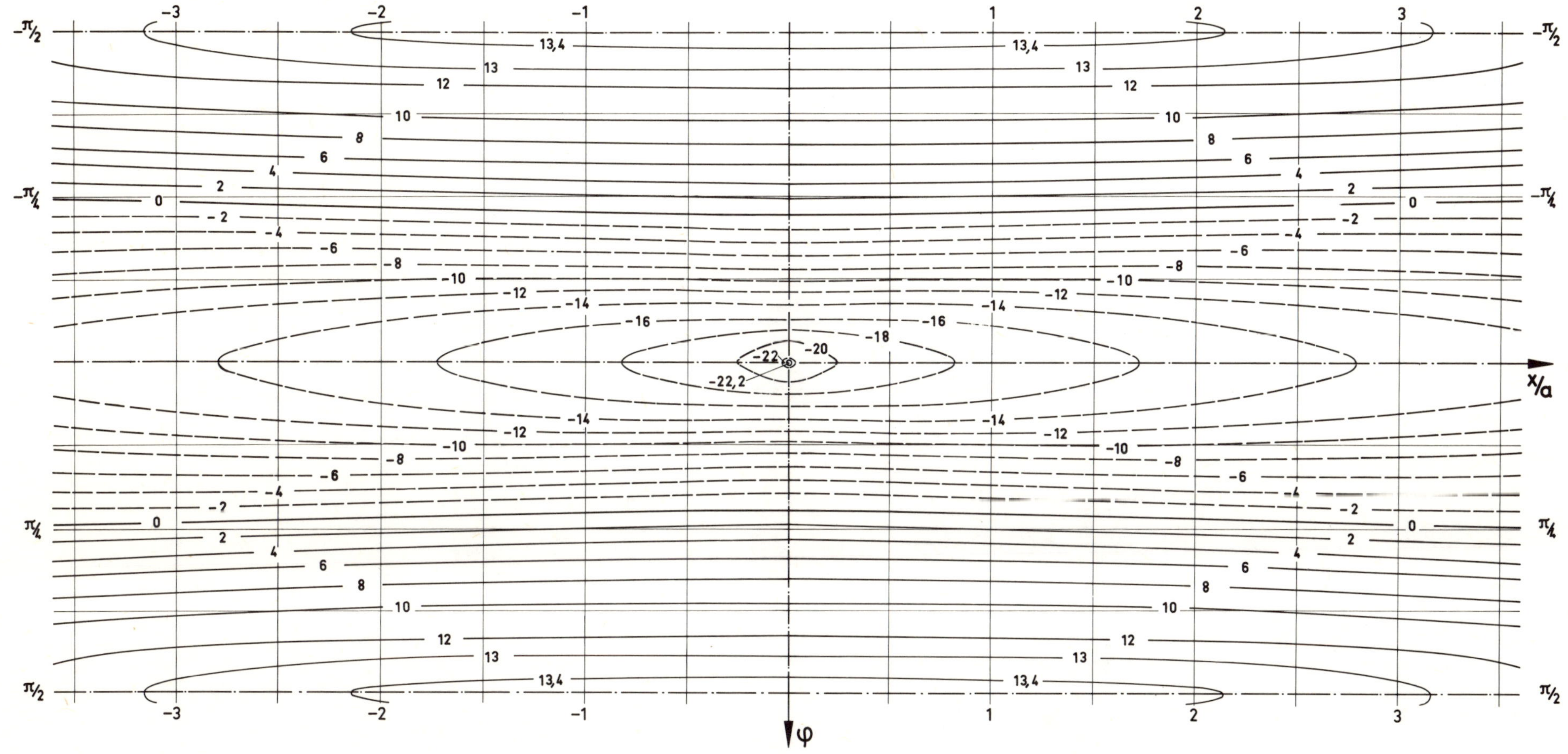

Einflußfläche (Zustandsfläche) der Verschiebung w

Influence surface (condition surface) for the displacement w

52	$100\,\pi\,K/a \cdot \chi_\varphi$
	$k = 1{,}0 \cdot 10^{-5}$

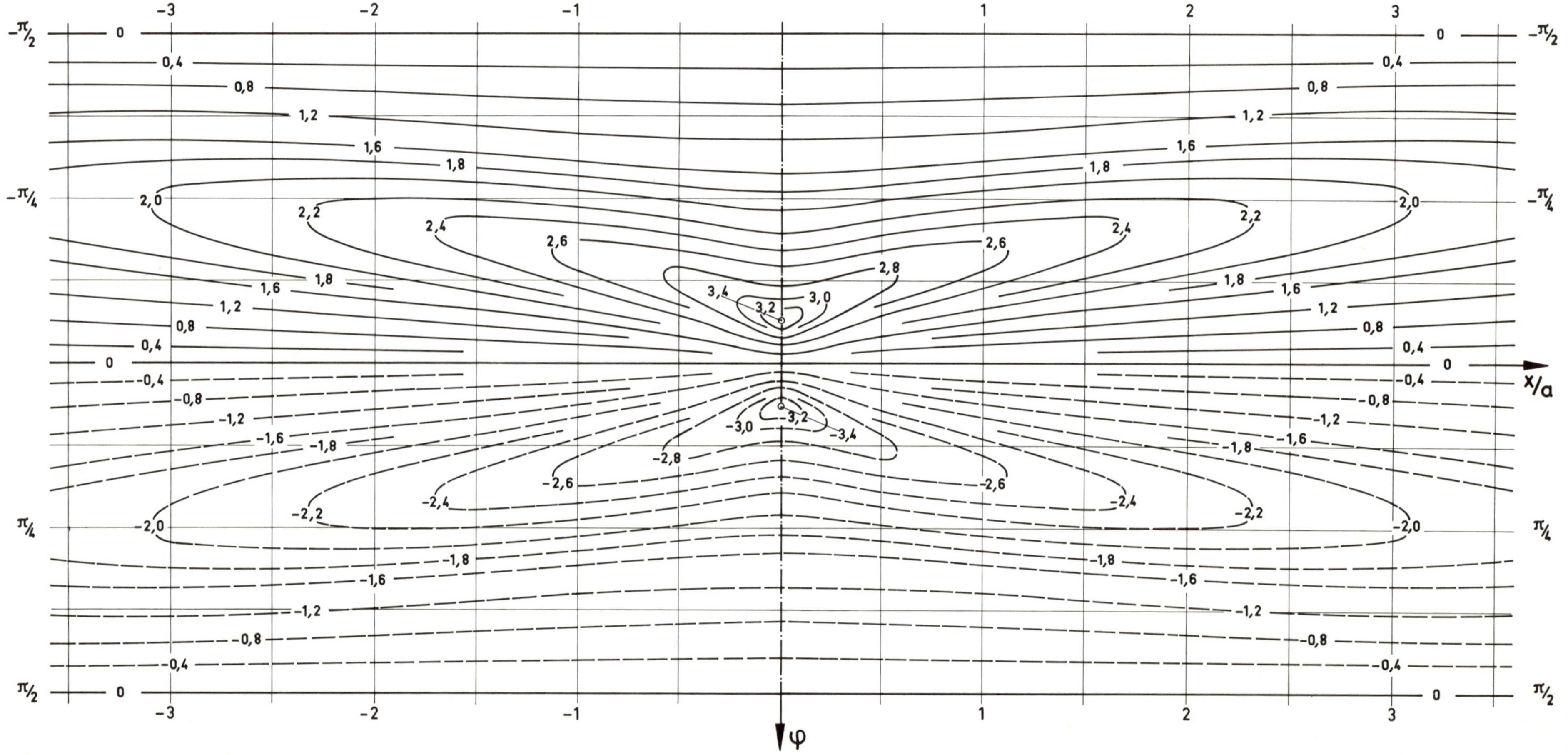

Einflußfläche (negative Zustandsfläche) der Verdrehung χ_φ

Influence surface (negative condition surface) for the rotation χ_φ

$\pi a \cdot n_x$	**53**
$k = 1{,}0 \cdot 10^{-5}$	

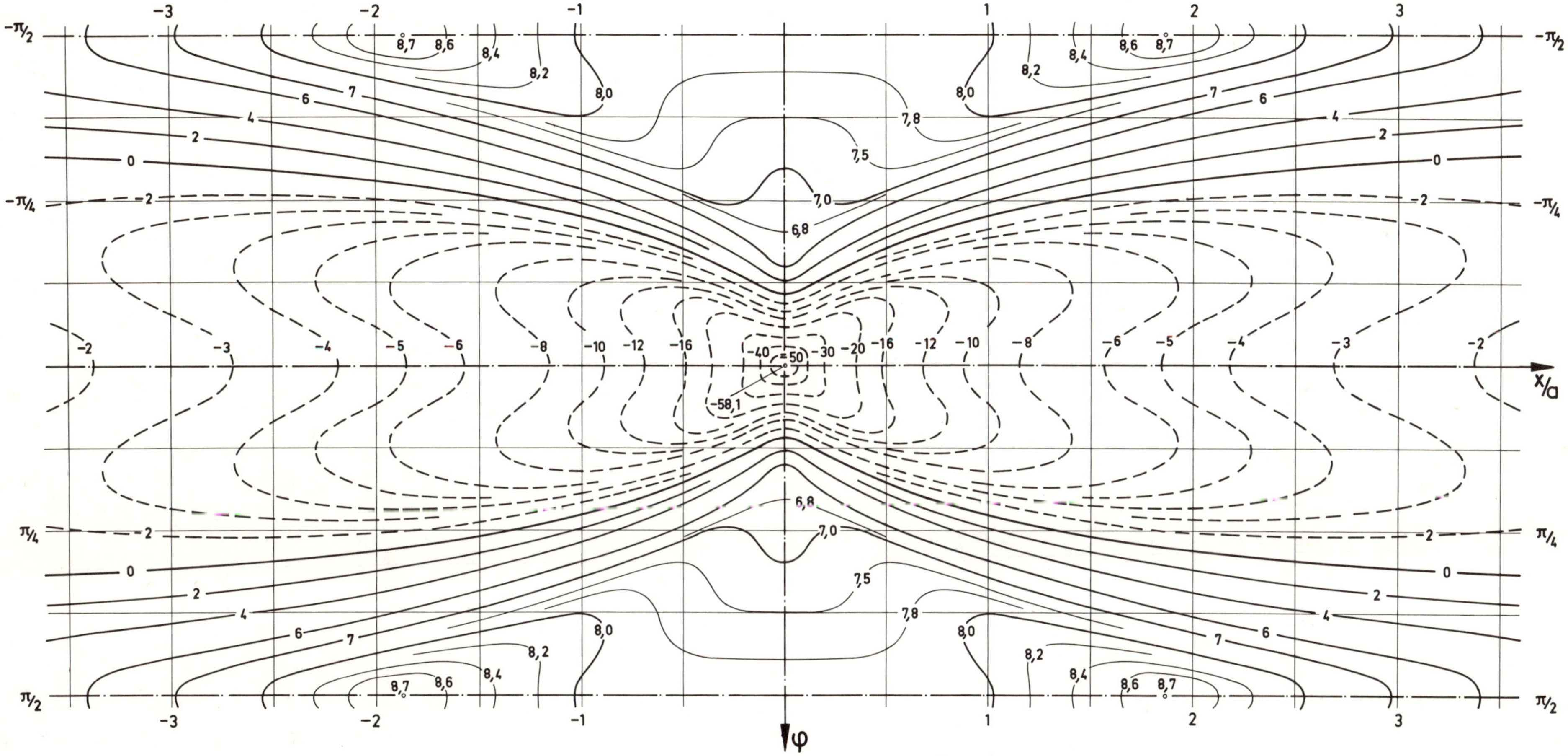

Einflußfläche (Zustandsfläche) der Normalkraft n_x

Influence surface (condition surface) for the normal force n_x

54	$\pi a \cdot n_{\varphi}$
	$k = 1{,}0 \cdot 10^{-5}$

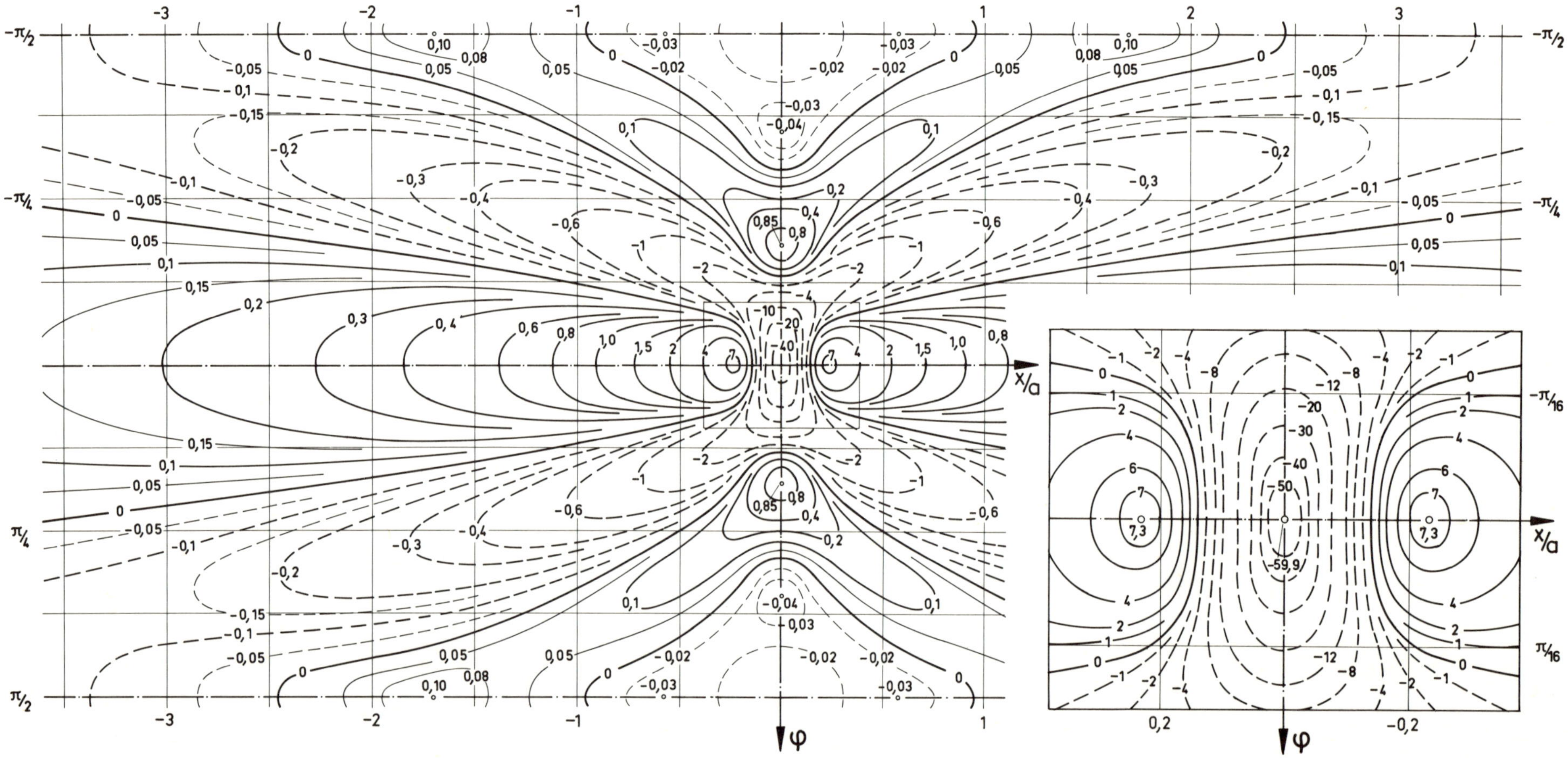

Einflußfläche (Zustandsfläche) der Normalkraft n_{φ}

Influence surface (condition surface) for the normal force n_{φ}

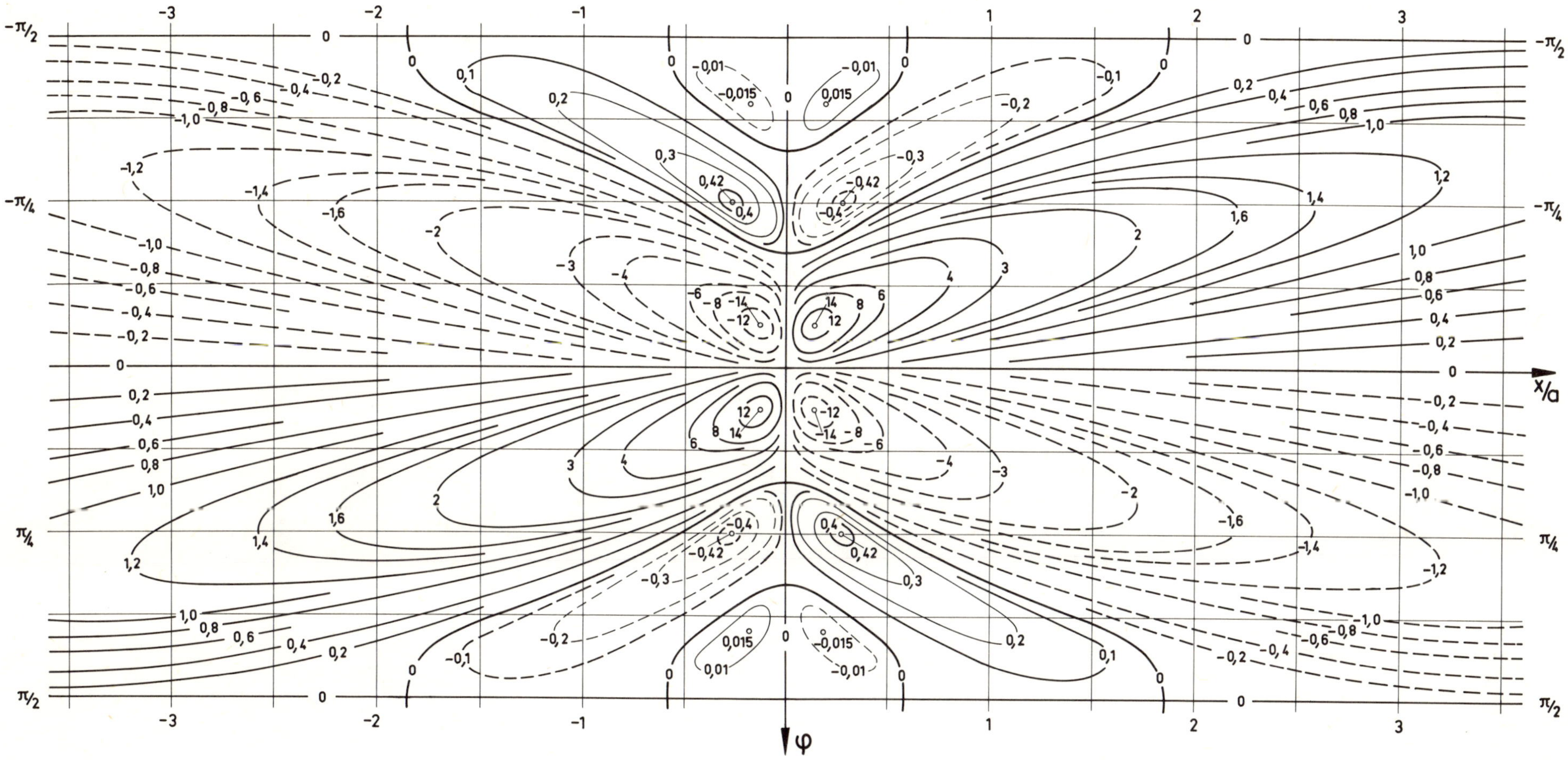

Einflußfläche (Zustandsfläche) der Randschubkraft $\bar{n}_{x\varphi}$

Influence surface (condition surface) for the boundary shearing force $\bar{n}_{x\varphi}$

56 | $100\,\pi \cdot m_x$ / $k = 1{,}0 \cdot 10^{-5}$

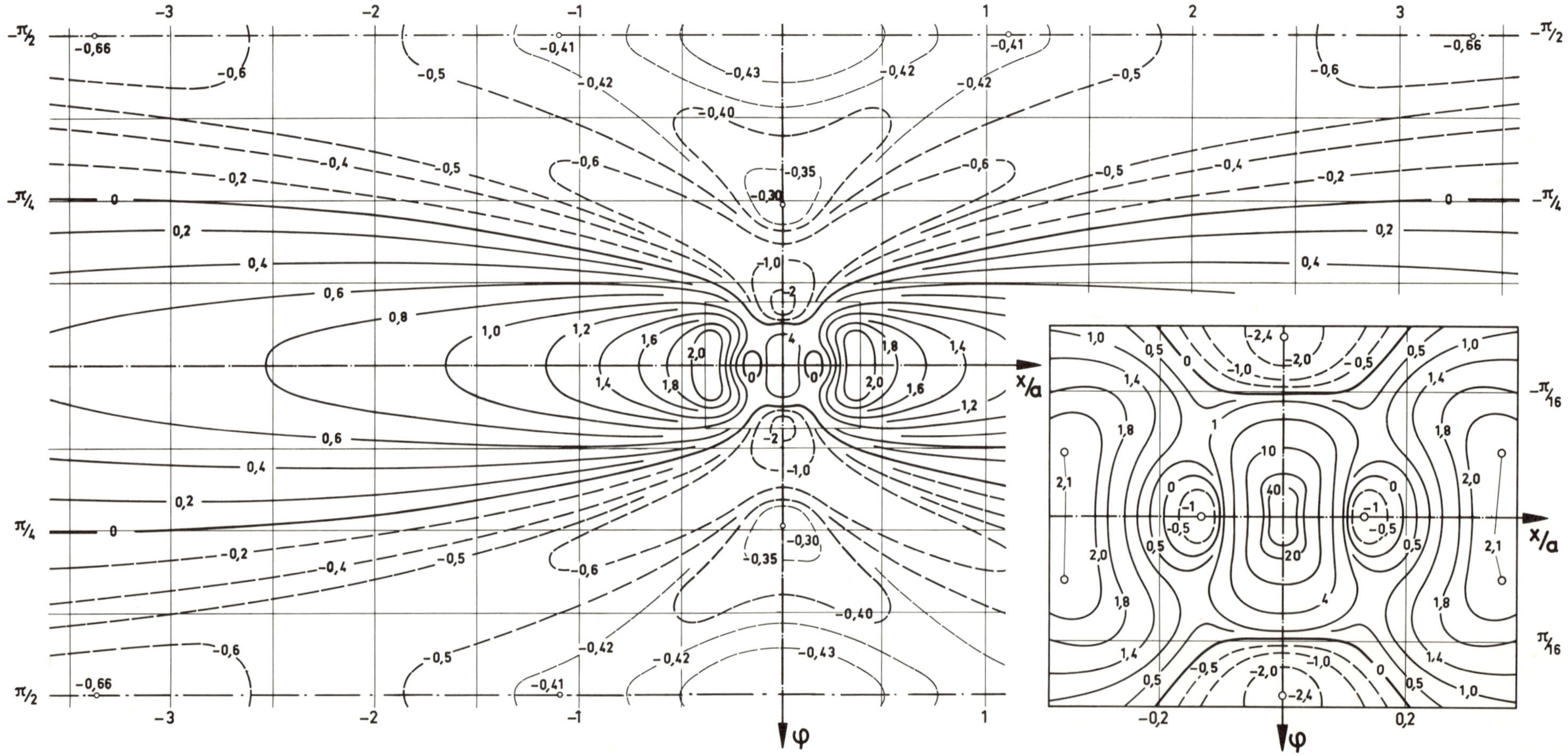

Einflußfläche (Zustandsfläche) des Biegemomentes m_x

Influence surface (condition surface) for the bending moment m_x

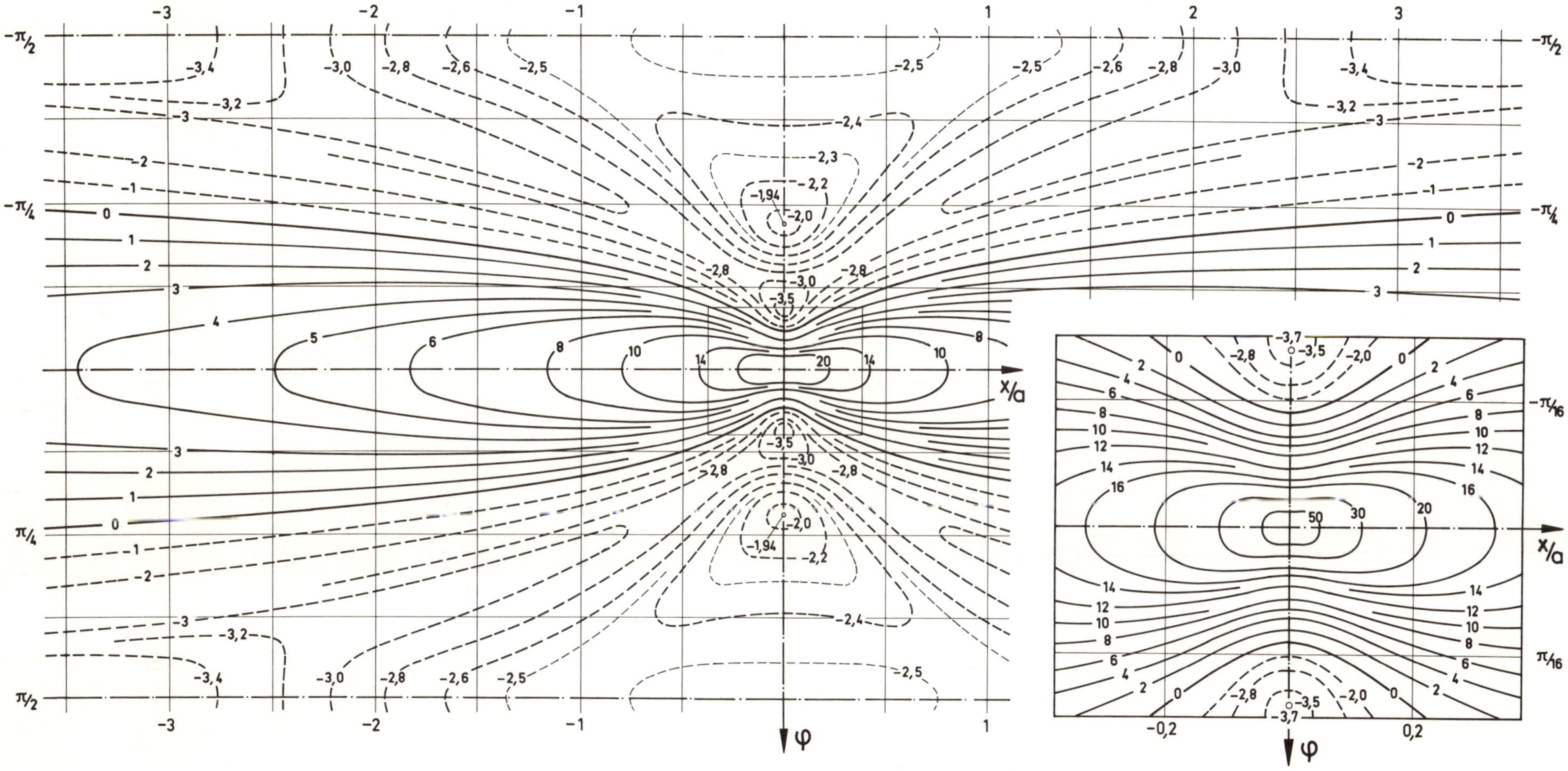

Einflußfläche (Zustandsfläche) des Biegemomentes m_φ

Influence surface (condition surface) for the bending moment m_φ

58	$100\,\pi \cdot m_{x\varphi}$
	$k = 1{,}0 \cdot 10^{-5}$

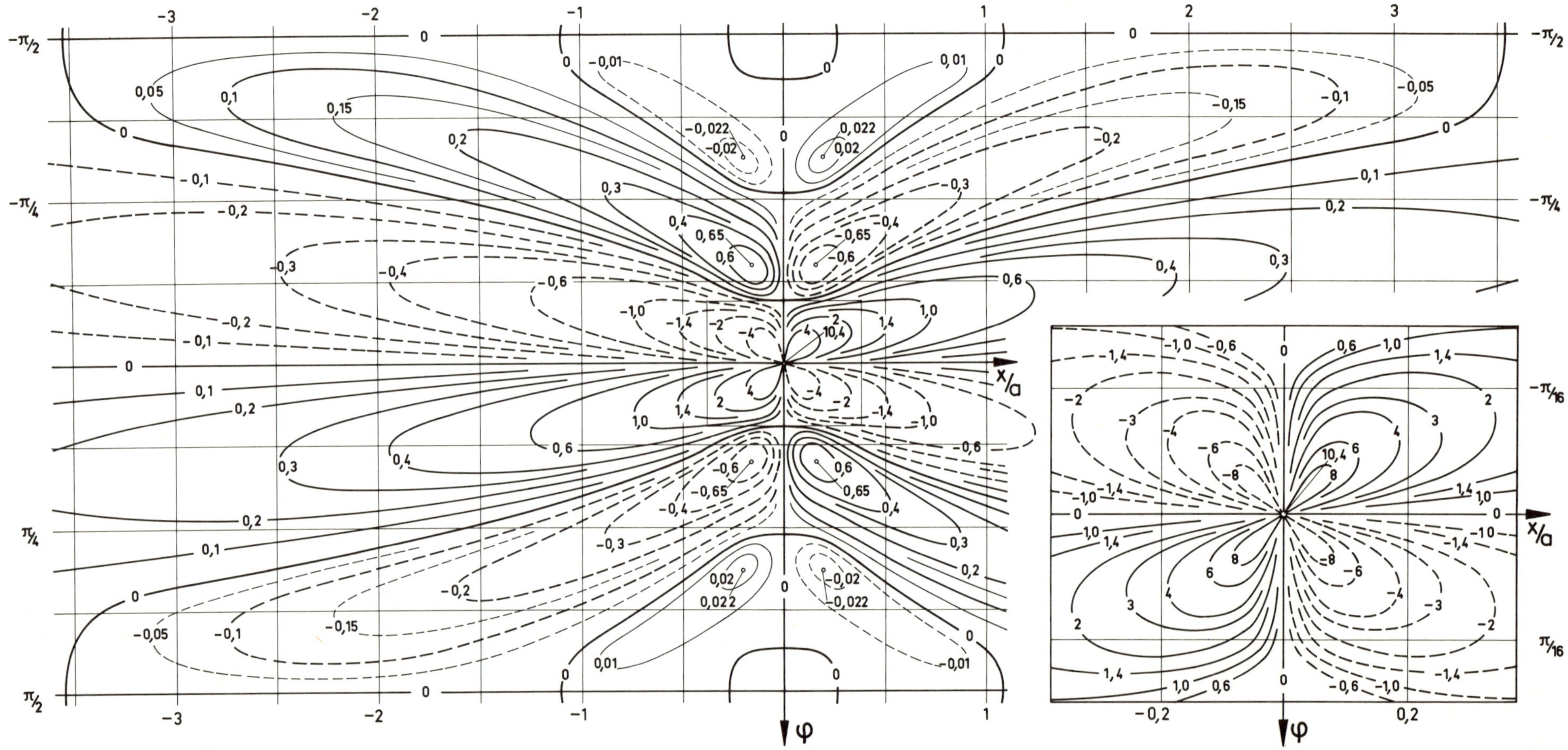

Einflußfläche (Zustandsfläche) des Drillmomentes $m_{x\varphi}$

Influence surface (condition surface) for the twisting moment $m_{x\varphi}$

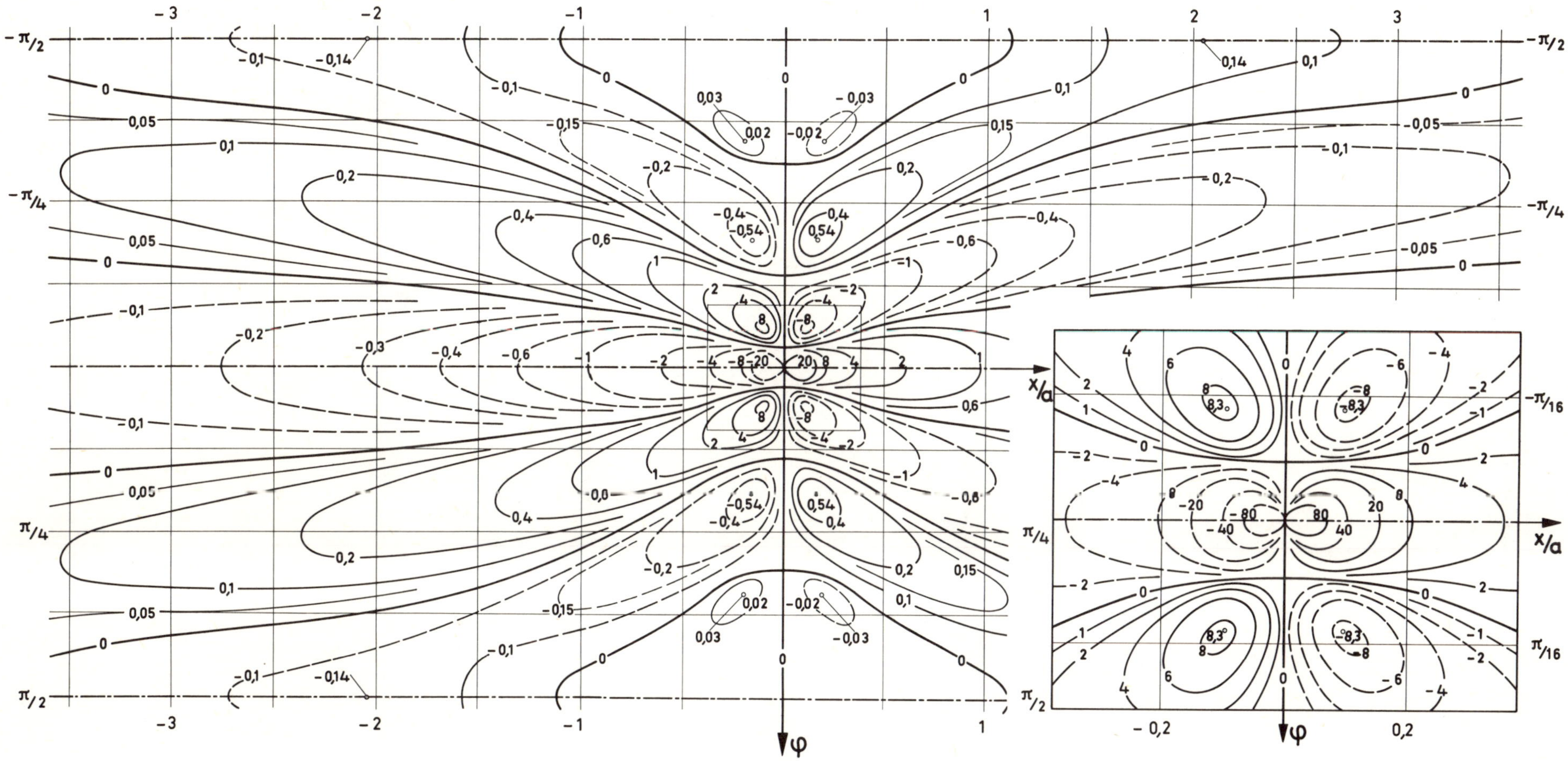

Einflußfläche (negative Zustandsfläche) der Randquerkraft $\bar{q}_x$

Influence surface (negative condition surface) for the transverse boundary force $\bar{q}_x$

60	$10\,\pi\,a \cdot \bar{q}_{\varphi}$
	$k = 1{,}0 \cdot 10^{-5}$

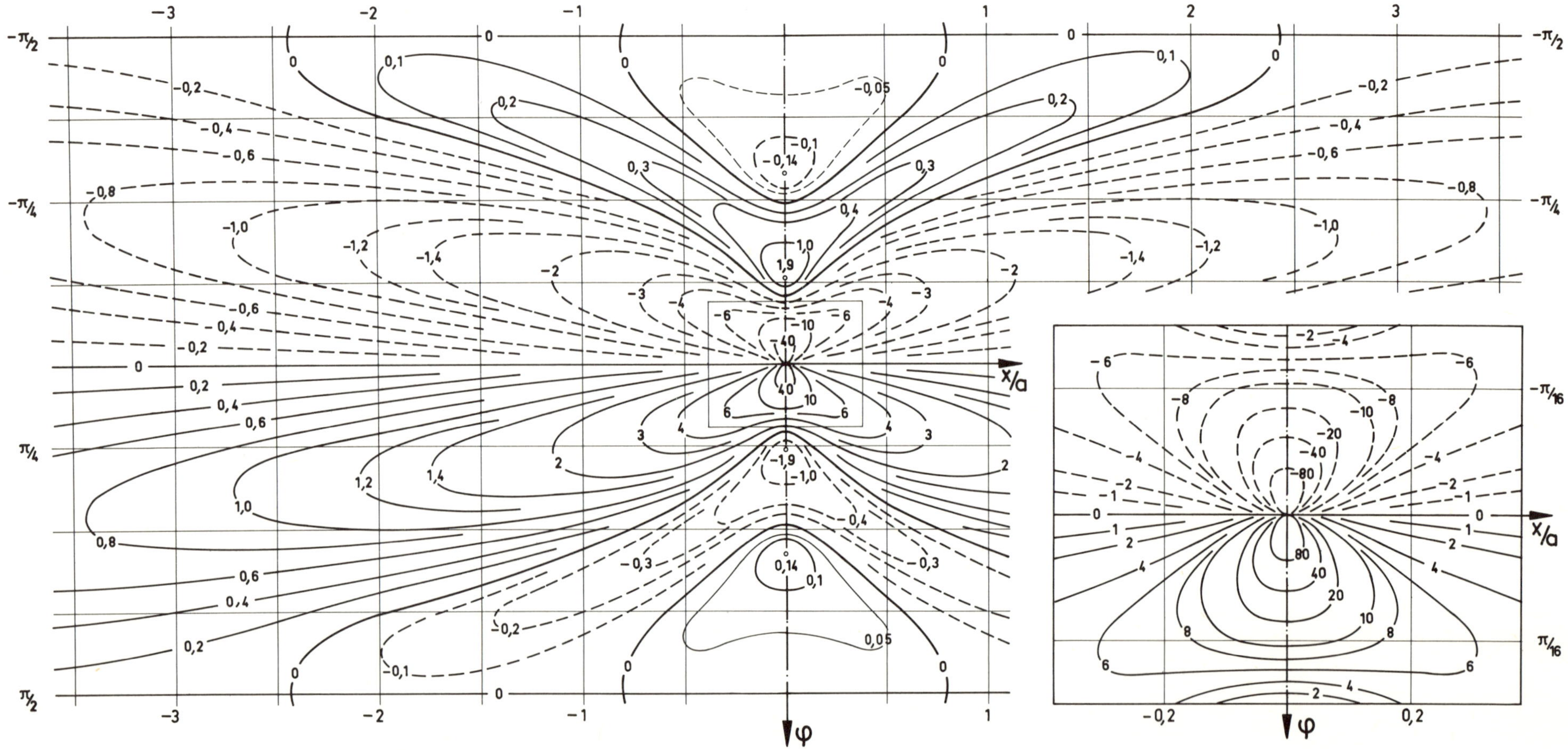

Einflußfläche (negative Zustandsfläche) der Randquerkraft $\bar{q}_{\varphi}$

Influence surface (negative condition surface) for the transverse boundary force $\bar{q}_{\varphi}$

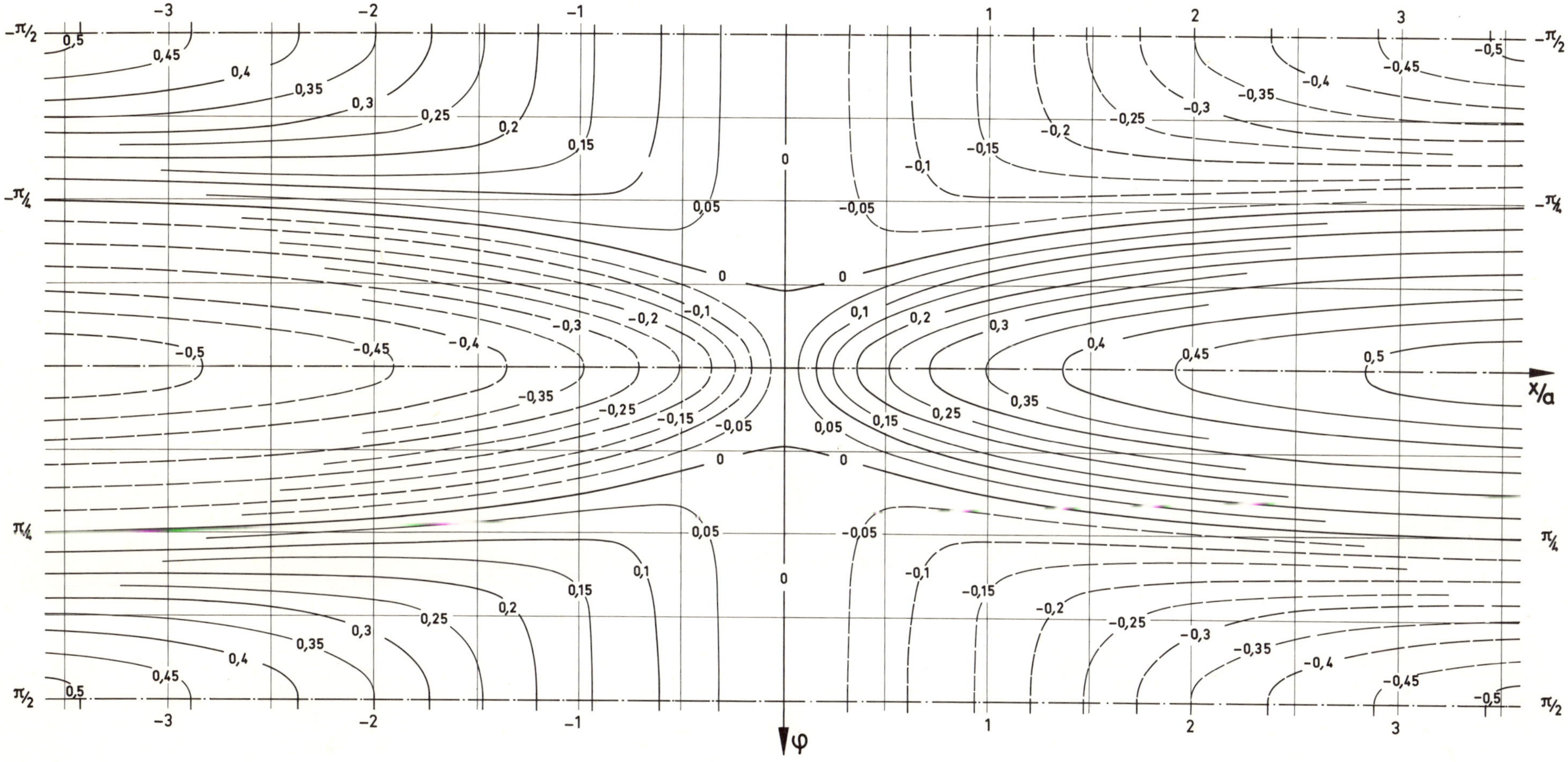

Einflußfläche (negative Zustandsfläche) der Verschiebung u

Influence surface (negative condition surface) for the displacement u

62	$1000\,\pi\,K/a^2 \cdot v$
	$k = 2{,}5 \cdot 10^{-5}$

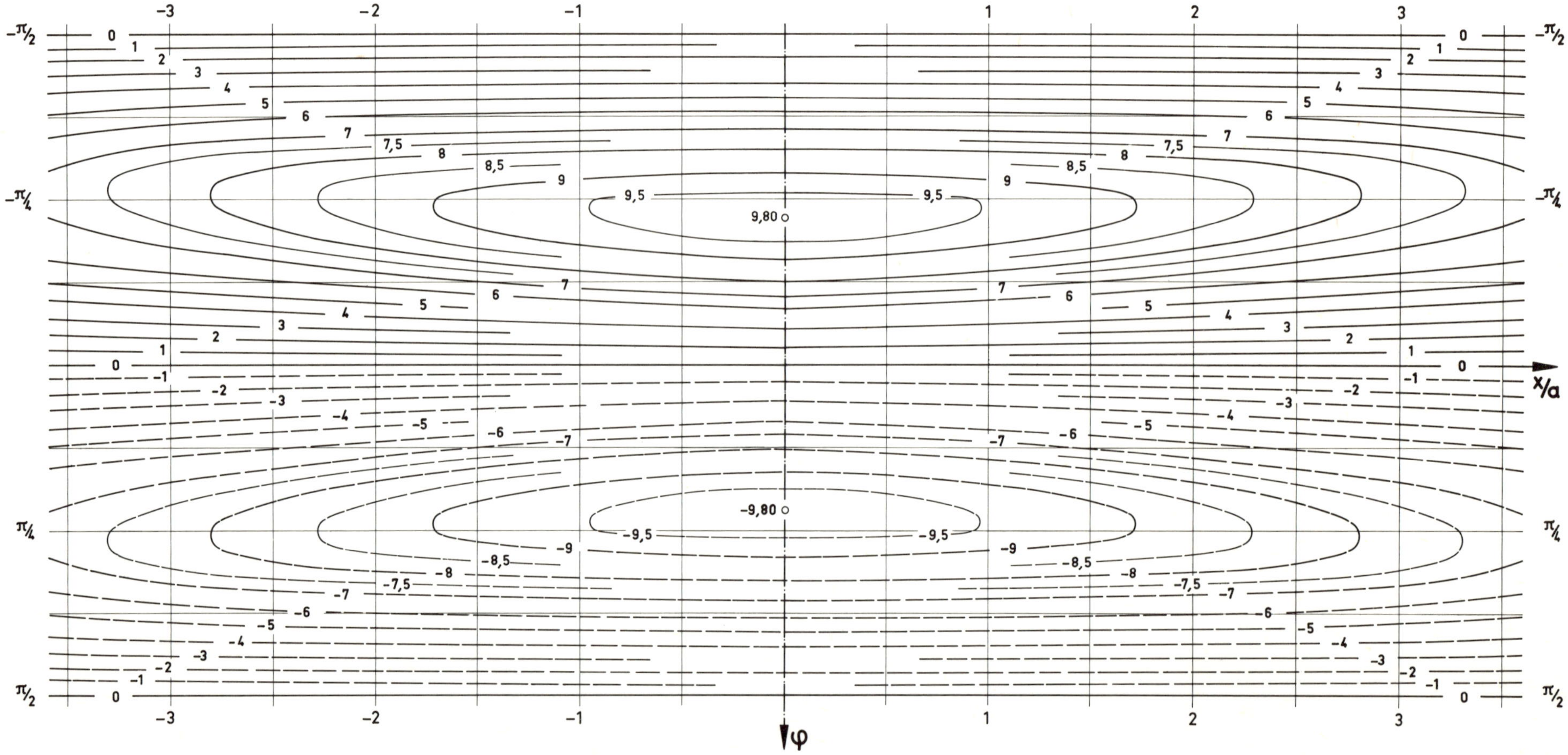

Einflußfläche (negative Zustandsfläche) der Verschiebung v

Influence surface (negative condition surface) for the displacement v

$1000\,\pi\, K/a^2 \cdot w$	**63**
$k = 2{,}5 \cdot 10^{-5}$	

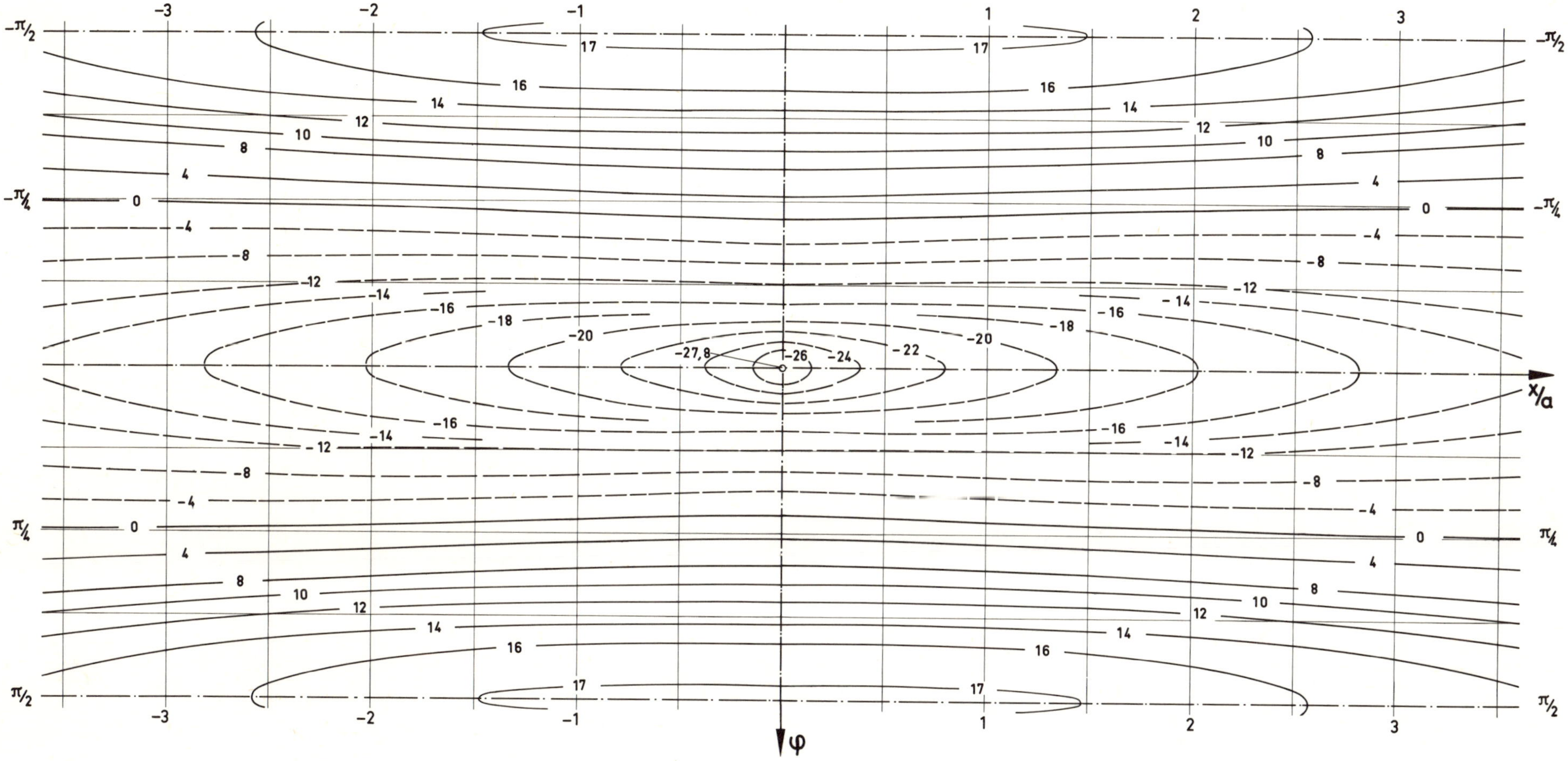

Einflußfläche (Zustandsfläche) der Verschiebung w

Influence surface (condition surface) for the displacement w

64	$100\,\pi\,K/a \cdot \chi_\varphi$
	$k = 2{,}5 \cdot 10^{-5}$

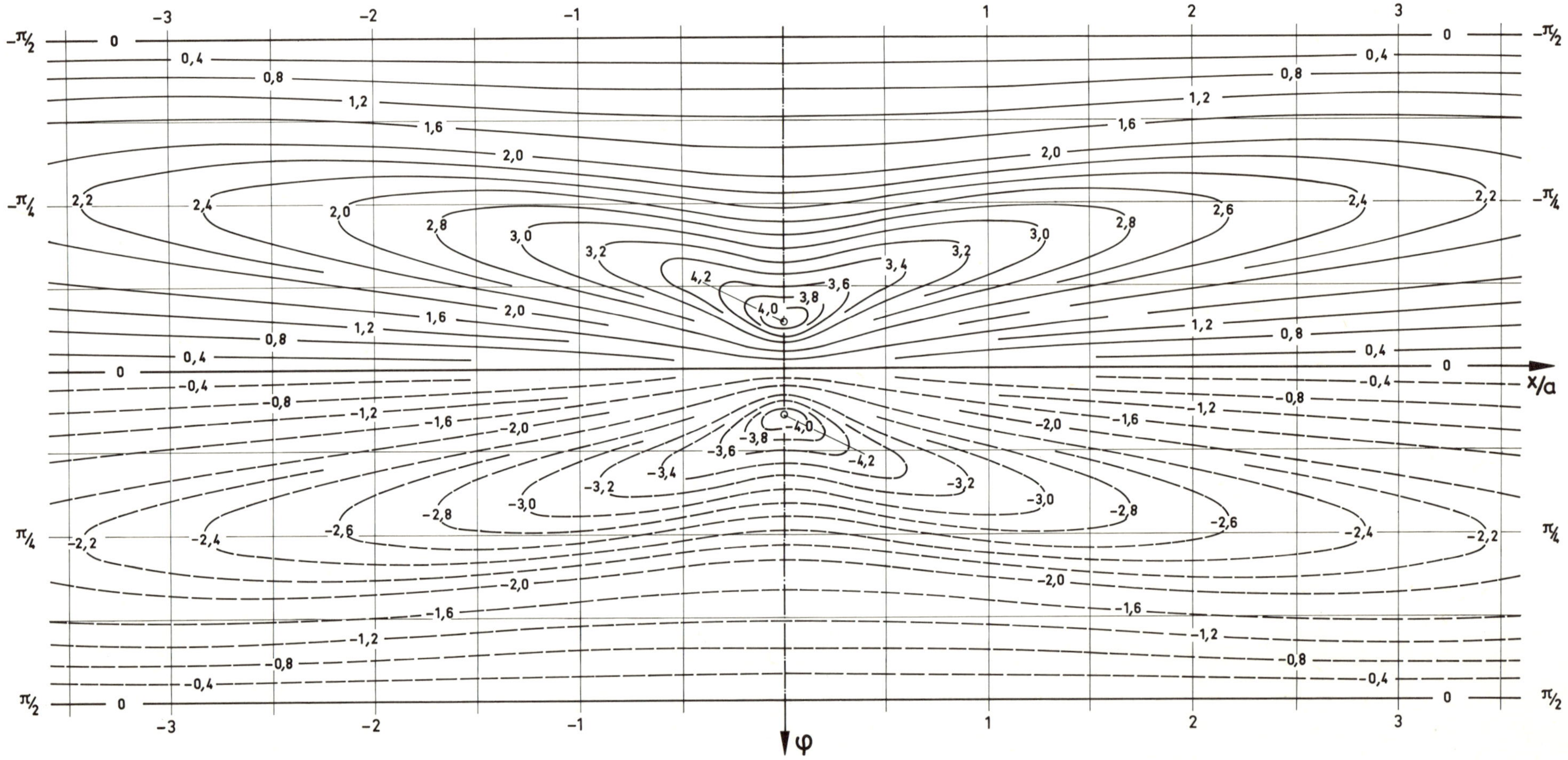

Einflußfläche (negative Zustandsfläche) der Verdrehung χ_φ

Influence surface (negative condition surface) for the rotation χ_φ

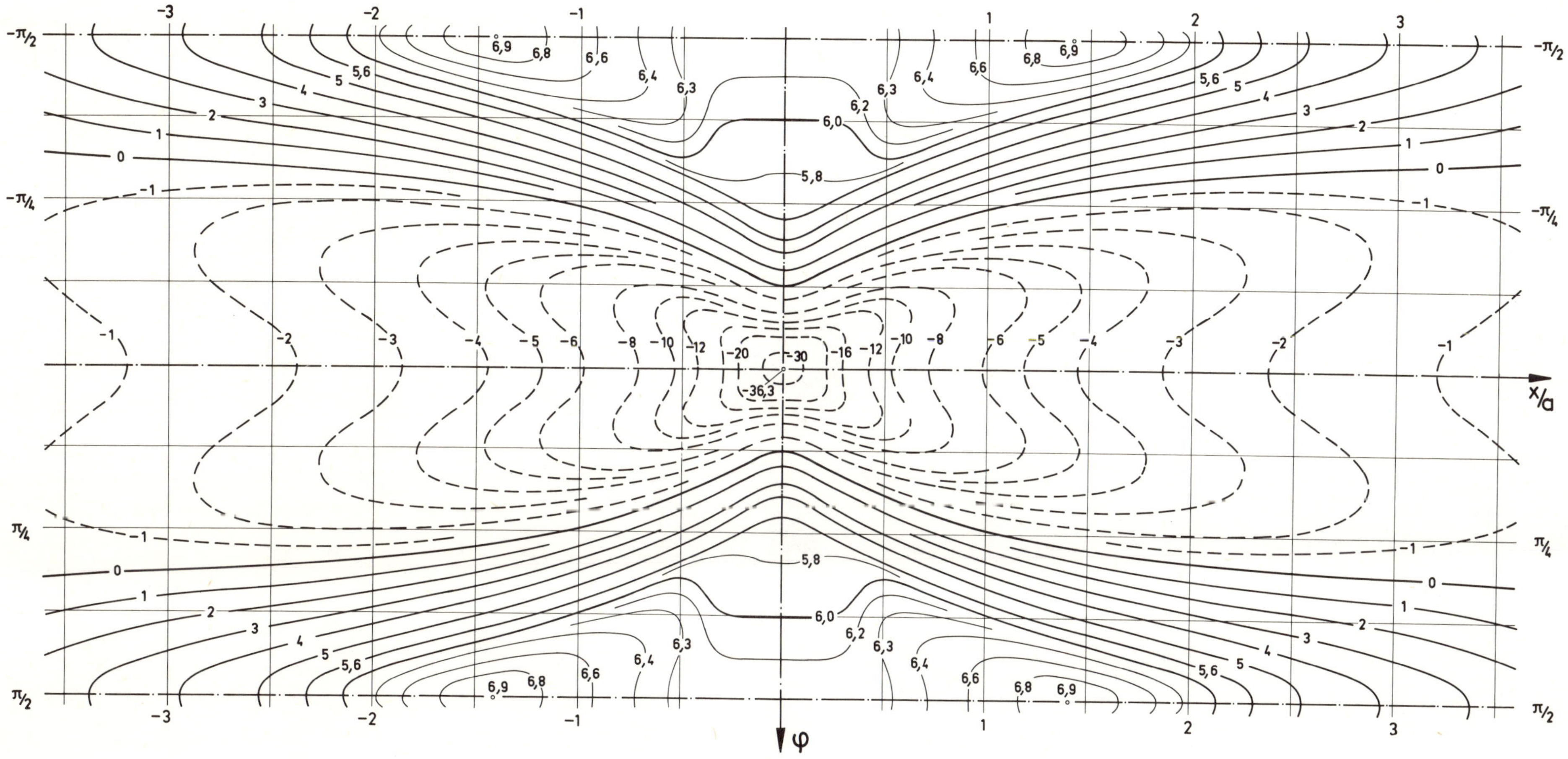

Einflußfläche (Zustandsfläche) der Normalkraft n_x

Influence surface (condition surface) for the normal force n_x

66	$\pi a \cdot n_\varphi$
	$k = 2{,}5 \cdot 10^{-5}$

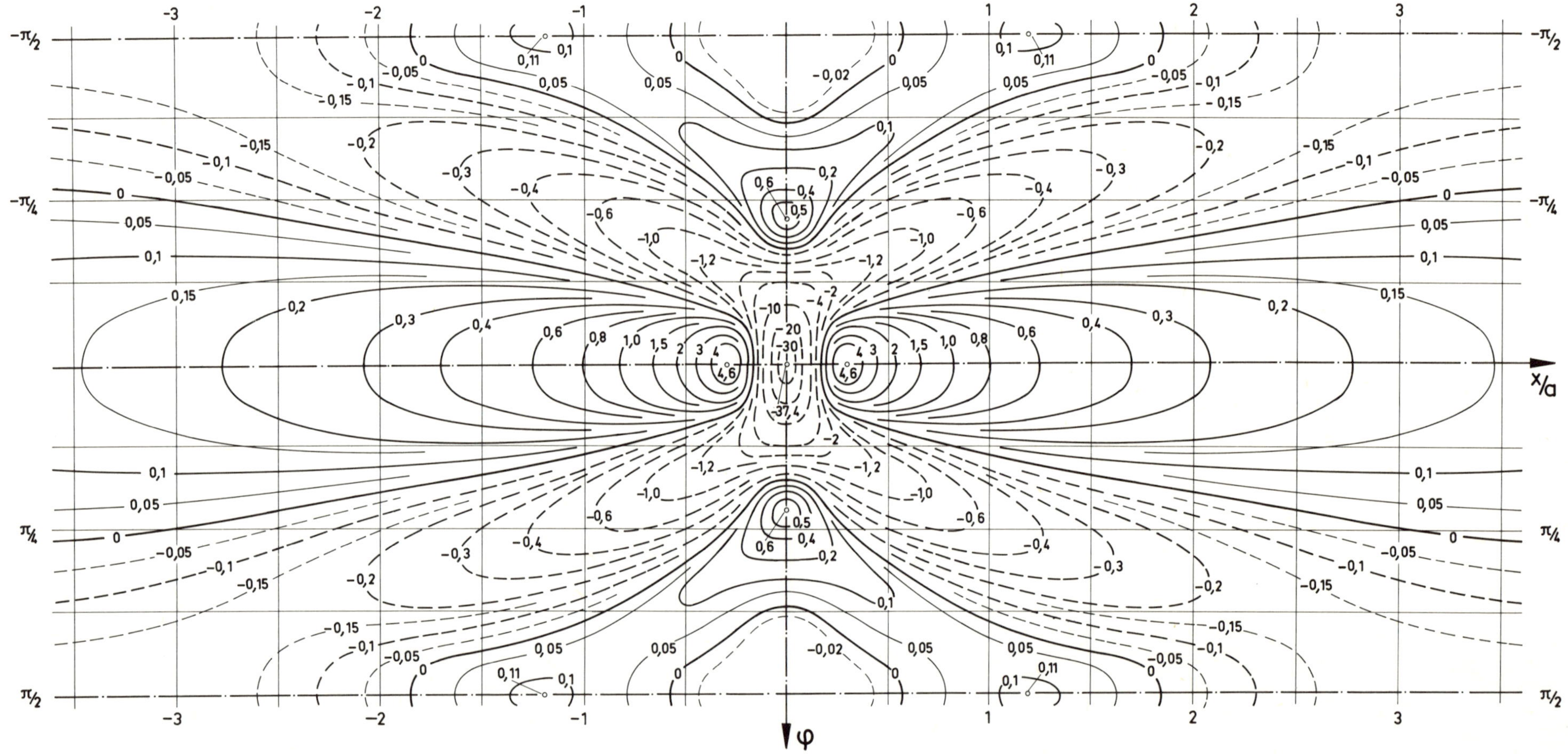

Einflußfläche (Zustandsfläche) der Normalkraft n_φ

Influence surface (condition surface) for the normal force n_φ

$\pi a \cdot \bar{n}_{x\varphi}$	**67**
$k = 2{,}5 \cdot 10^{-5}$	

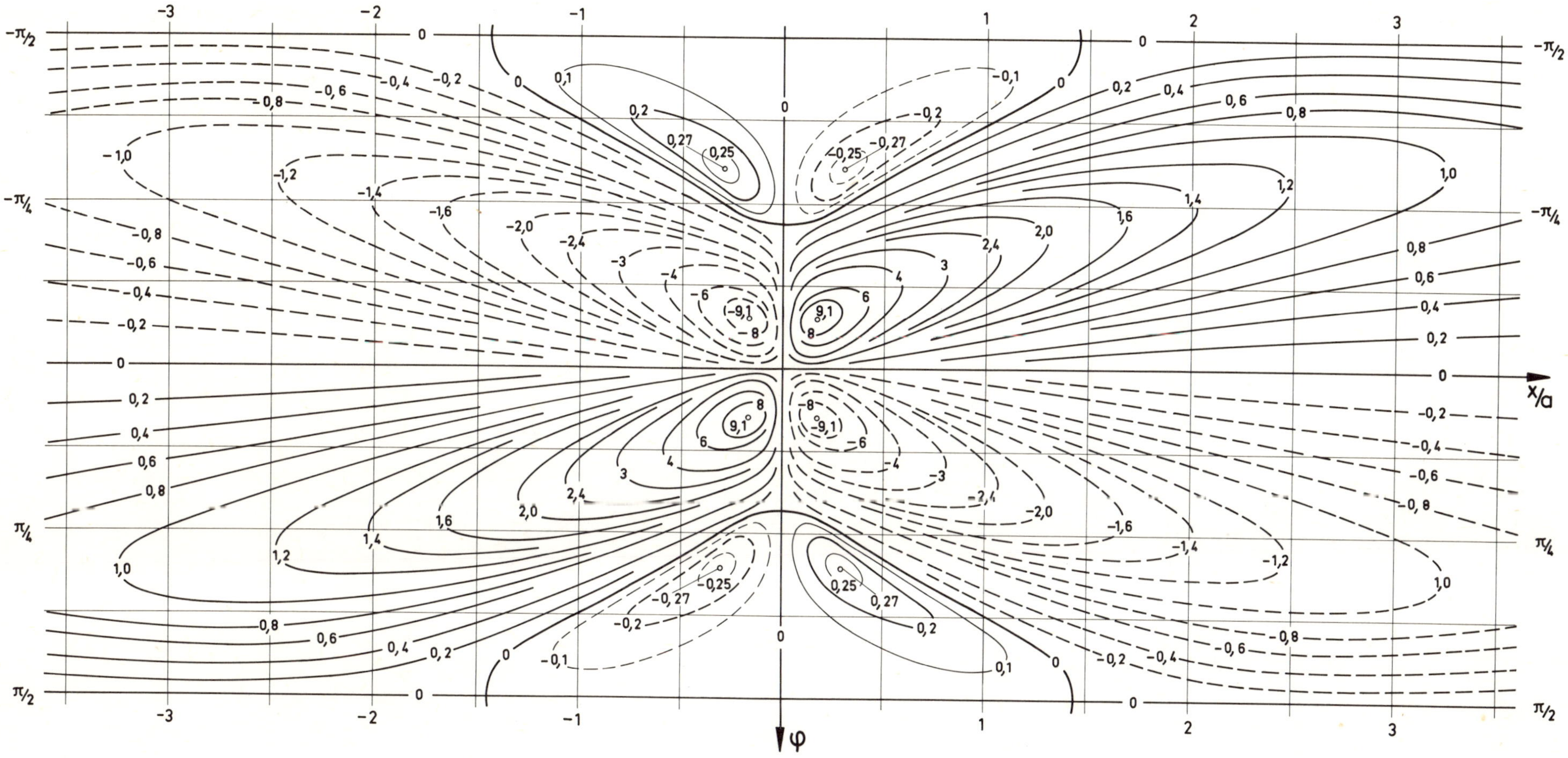

Einflußfläche (Zustandsfläche) der Randschubkraft $\bar{n}_{x\varphi}$

Influence surface (condition surface) for the boundary shearing force $\bar{n}_{x\varphi}$

68 | $100\,\pi \cdot m_x$ | $k = 2{,}5 \cdot 10^{-5}$

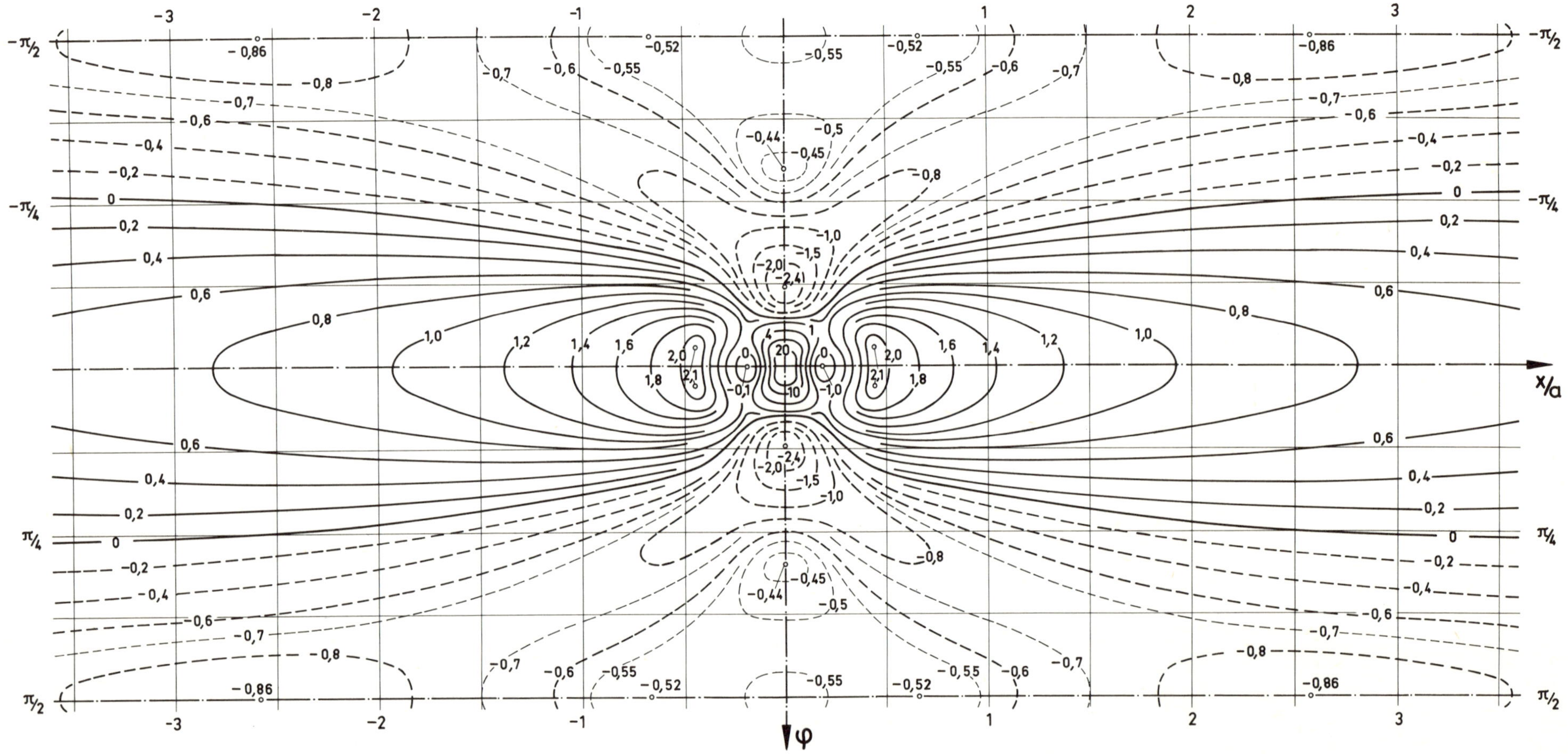

Einflußfläche (Zustandsfläche) des Biegemomentes m_x

Influence surface (condition surface) for the bending moment m_x

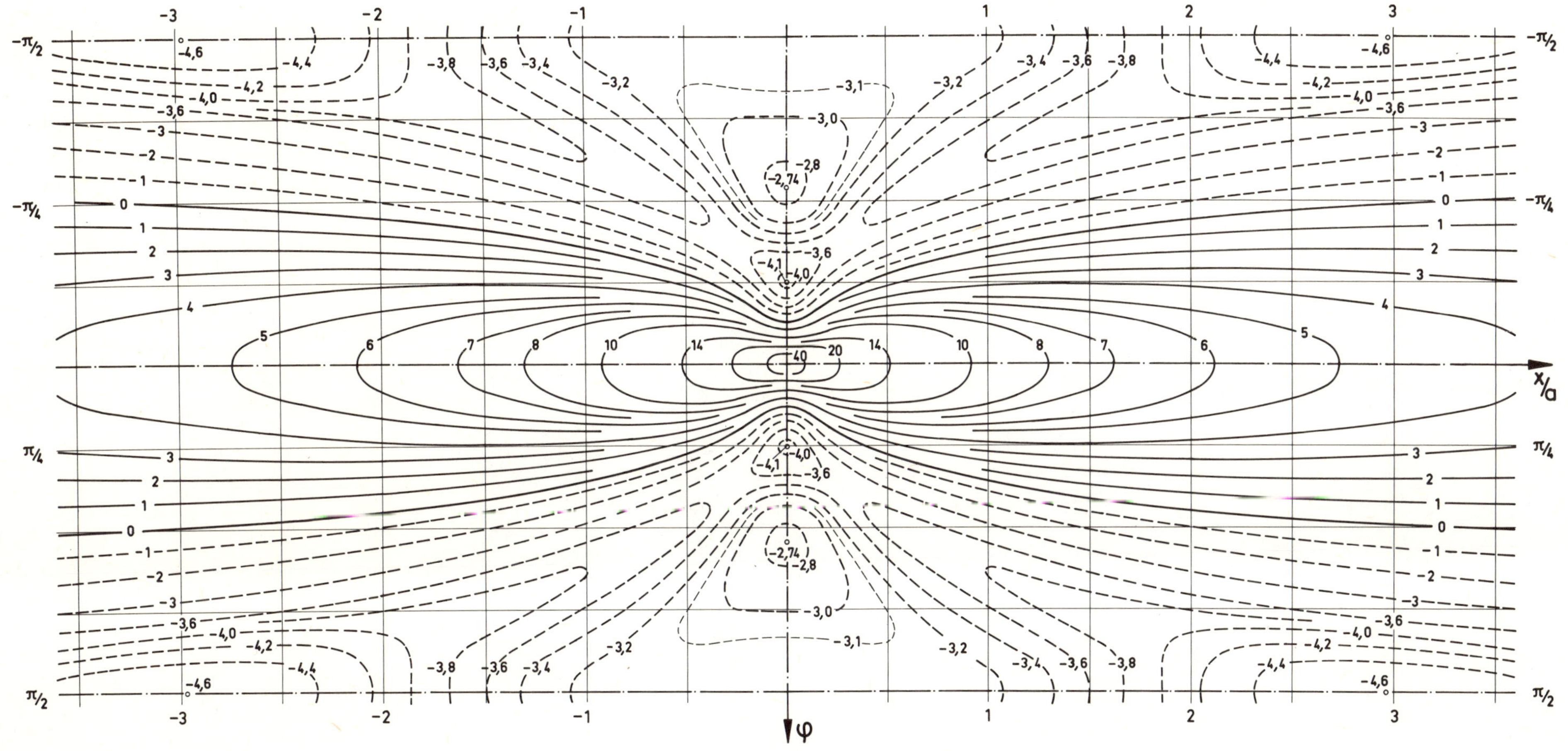

Einflußfläche (Zustandsfläche) des Biegemomentes m_φ

Influence surface (condition surface) for the bending moment m_φ

70	$100\,\pi \cdot m_{x\varphi}$
	$k = 2{,}5 \cdot 10^{-5}$

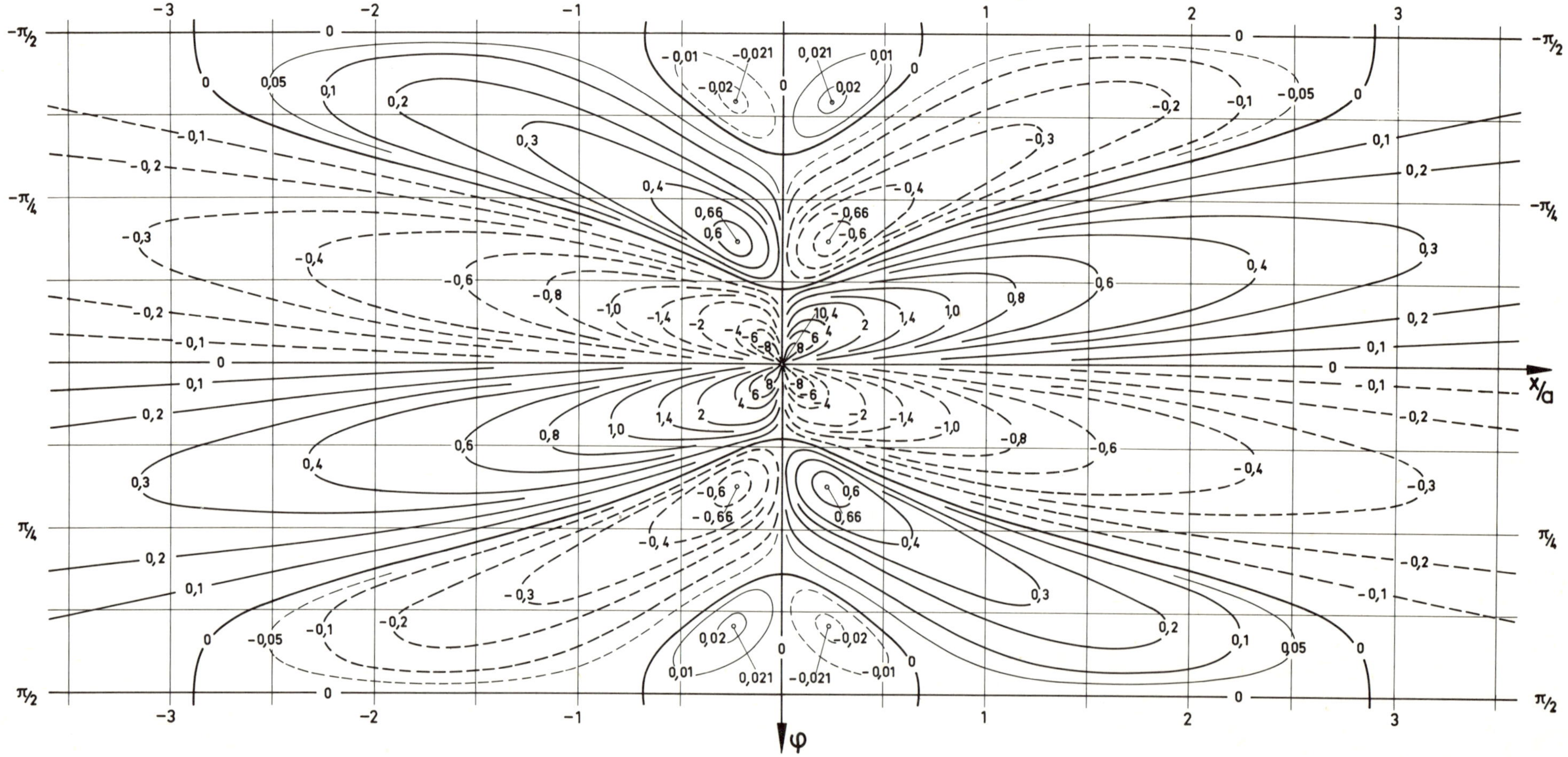

Einflußfläche (Zustandsfläche) des Drillmomentes $m_{x\varphi}$

Influence surface (condition surface) for the twisting moment $m_{x\varphi}$

$10\,\pi\, a \cdot \bar{q}_x$	**71**
$k = 2{,}5 \cdot 10^{-5}$	

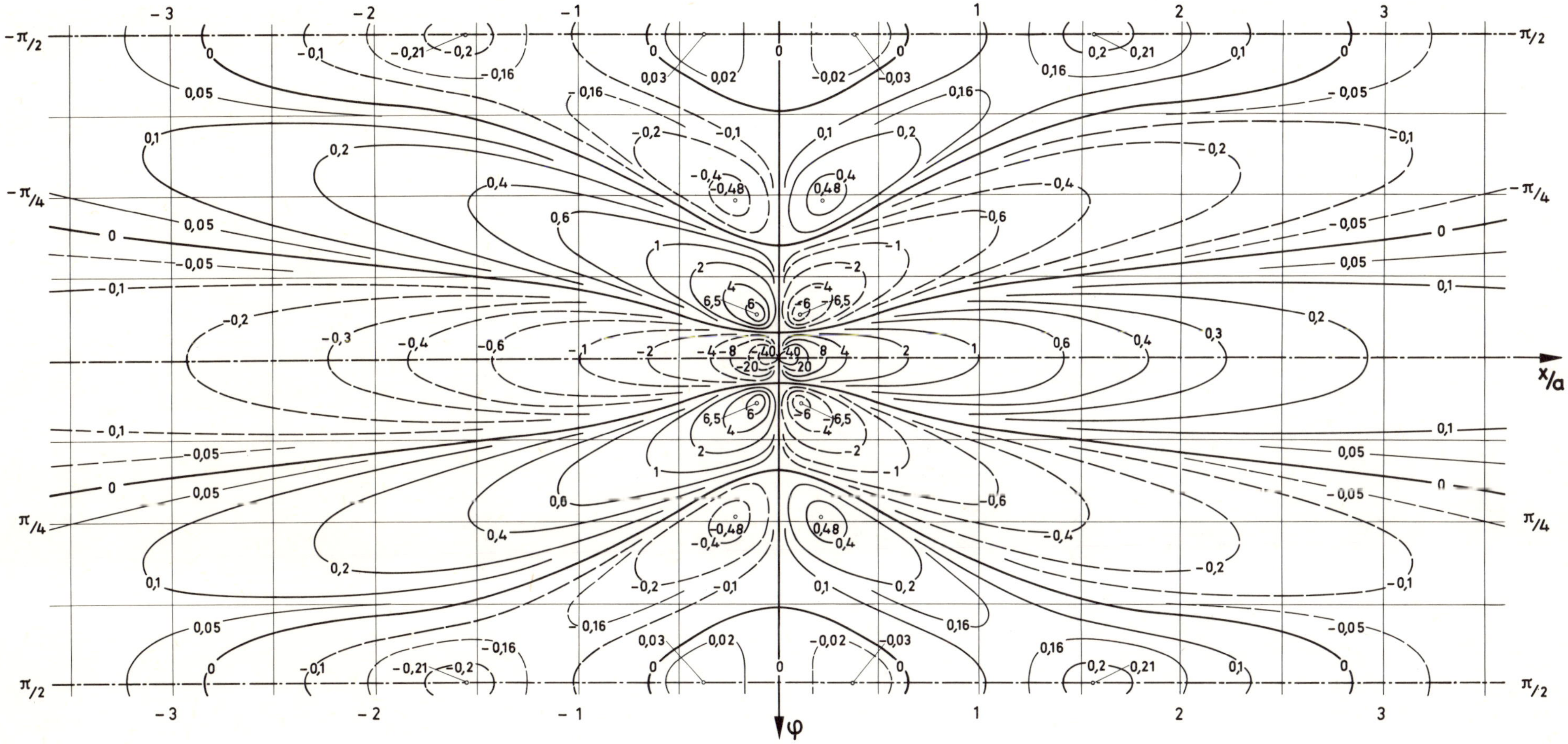

Einflußfläche (negative Zustandsfläche) der Randquerkraft $\bar{q}_x$

Influence surface (negative condition surface) for the transverse boundary force $\bar{q}_x$

72 | $10\,\pi\, a \cdot \bar{q}_{\varphi}$ | $k = 2{,}5 \cdot 10^{-5}$

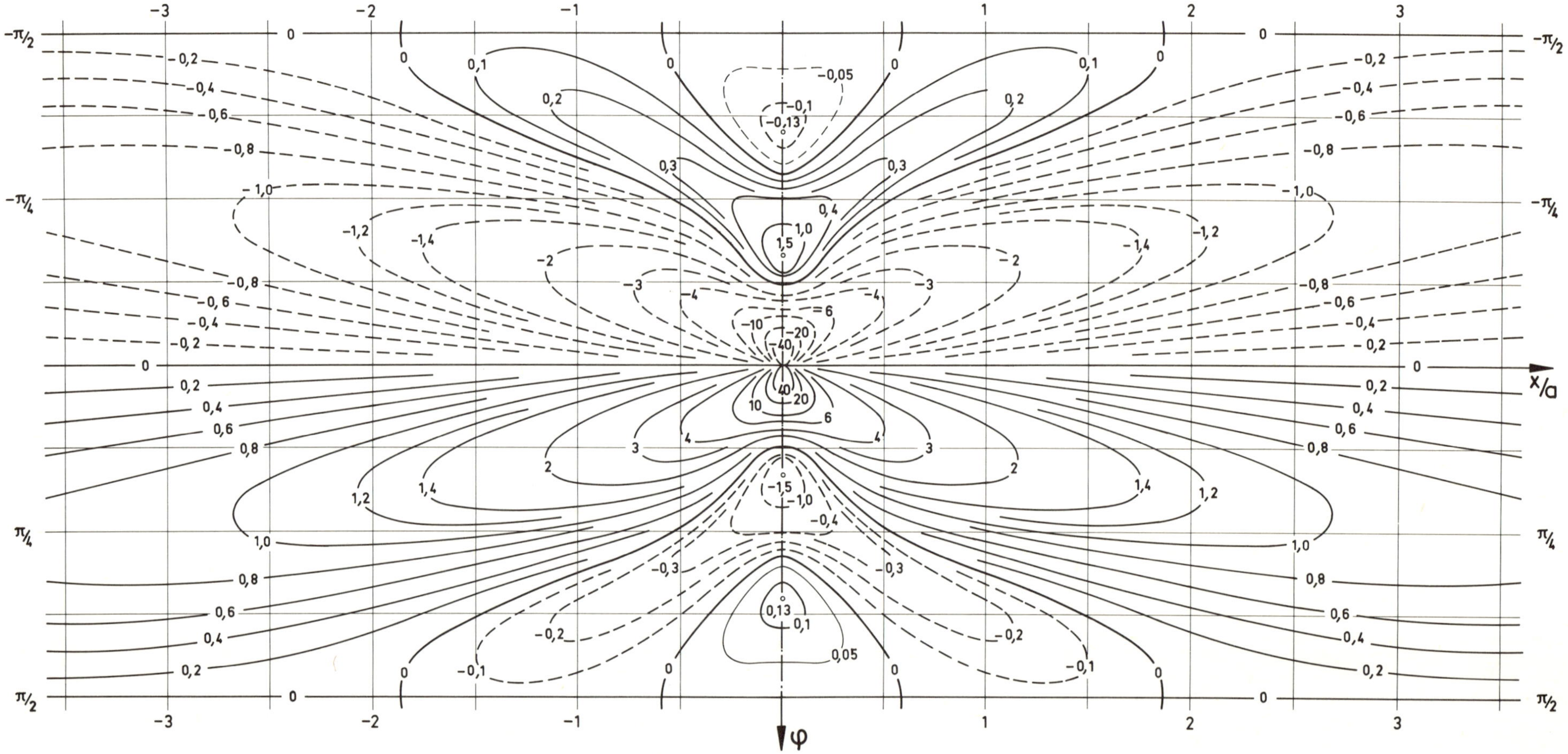

Einflußfläche (negative Zustandsfläche) der Randquerkraft $\bar{q}_{\varphi}$

Influence surface (negative condition surface) for the transverse boundary force $\bar{q}_{\varphi}$

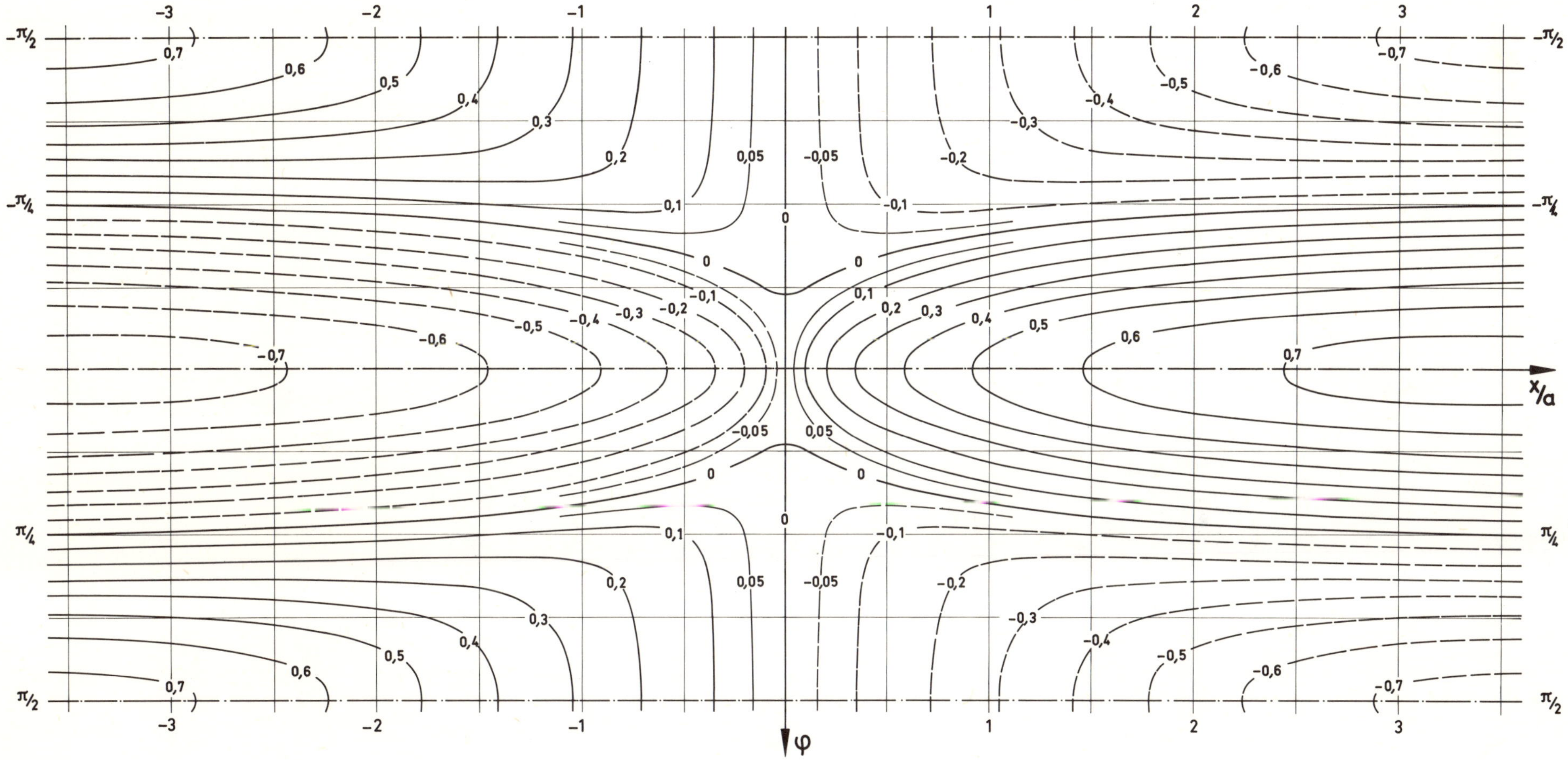

Einflußfläche (negative Zustandsfläche) der Verschiebung *u*

Influence surface (negative condition surface) for the displacement *u*

74	$1000\,\pi\,K/a^2 \cdot v$
	$k = 5{,}0 \cdot 10^{-5}$

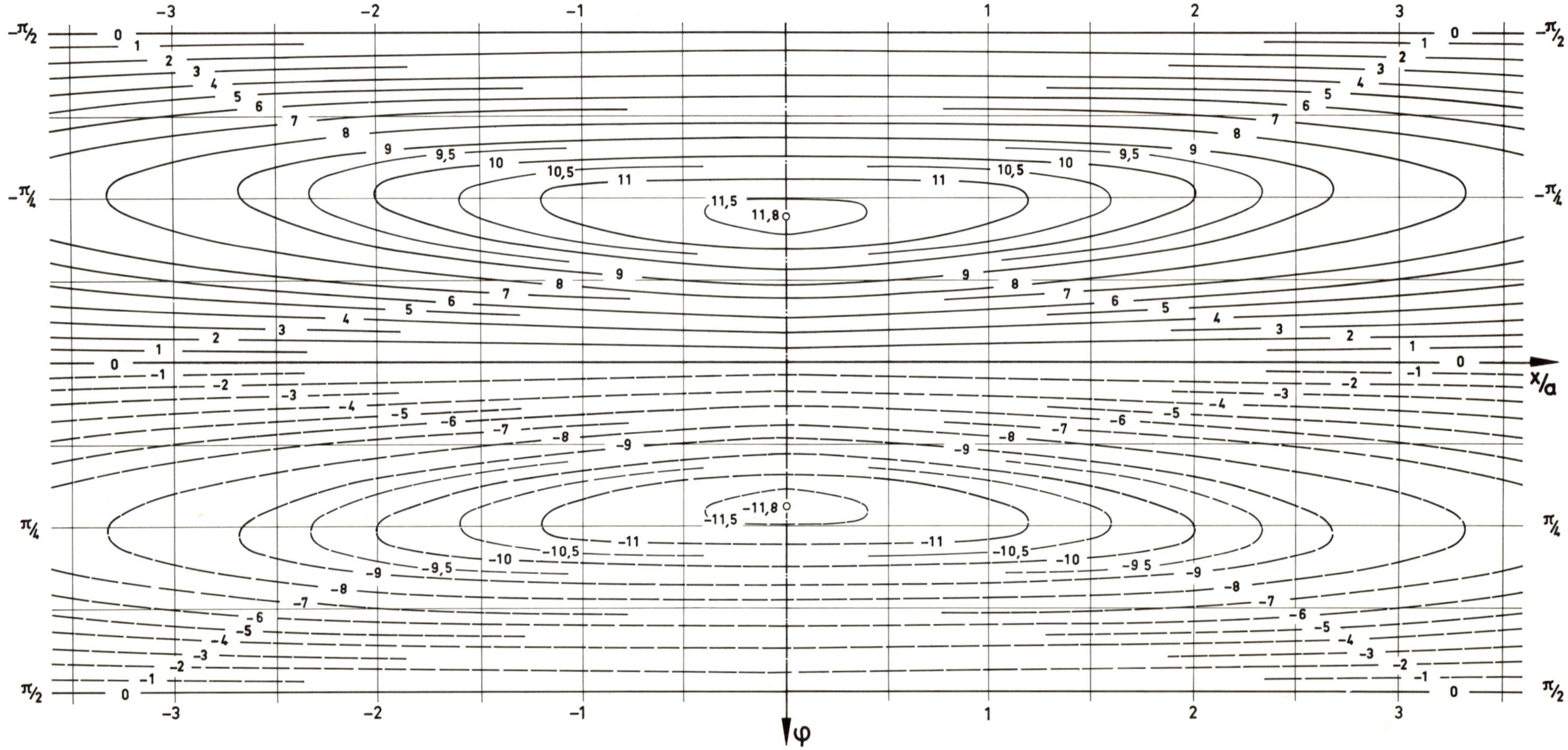

Einflußfläche (negative Zustandsfläche) der Verschiebung v

Influence surface (negative condition surface) for the displacement v

$1000\,\pi\,K/a^2 \cdot w$	**75**
$k = 5{,}0 \cdot 10^{-5}$	

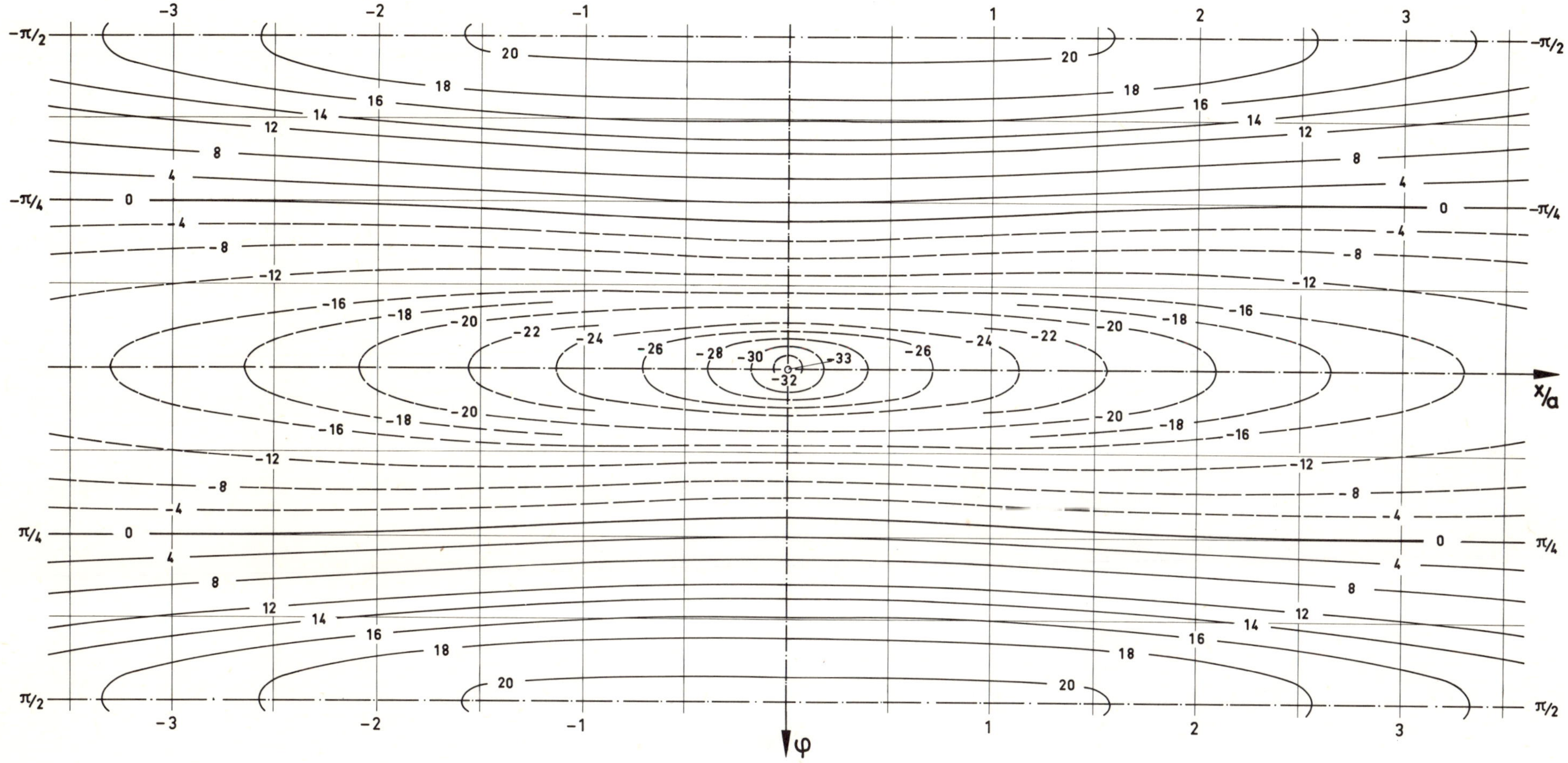

Einflußfläche (Zustandsfläche) der Verschiebung w

Influence surface (condition surface) for the displacement w

76	$100\,\pi\, K/a \cdot \chi_{\varphi}$
	$k = 5{,}0 \cdot 10^{-5}$

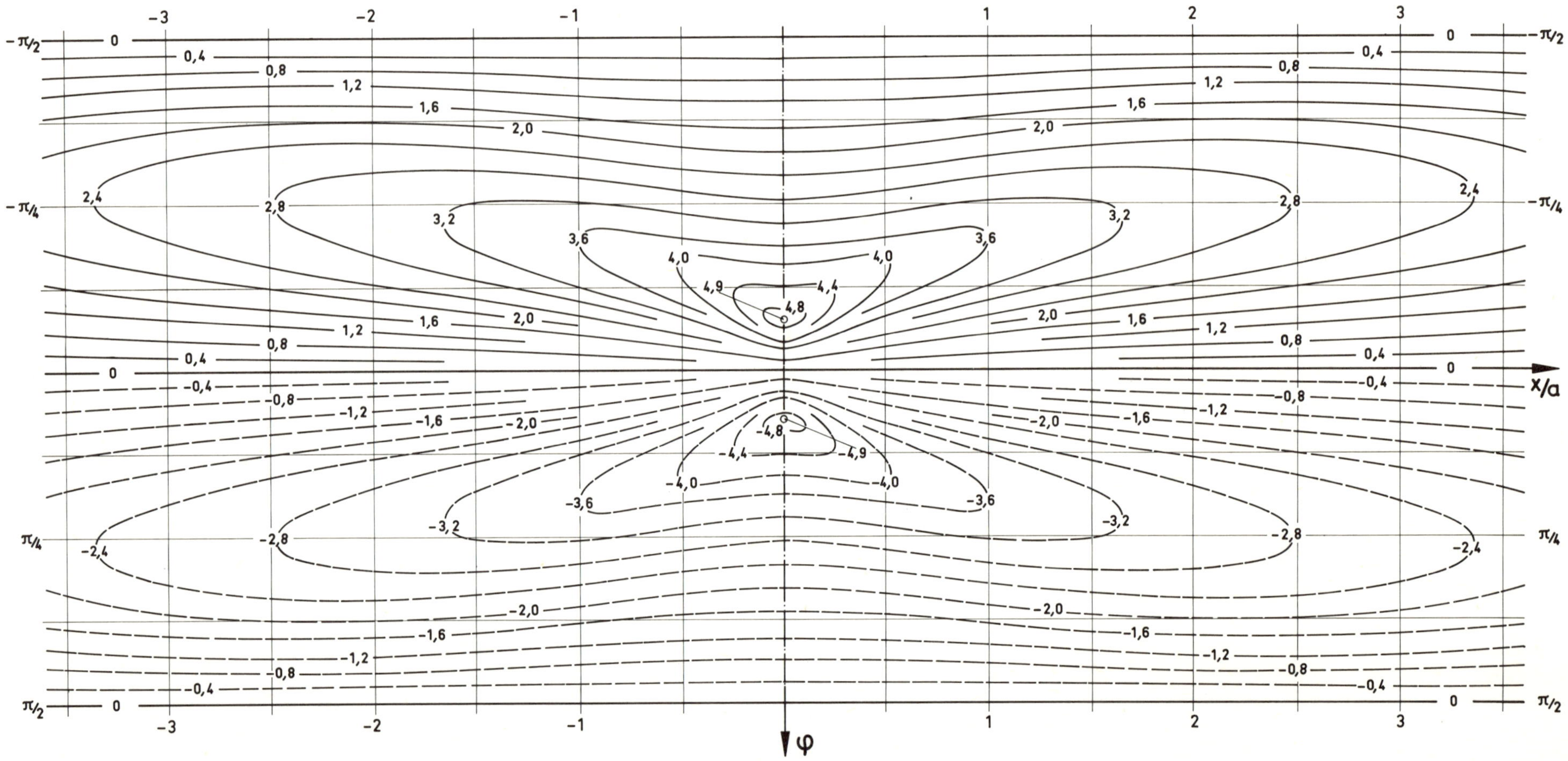

Einflußfläche (negative Zustandsfläche) der Verdrehung χ_{φ}

Influence surface (negative condition surface) for the rotation χ_{φ}

$\pi a \cdot n_x$	**77**
$k = 5{,}0 \cdot 10^{-5}$	

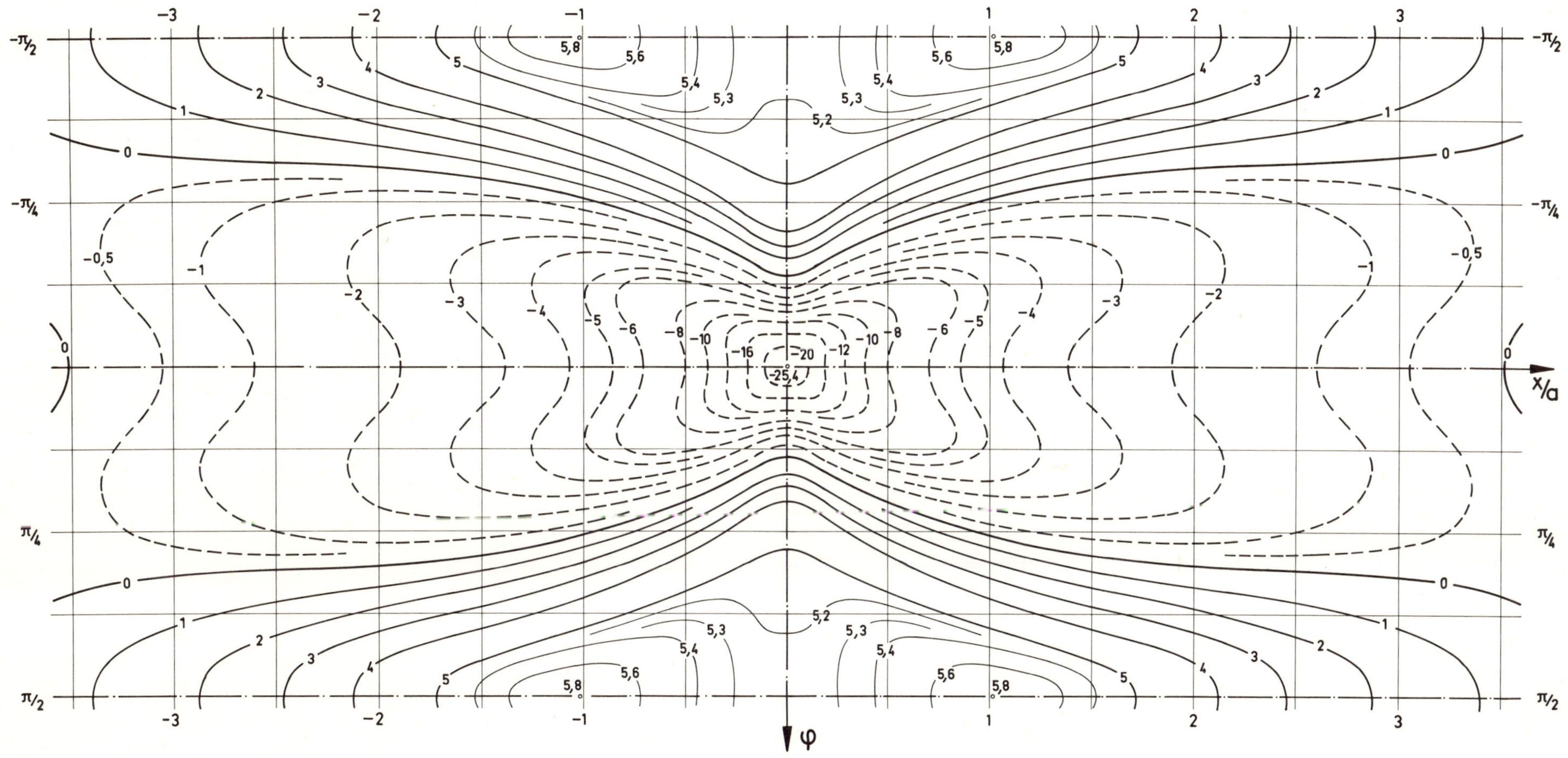

Einflußfläche (Zustandsfläche) der Normalkraft n_x

Influence surface (condition surface) for the normal force n_x

78	$\pi a \cdot n_\varphi$
	$k = 5{,}0 \cdot 10^{-5}$

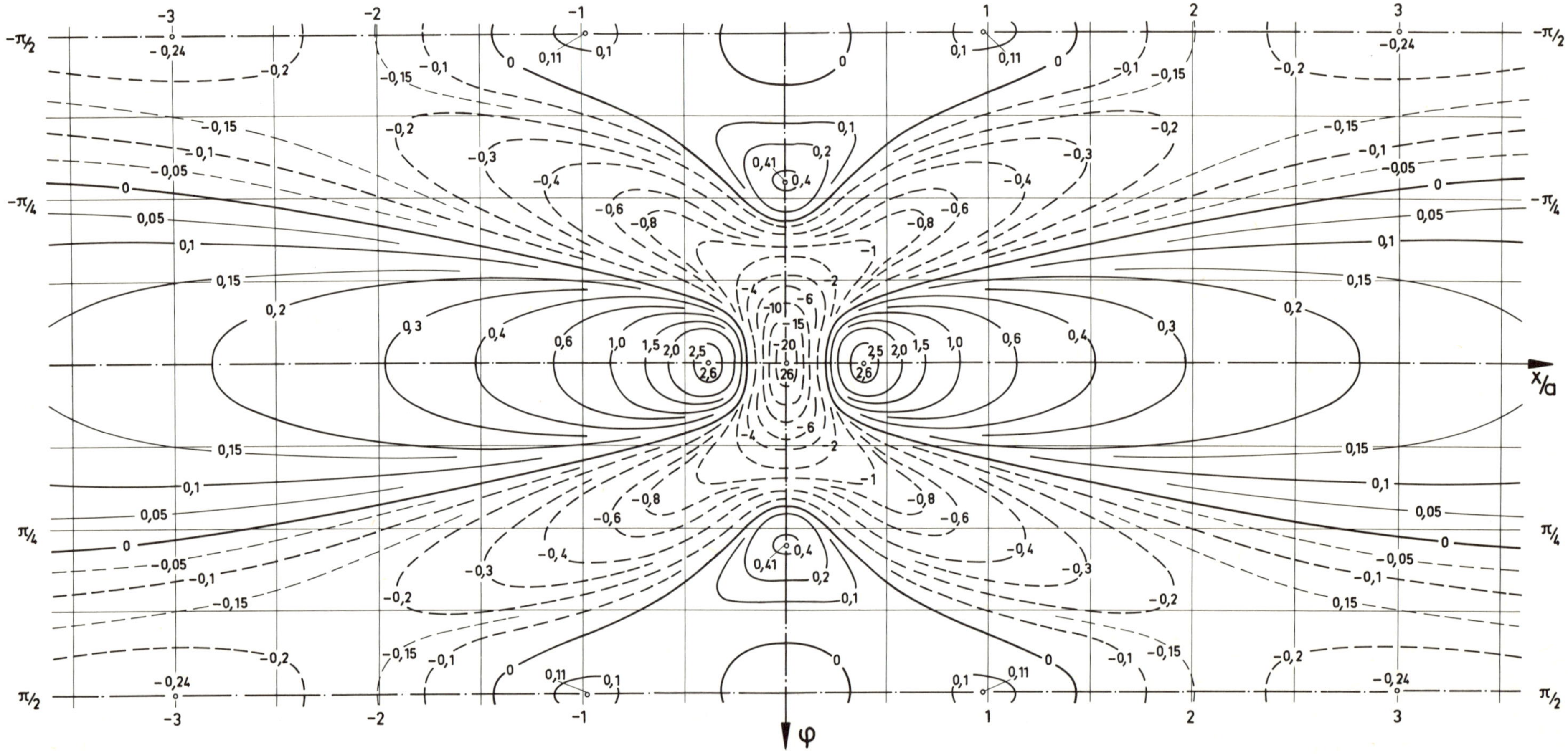

Einflußfläche (Zustandsfläche) der Normalkraft n_φ

Influence surface (condition surface) for the normal force n_φ

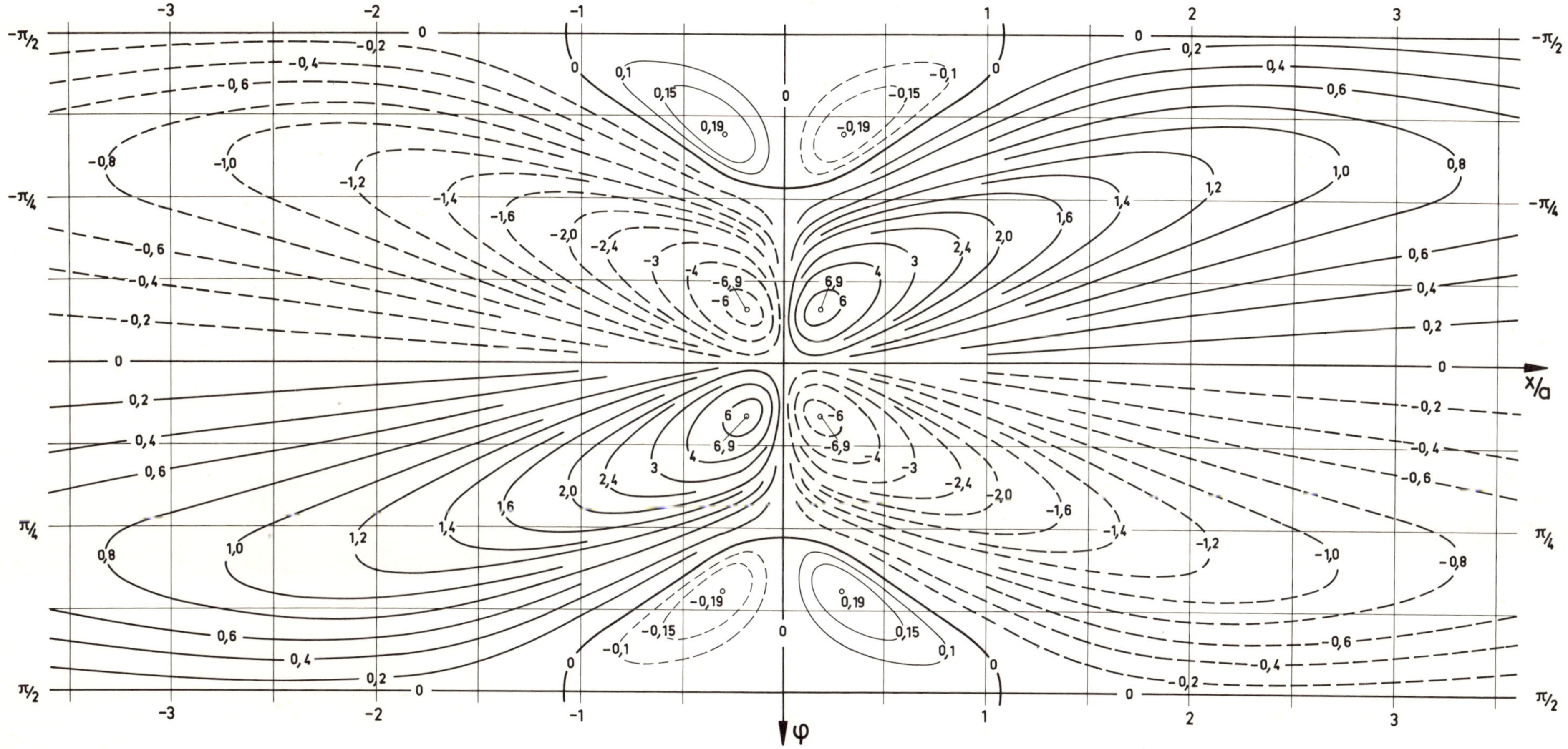

Einflußfläche (Zustandsfläche) der Randschubkraft $\bar{n}_{x\varphi}$

Influence surface (condition surface) for the boundary shearing force $\bar{n}_{x\varphi}$

80 | $100\,\pi \cdot m_x$ | $k = 5{,}0 \cdot 10^{-5}$

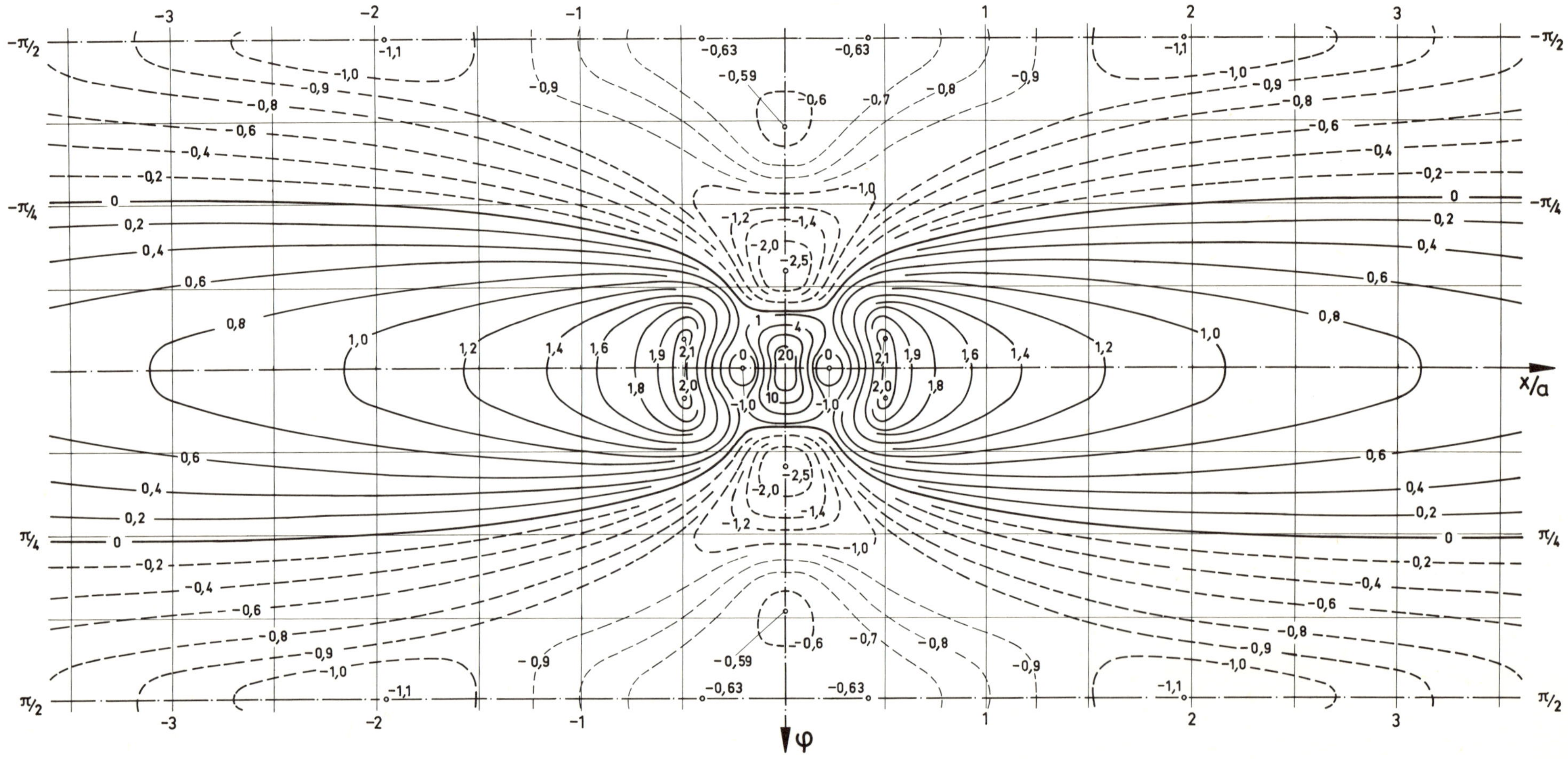

Einflußfläche (Zustandsfläche) des Biegemomentes m_x

Influence surface (condition surface) for the bending moment m_x

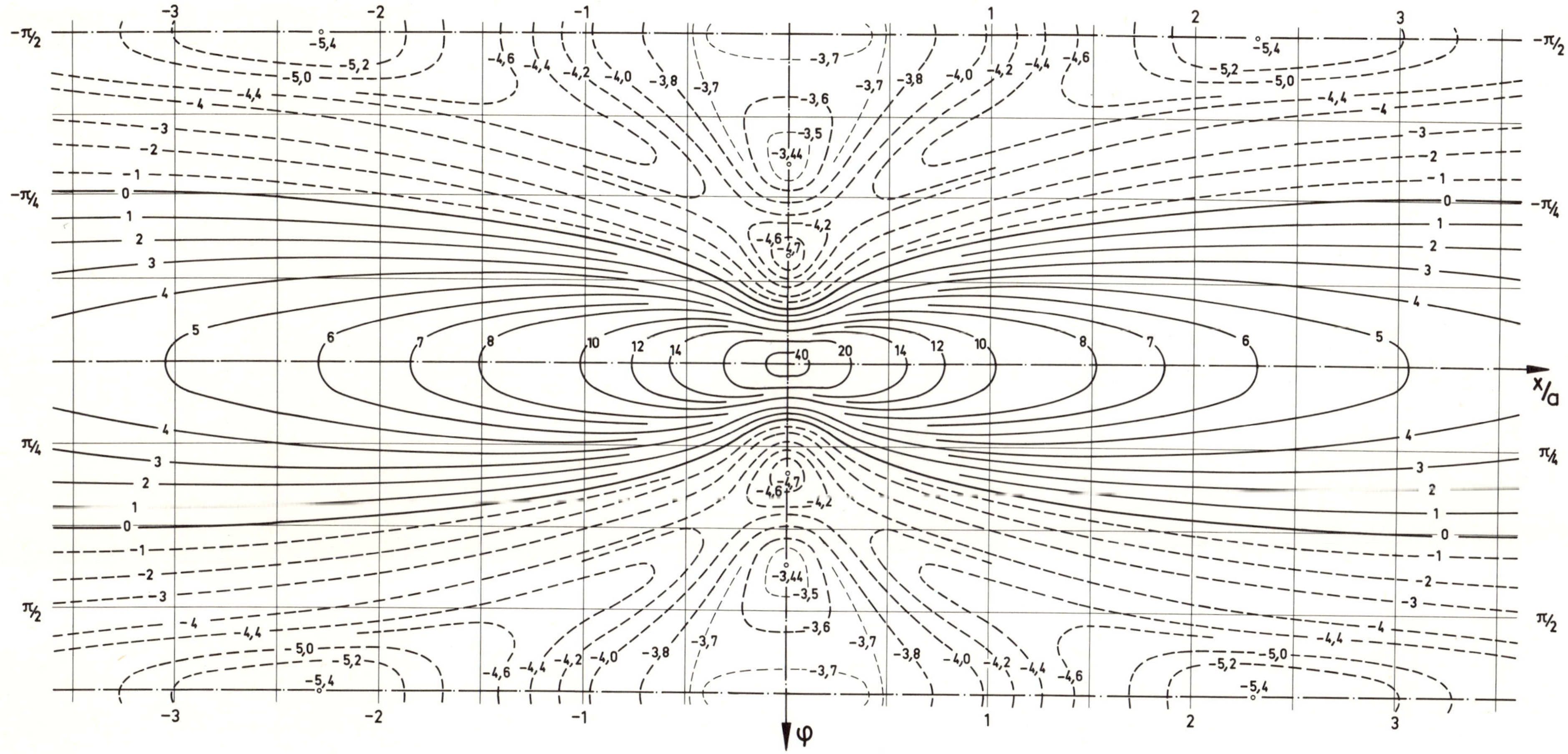

Einflußfläche (Zustandsfläche) des Biegemomentes m_φ

Influence surface (condition surface) for the bending moment m_φ

82	$100\,\pi \cdot m_{x\varphi}$
	$k = 5{,}0 \cdot 10^{-5}$

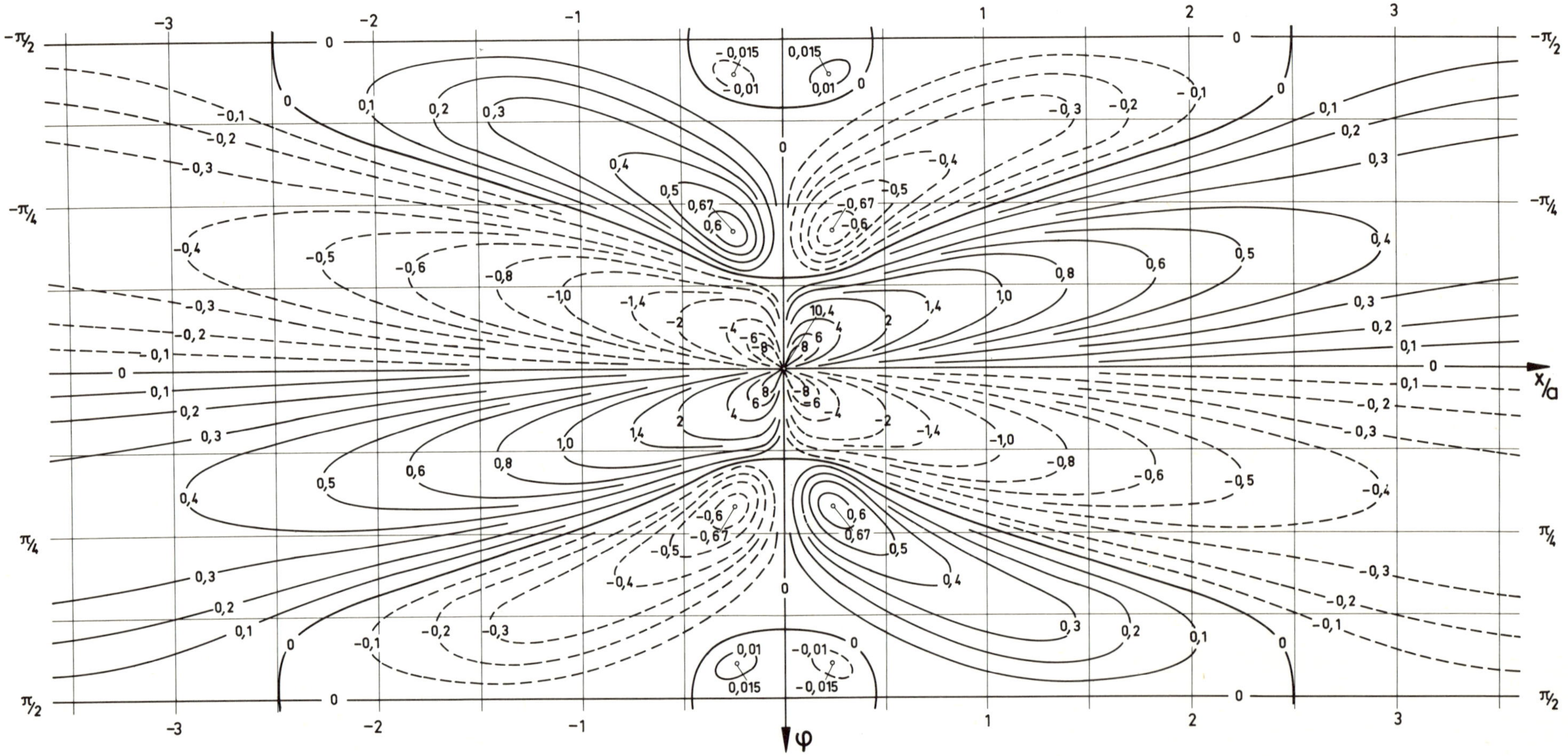

Einflußfläche (Zustandsfläche) des Drillmomentes $m_{x\varphi}$

Influence surface (condition surface) for the twisting moment $m_{x\varphi}$

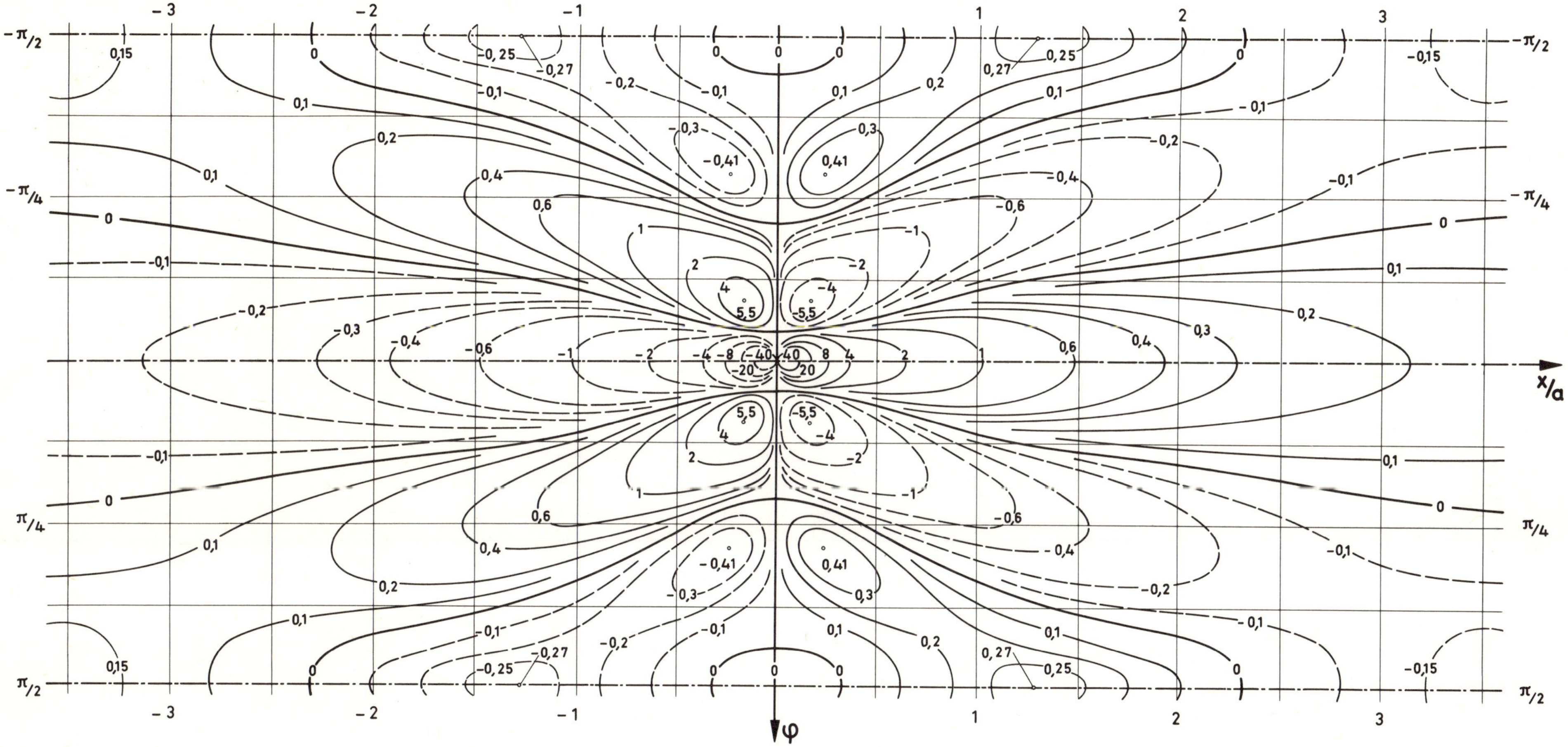

Einflußfläche (negative Zustandsfläche) der Randquerkraft $\bar{q}_x$

Influence surface (negative condition surface) for the transverse boundary force $\bar{q}_x$

84 | $10\,\pi\, a \cdot \bar{q}_{\varphi}$ | $k = 5{,}0 \cdot 10^{-5}$

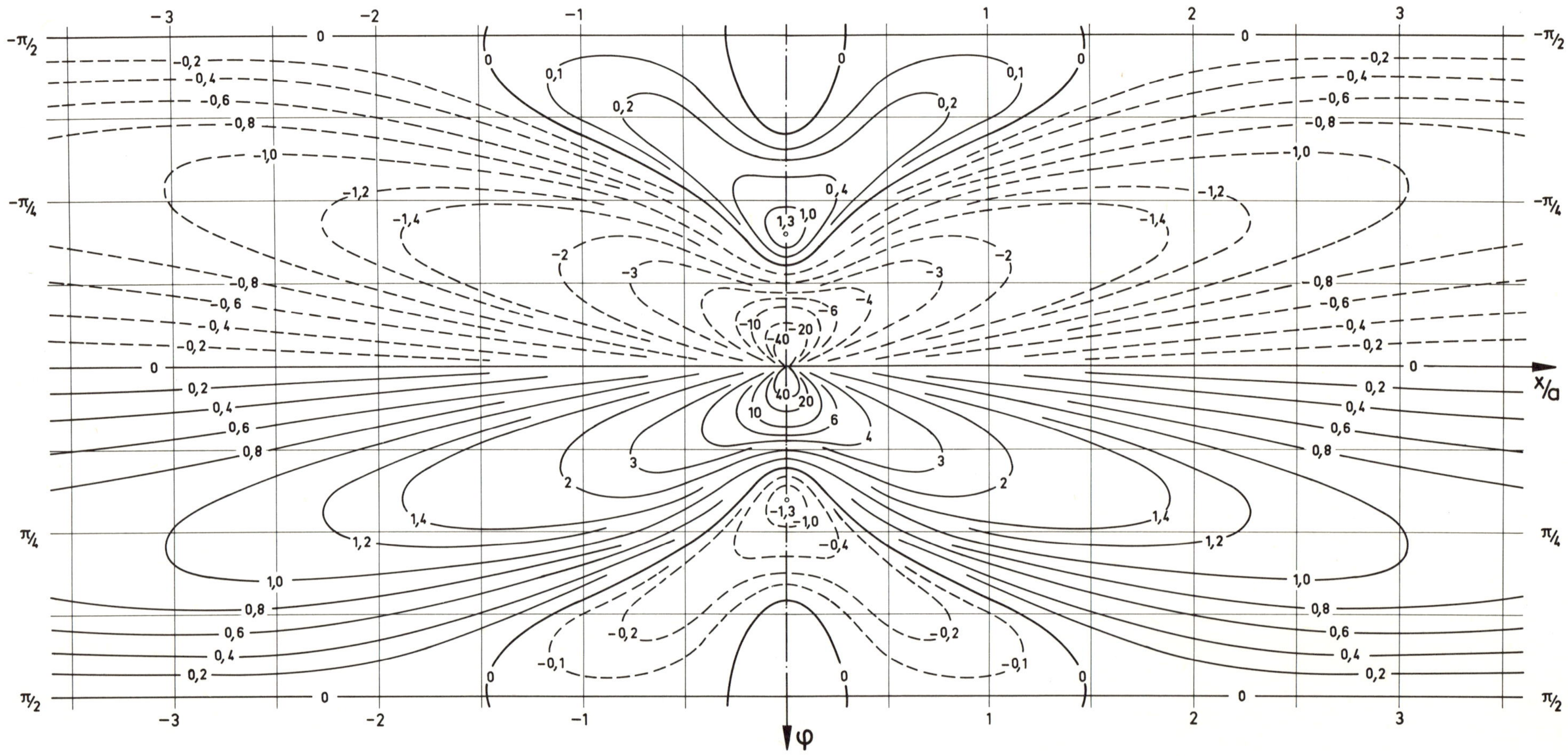

Einflußfläche (negative Zustandsfläche) der Randquerkraft $\bar{q}_{\varphi}$

Influence surface (negative condition surface) for the transverse boundary force $\bar{q}_{\varphi}$

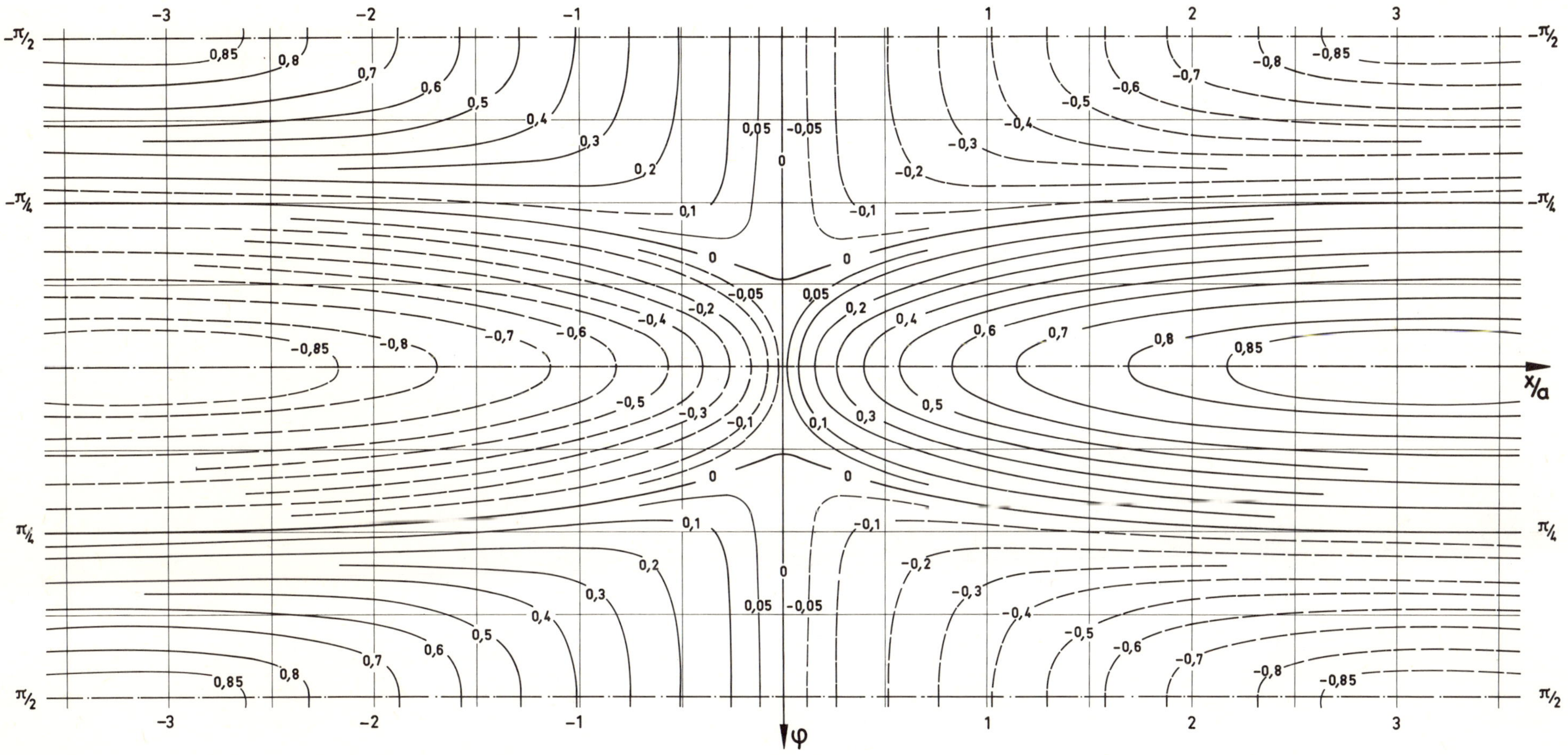

Einflußfläche (negative Zustandsfläche) der Verschiebung u

Influence surface (negative condition surface) for the displacement u

86 | $1000\,\pi\,K/a^2 \cdot v$ | $k = 7{,}5 \cdot 10^{-5}$

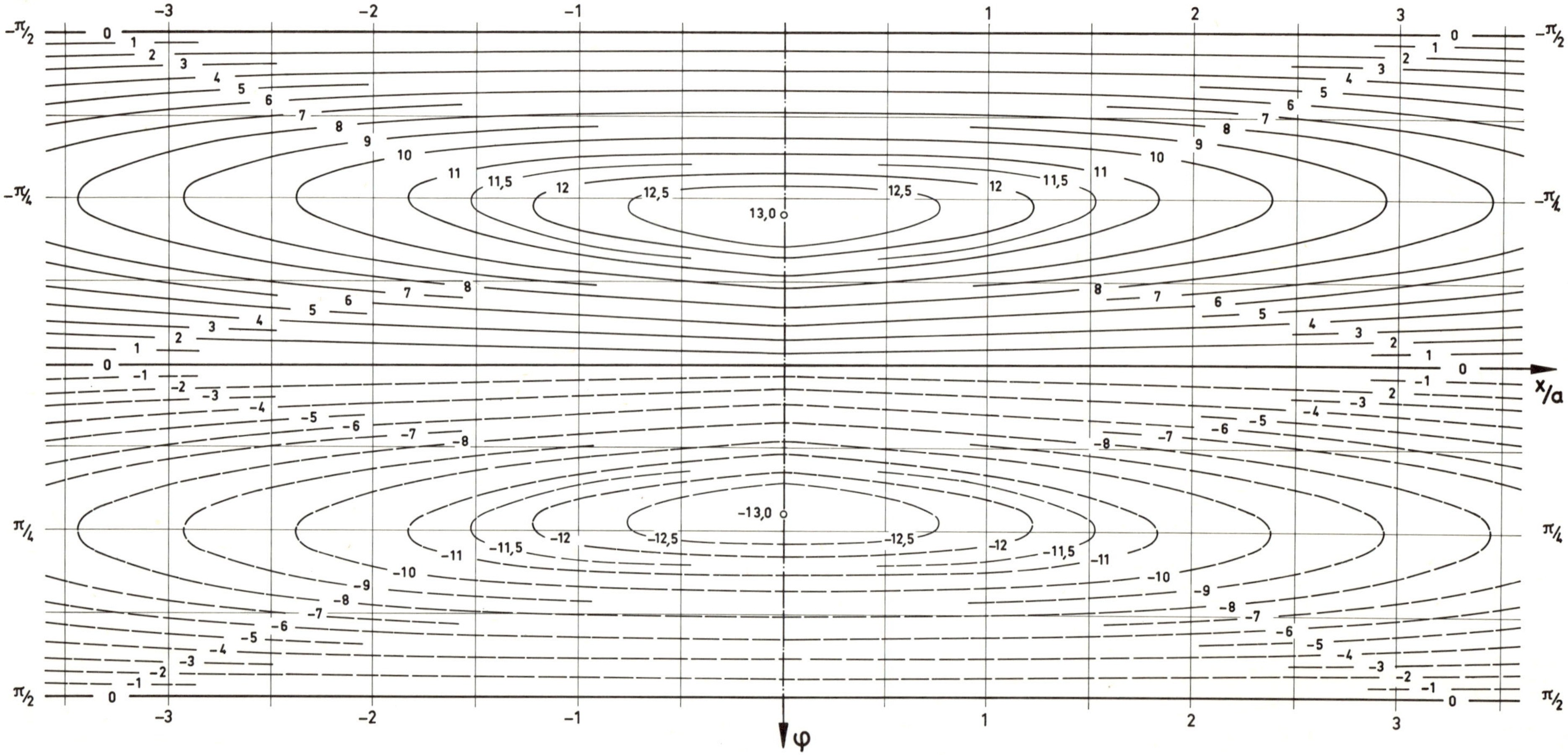

Einflußfläche (negative Zustandsfläche) der Verschiebung v

Influence surface (negative condition surface) for the displacement v

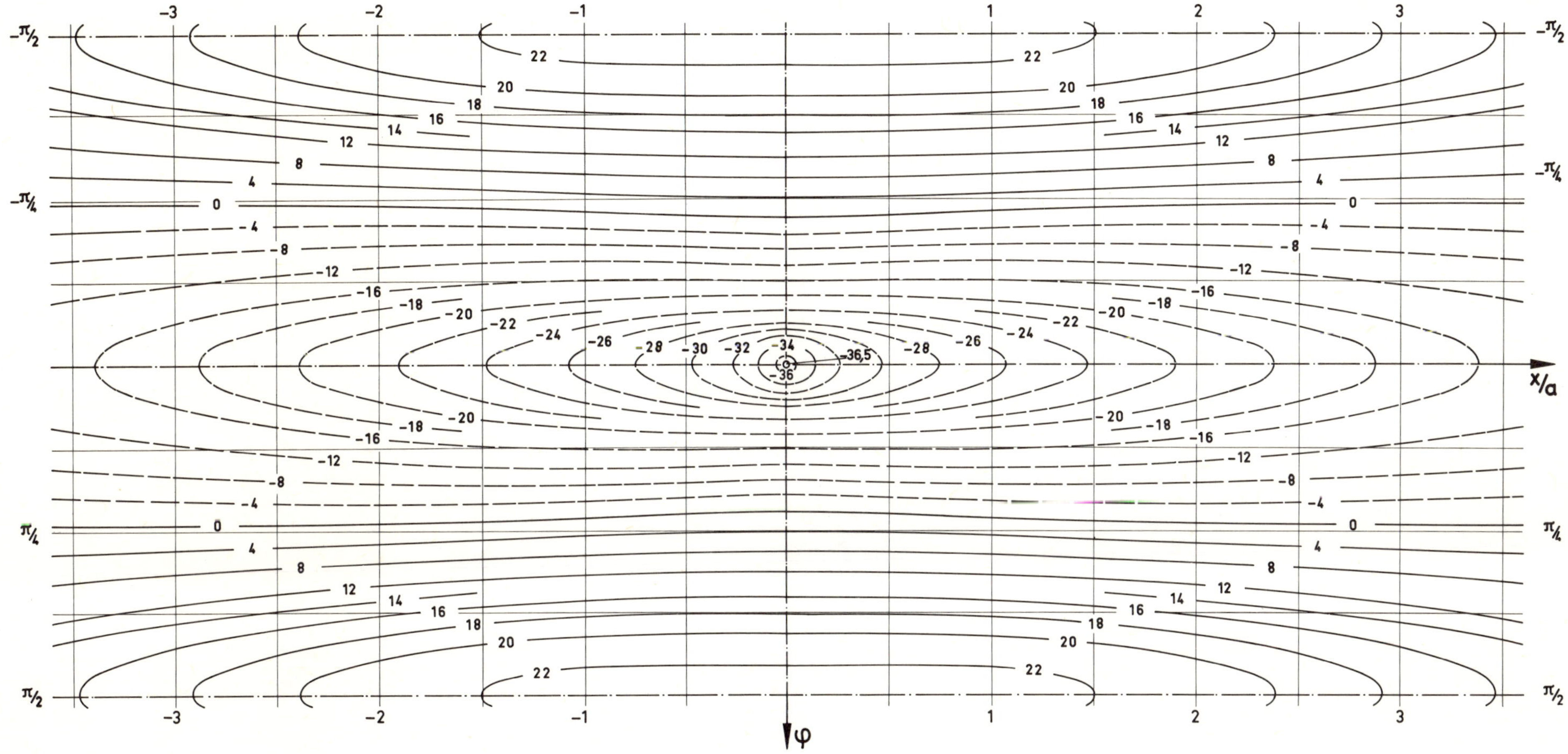

Einflußfläche (Zustandsfläche) der Verschiebung w

Influence surface (condition surface) for the displacement w

88	$100\,\pi\,K/a \cdot \chi_{\varphi}$ $k = 7{,}5 \cdot 10^{-5}$

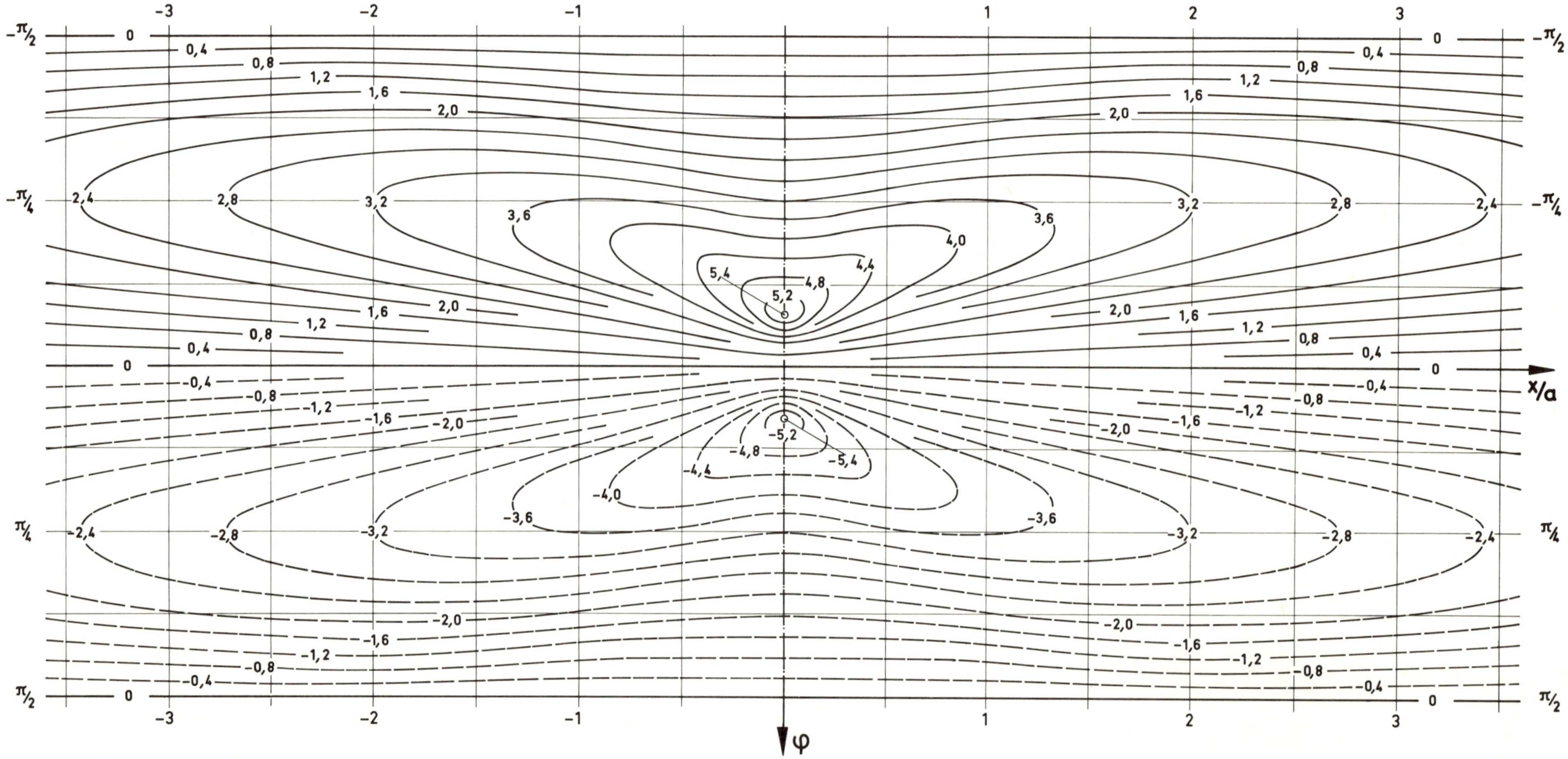

Einflußfläche (negative Zustandsfläche) der Verdrehung χ_{φ}

Influence surface (negative condition surface) for the rotation χ_{φ}

$\pi a \cdot n_x$	**89**
$k = 7{,}5 \cdot 10^{-5}$	

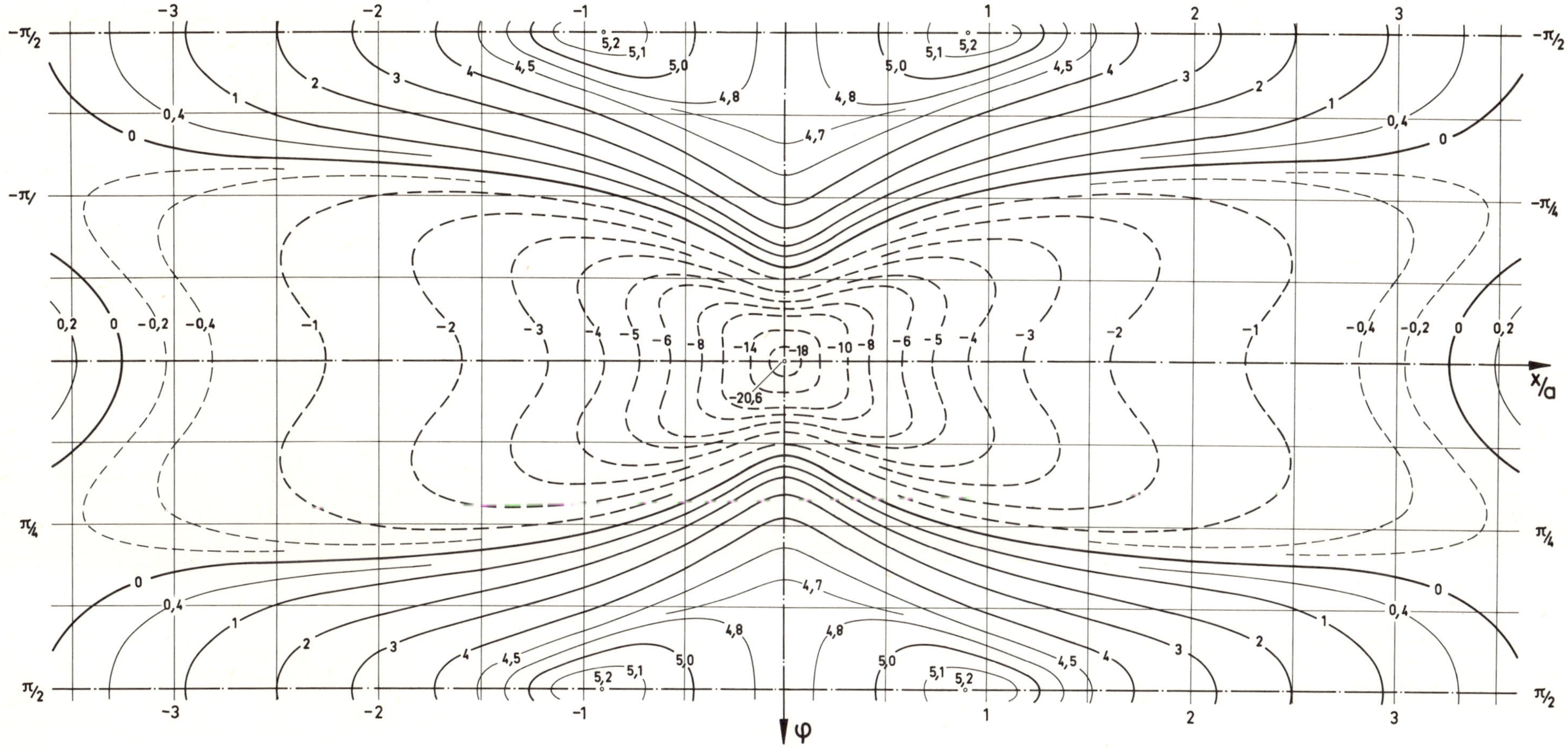

Einflußfläche (Zustandsfläche) der Normalkraft n_x

Influence surface (condition surface) for the normal force n_x

90	$\pi a \cdot n_\varphi$
	$k = 7{,}5 \cdot 10^{-5}$

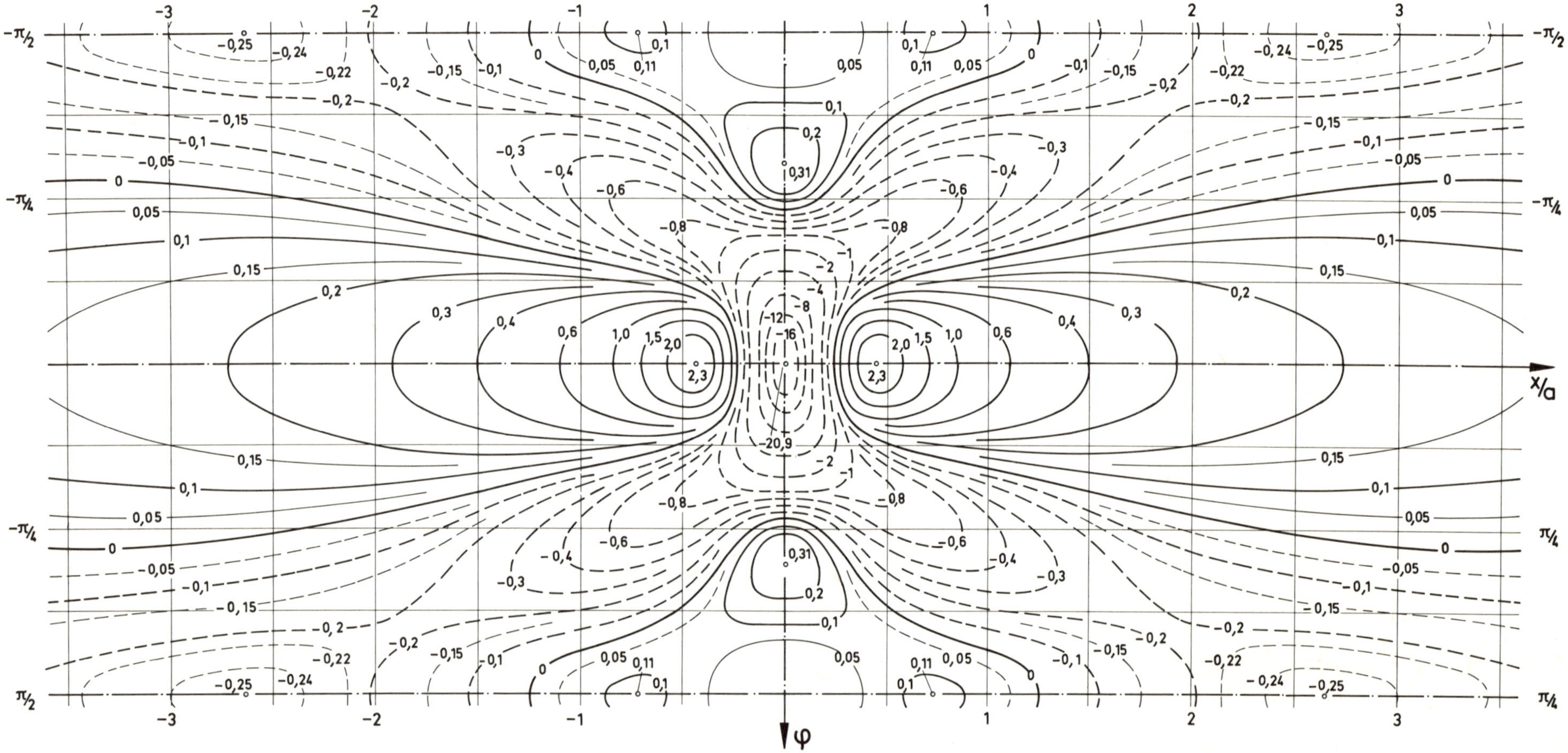

Einflußfläche (Zustandsfläche) der Normalkraft n_φ

Influence surface (condition surface) for the normal force n_φ

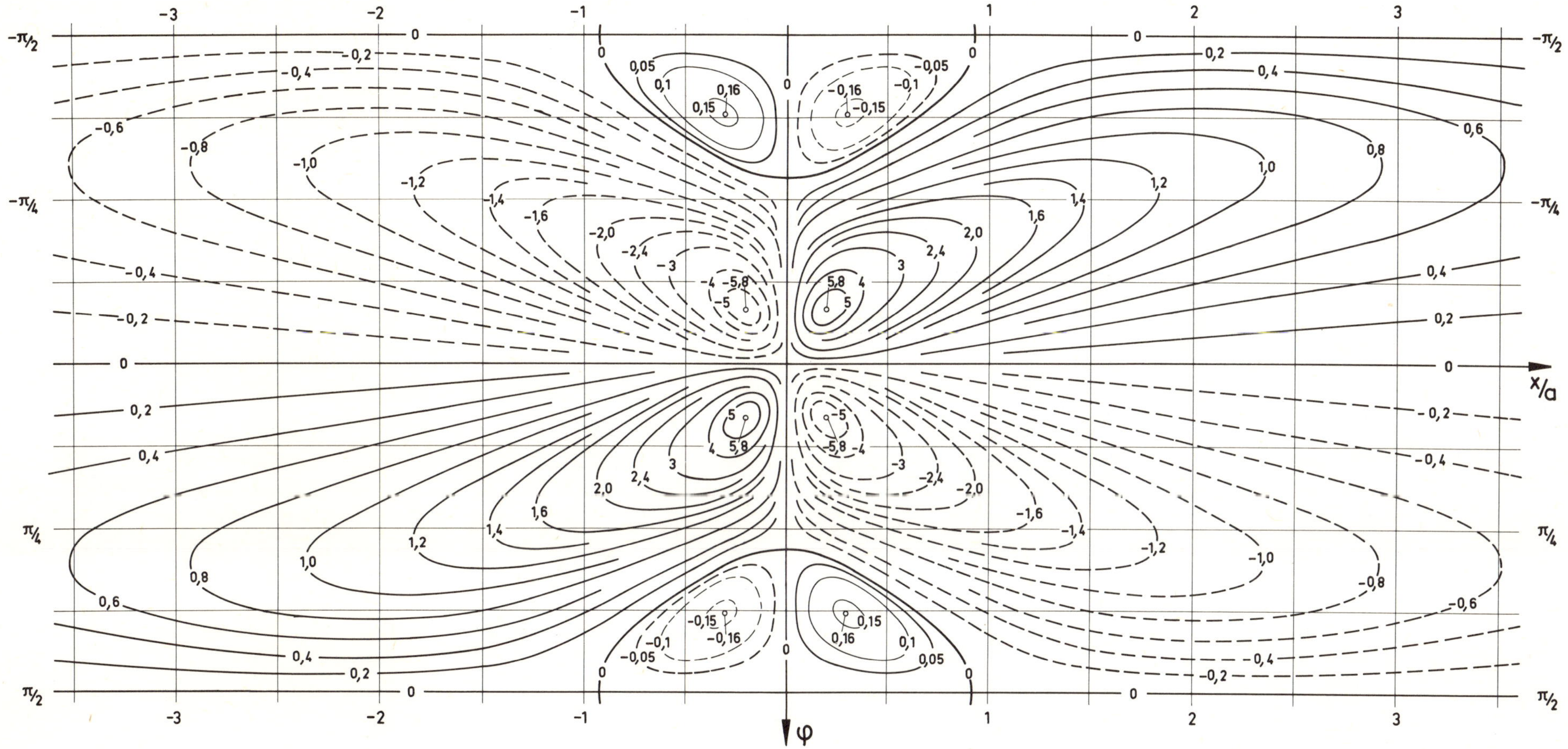

Einflußfläche (Zustandsfläche) der Randschubkraft $\bar{n}_{x\varphi}$

Influence surface (condition surface) for the boundary shearing force $\bar{n}_{x\varphi}$

92	$100\,\pi \cdot m_x$
	$k = 7{,}5 \cdot 10^{-5}$

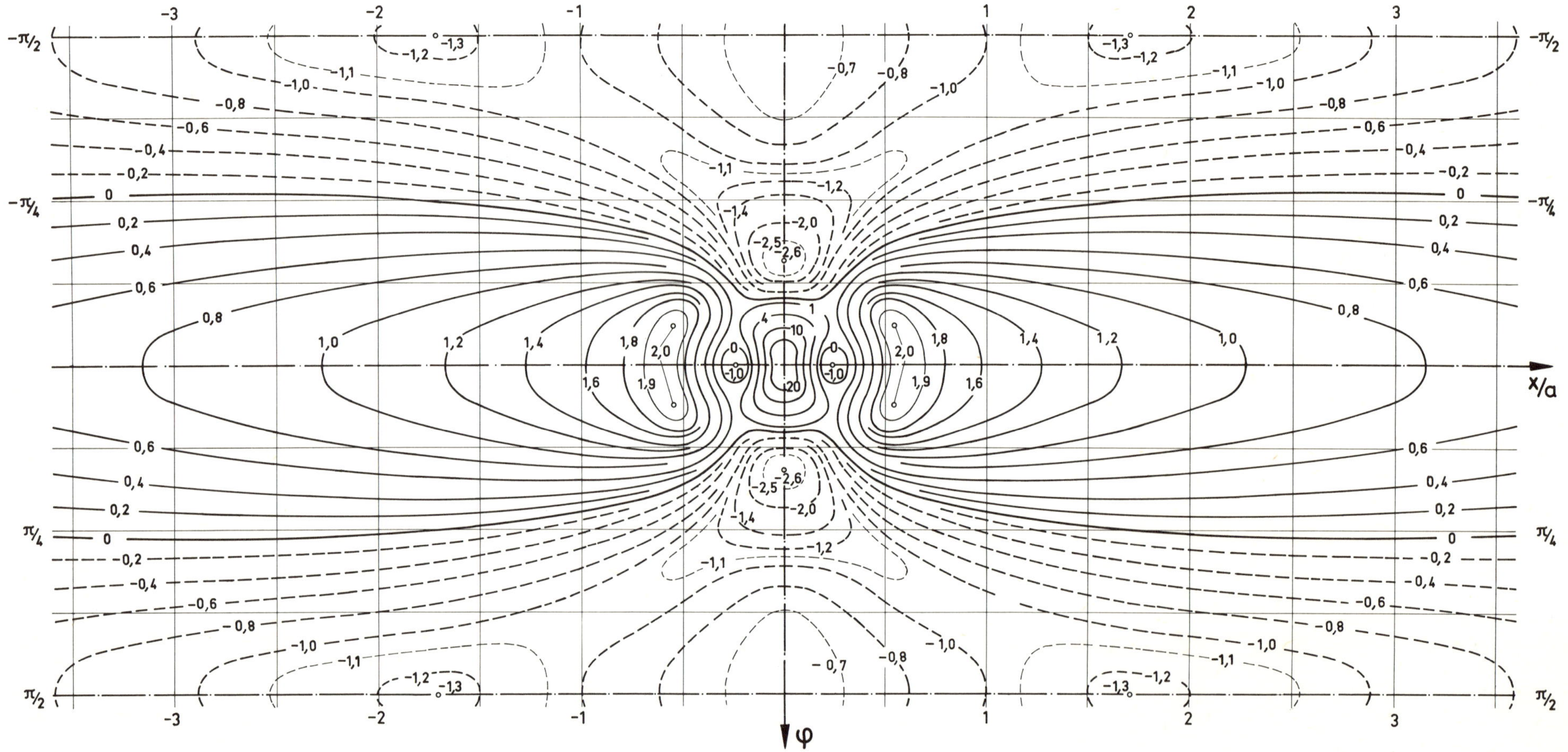

Einflußfläche (Zustandsfläche) des Biegemomentes m_x

Influence surface (condition surface) for the bending moment m_x

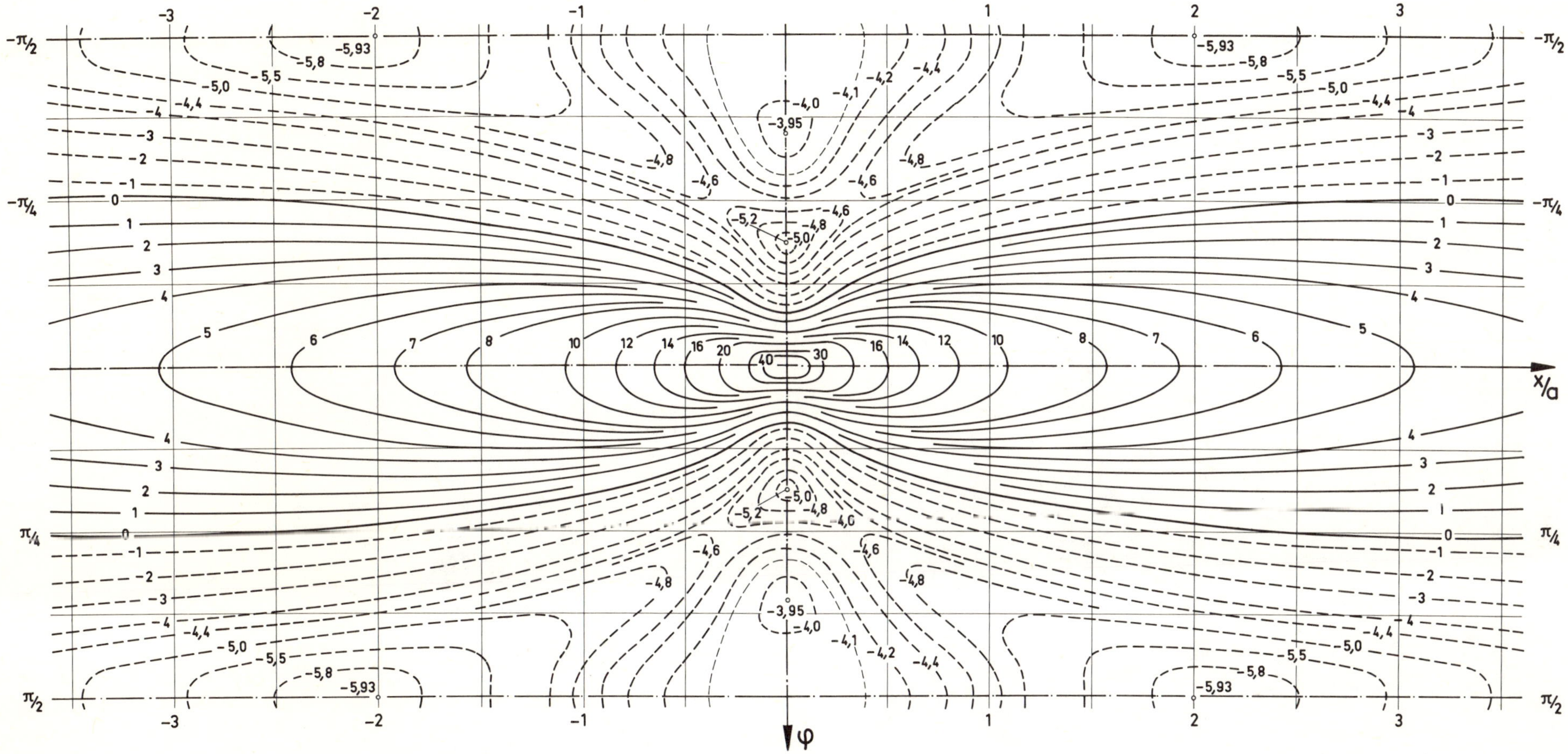

Einflußfläche (Zustandsfläche) des Biegemomentes m_φ

Influence surface (condition surface) for the bending moment m_φ

94	$100\,\pi \cdot m_{x\varphi}$
	$k = 7{,}5 \cdot 10^{-5}$

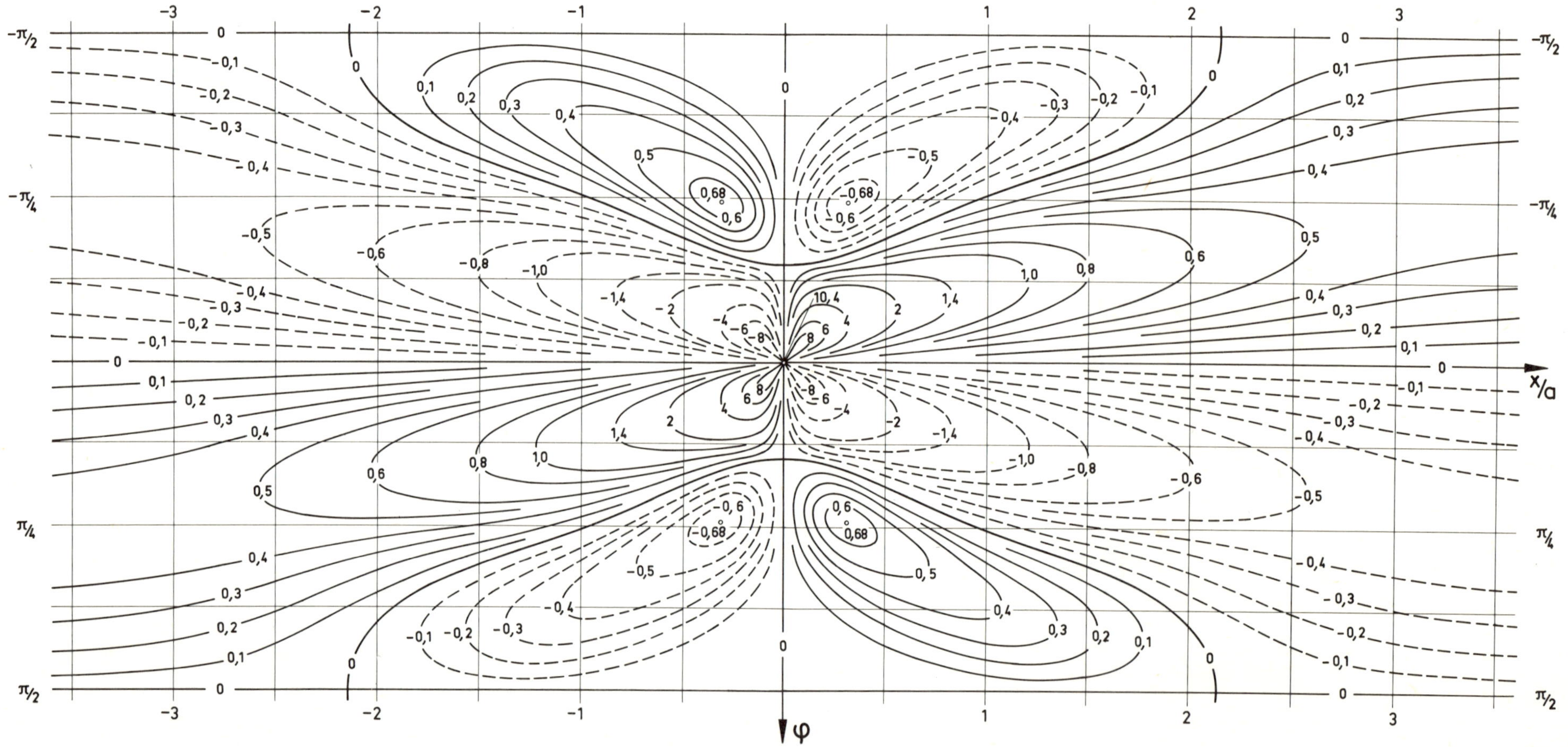

Einflußfläche (Zustandsfläche) des Drillmomentes $m_{x\varphi}$

Influence surface (condition surface) for the twisting moment $m_{x\varphi}$

$10\,\pi\,a \cdot \bar{q}_x$	**95**
$k = 7{,}5 \cdot 10^{-5}$	

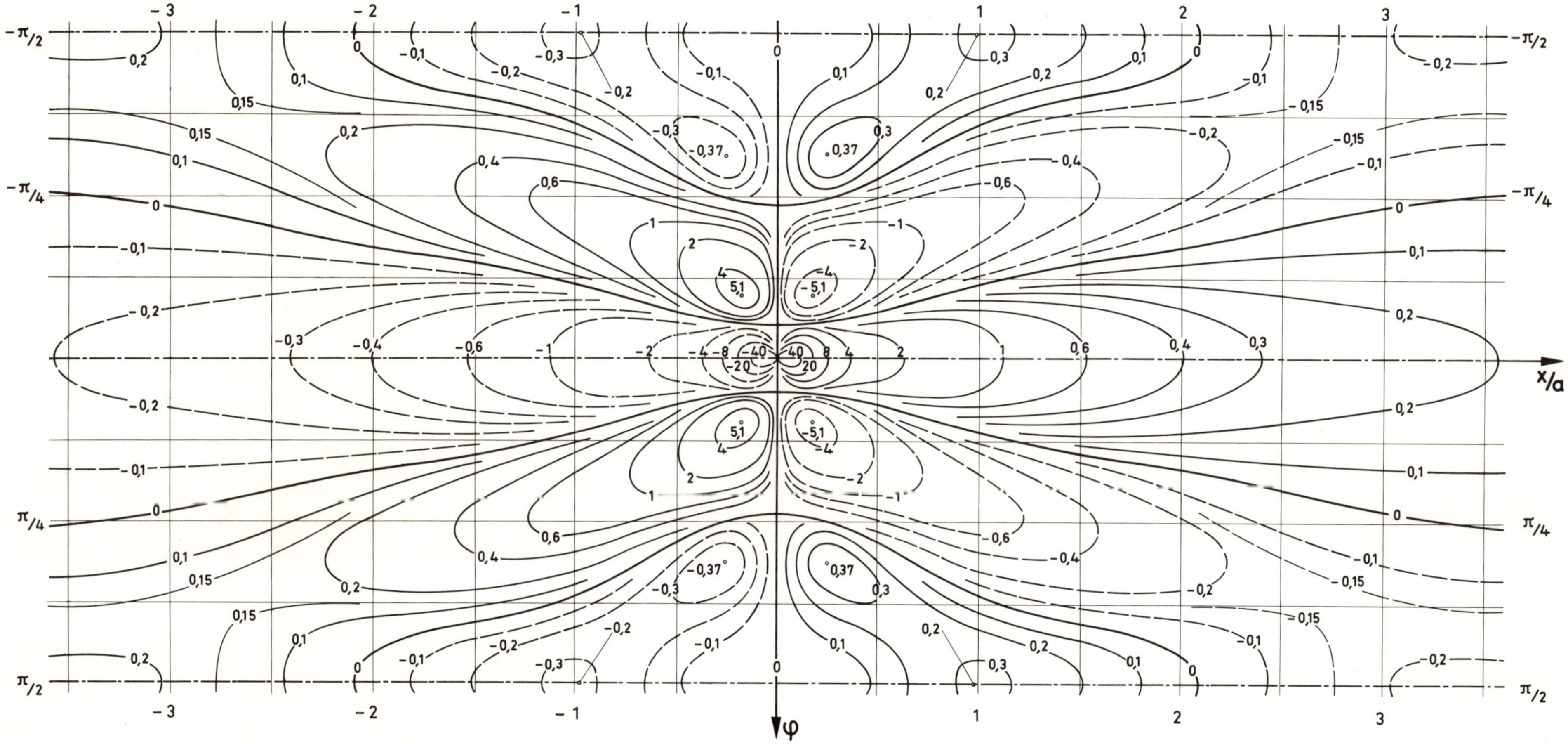

Einflußfläche (negative Zustandsfläche) der Randquerkraft $\bar{q}_x$

Influence surface (negative condition surface) for the transverse boundary force $\bar{q}_x$

96 | $10\,\pi\,a \cdot \bar{q}_{\varphi}$ | $k = 7{,}5 \cdot 10^{-5}$

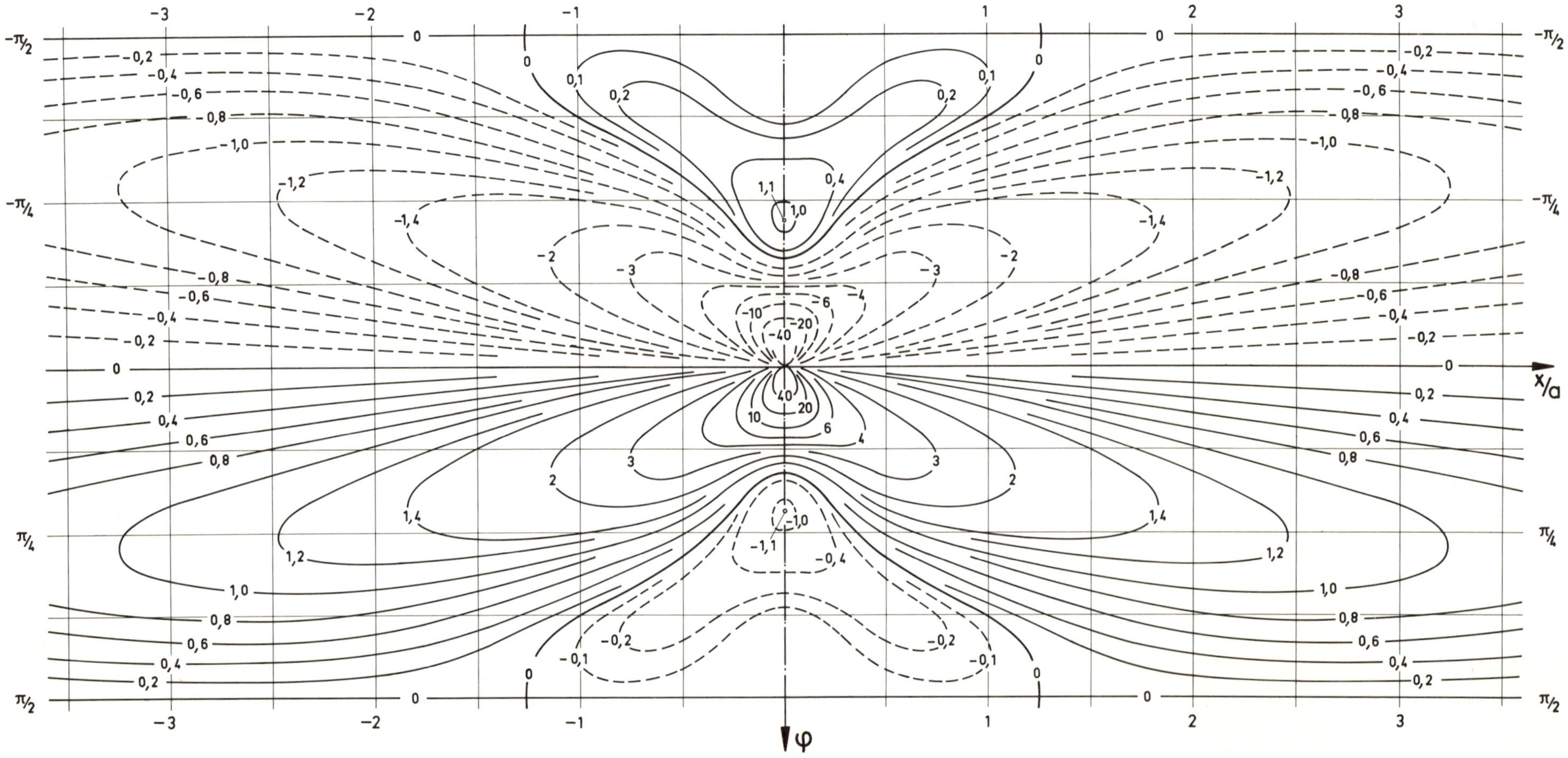

Einflußfläche (negative Zustandsfläche) der Randquerkraft $\bar{q}_{\varphi}$

Influence surface (negative condition surface) for the transverse boundary force $\bar{q}_{\varphi}$

$1000\,\pi\,K/a^2 \cdot u$	**97**
$k = 1{,}0 \cdot 10^{-4}$	

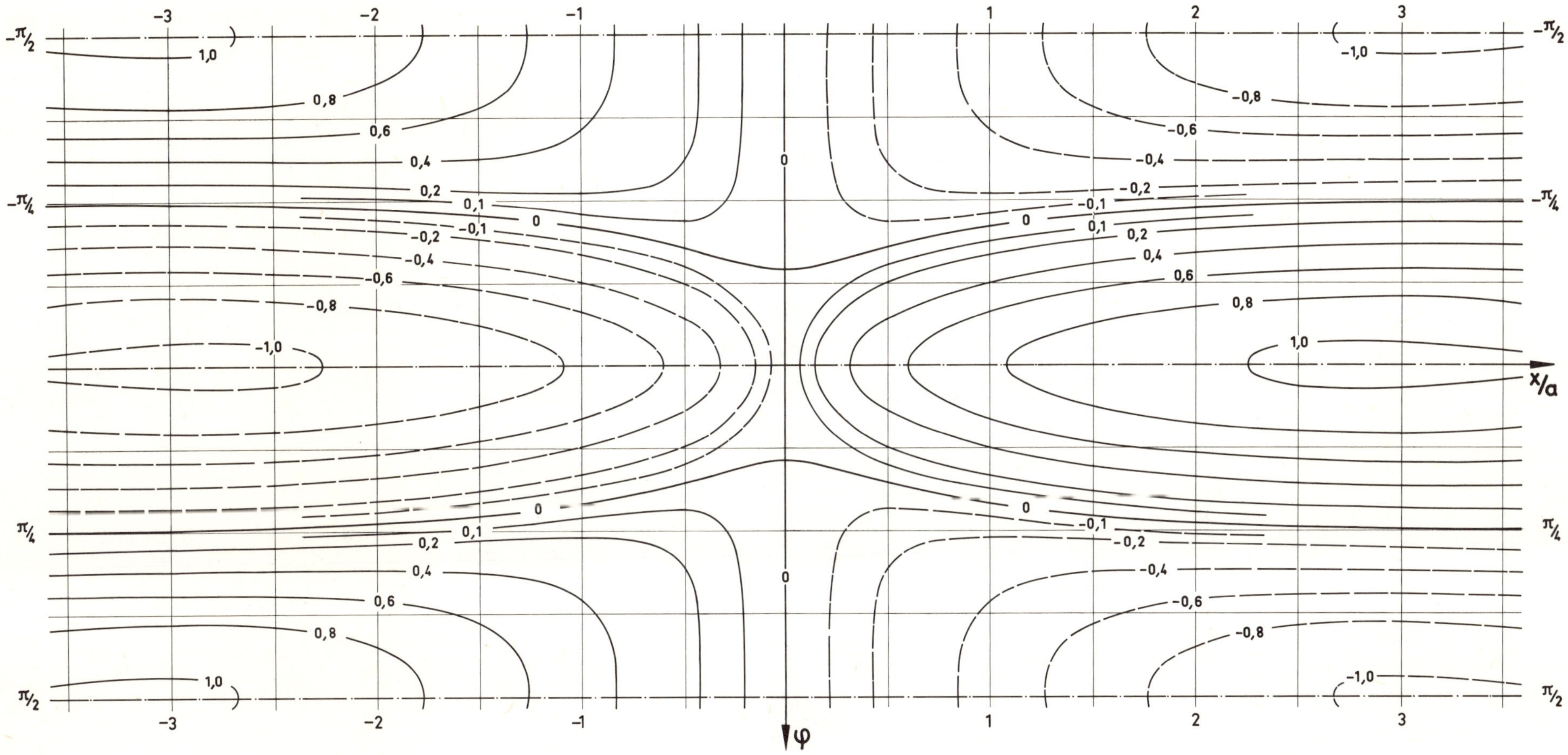

Einflußfläche (negative Zustandsfläche) der Verschiebung *u*

Influence surface (negative condition surface) for the displacement *u*

98	$1000\,\pi\,K/a^2 \cdot v$ $k = 1{,}0 \cdot 10^{-4}$

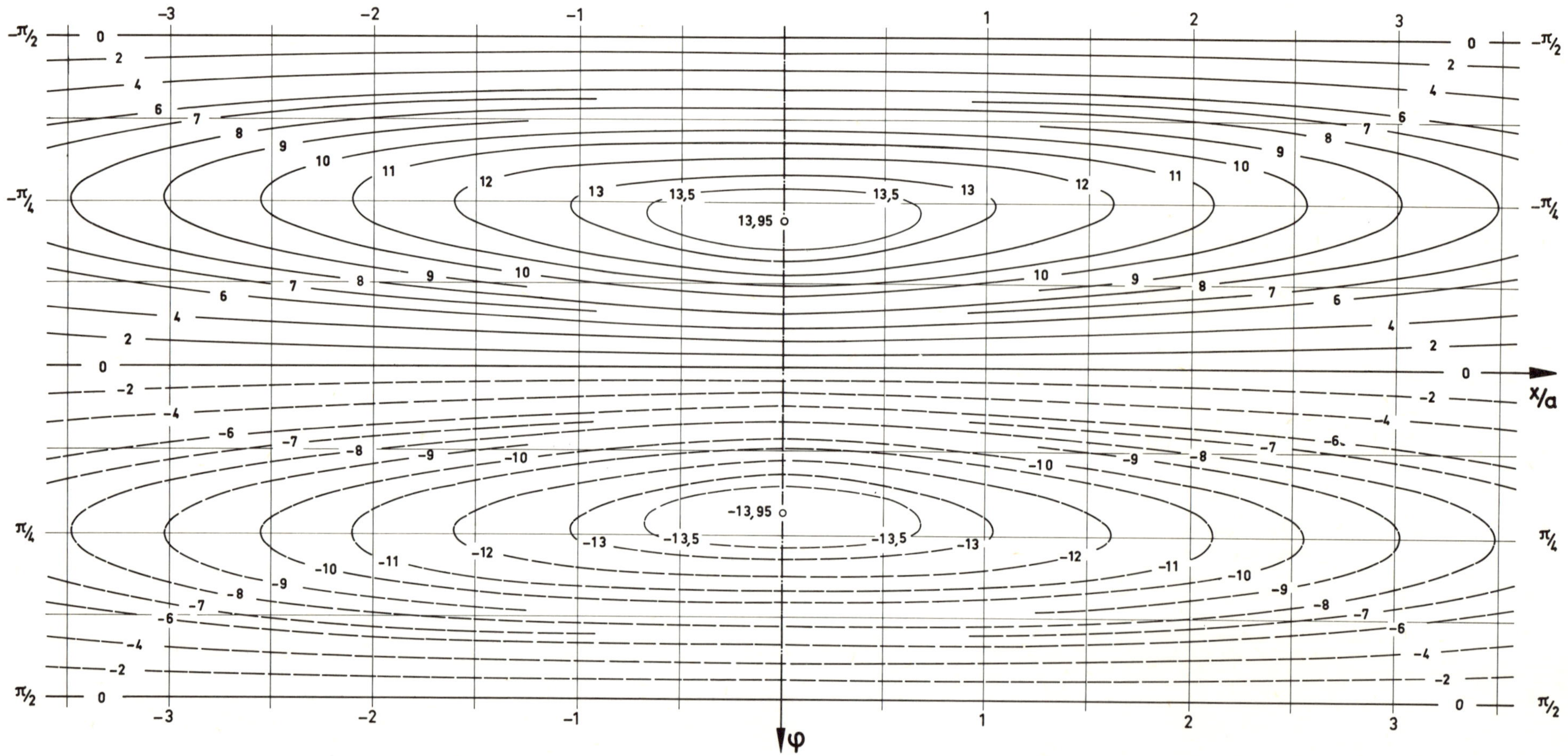

Einflußfläche (negative Zustandsfläche) der Verschiebung v

Influence surface (negative condition surface) for the displacement v

$1000\,\pi\,K/a^2 \cdot w$	**99**
$k = 1{,}0 \cdot 10^{-4}$	

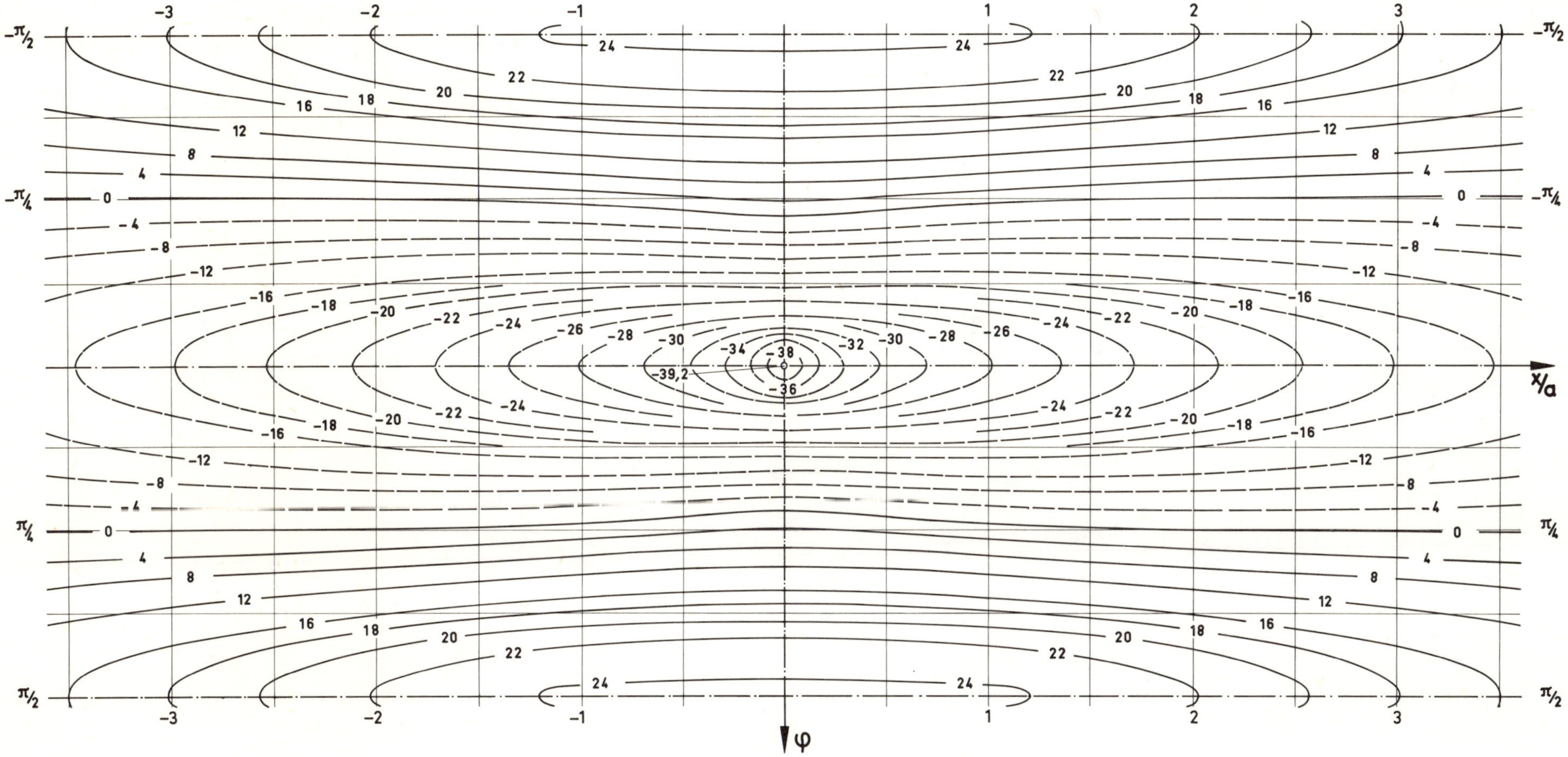

Einflußfläche (Zustandsfläche) der Verschiebung *w*

Influence surface (condition surface) for the displacement *w*

100	$100\,\pi\,K/a \cdot \chi_{\varphi}$
	$k = 1{,}0 \cdot 10^{-4}$

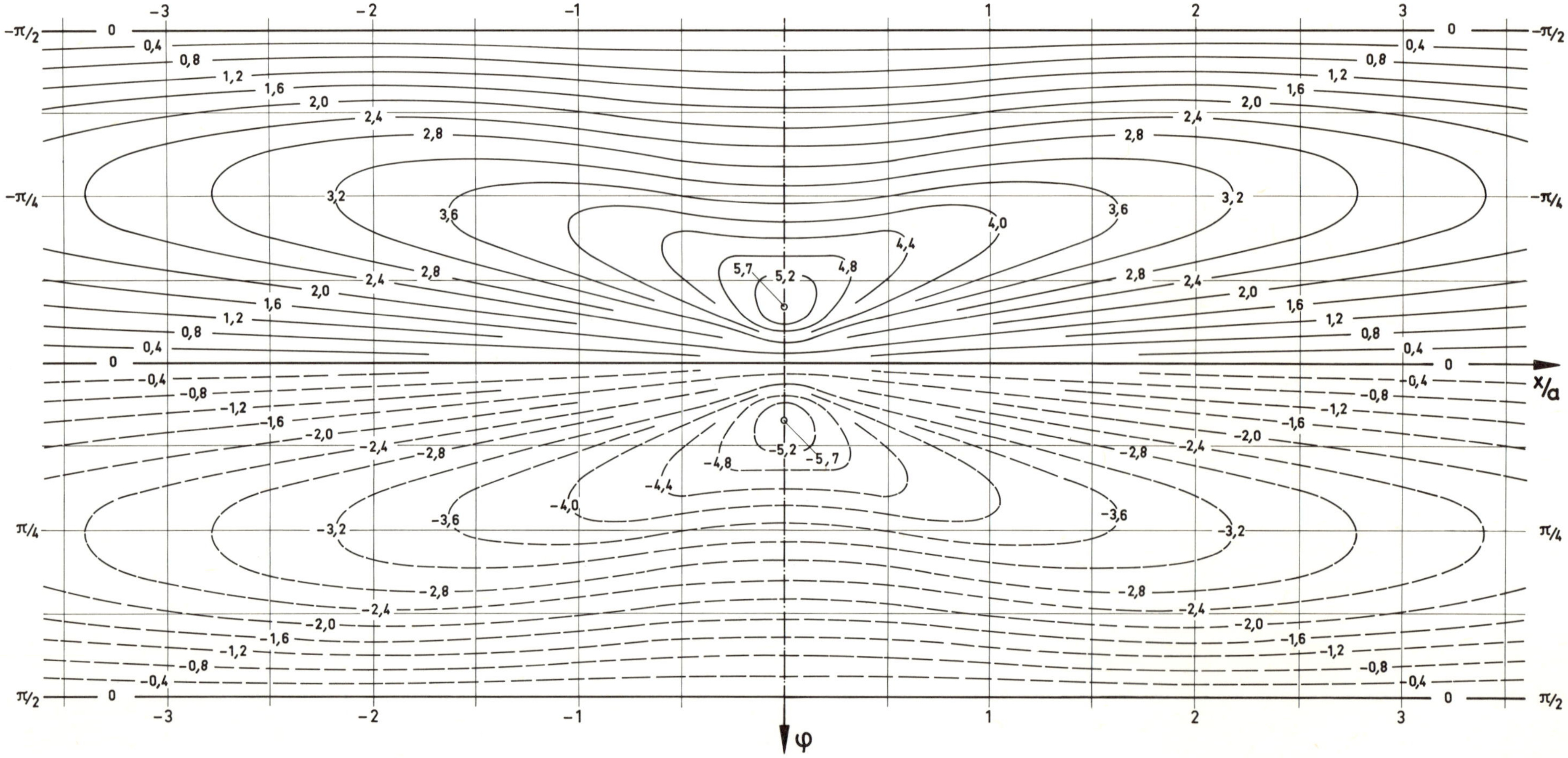

Einflußfläche (negative Zustandsfläche) der Verdrehung χ_{φ}

Influence surface (negative condition surface) for the rotation χ_{φ}

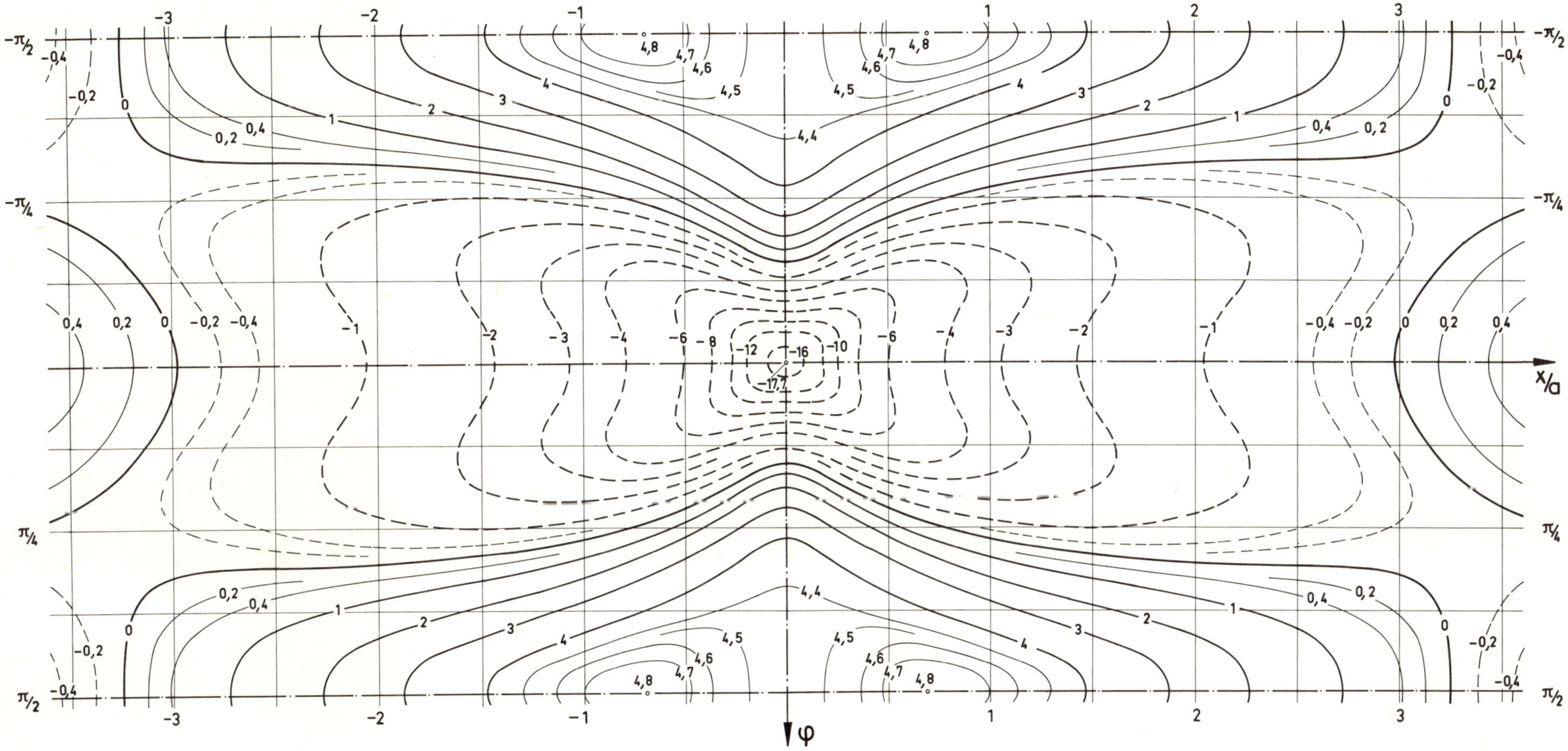

Einflußfläche (Zustandsfläche) der Normalkraft n_x

Influence surface (condition surface) for the normal force n_x

102	$\pi a \cdot n_\varphi$
	$k=1{,}0 \cdot 10^{-4}$

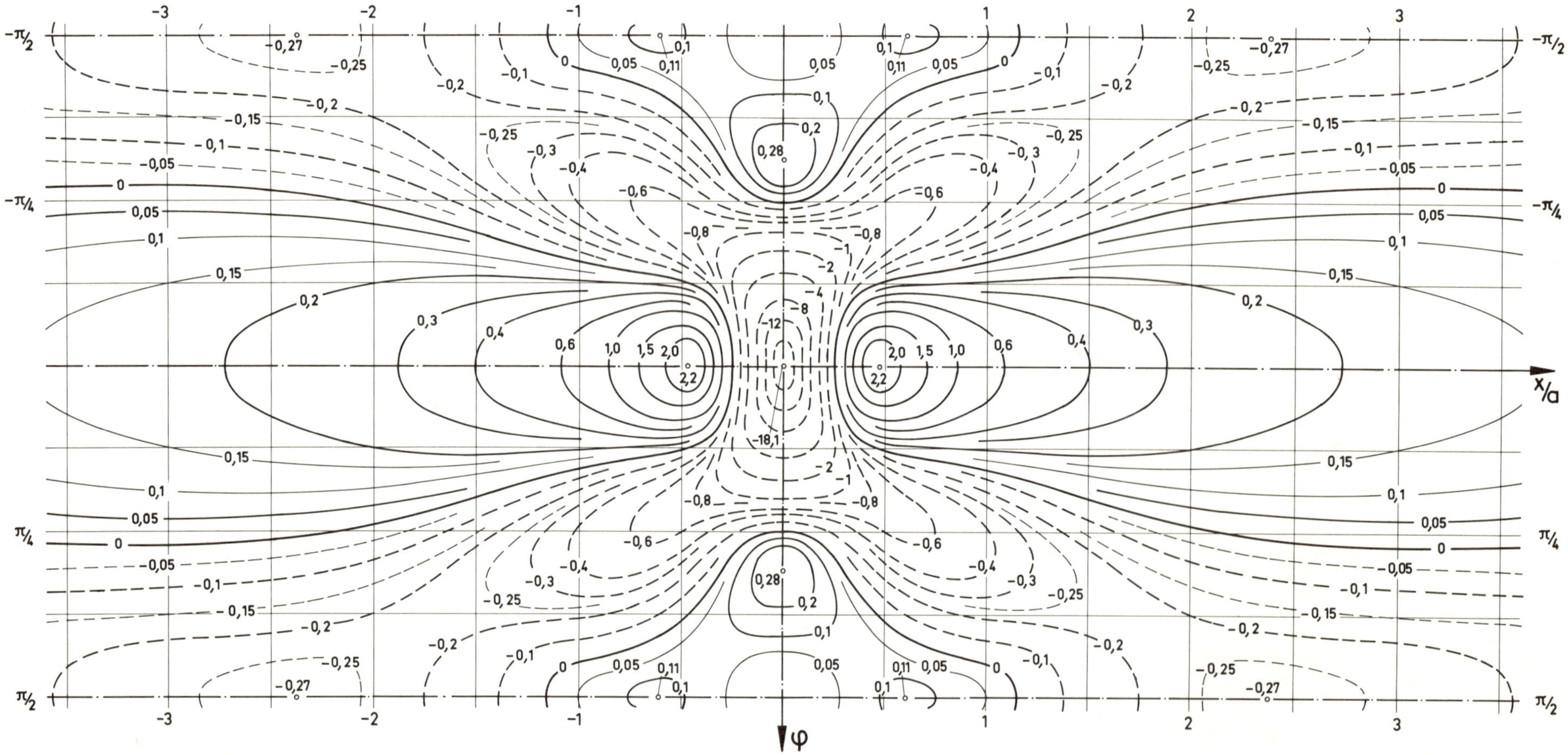

Einflußfläche (Zustandsfläche) der Normalkraft n_φ

Influence surface (condition surface) for the normal force n_φ

$\pi a \cdot \bar{n}_{x\varphi}$	**103**
$k=1{,}0 \cdot 10^{-4}$	

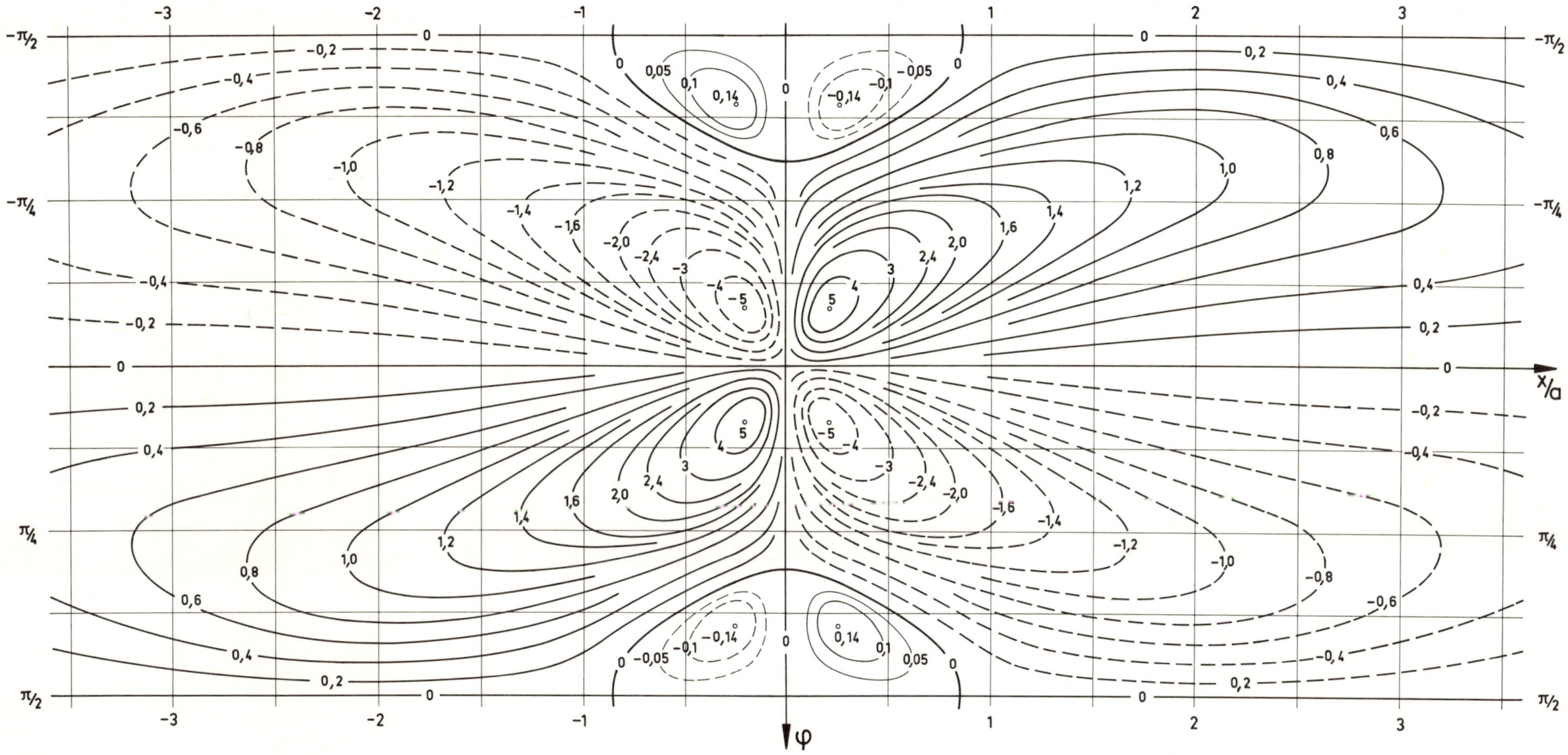

Einflußfläche (Zustandsfläche) der Randschubkraft $\bar{n}_{x\varphi}$

Influence surface (condition surface) for the boundary shearing force $\bar{n}_{x\varphi}$

104	$100\,\pi \cdot m_x$
	$k = 1{,}0 \cdot 10^{-4}$

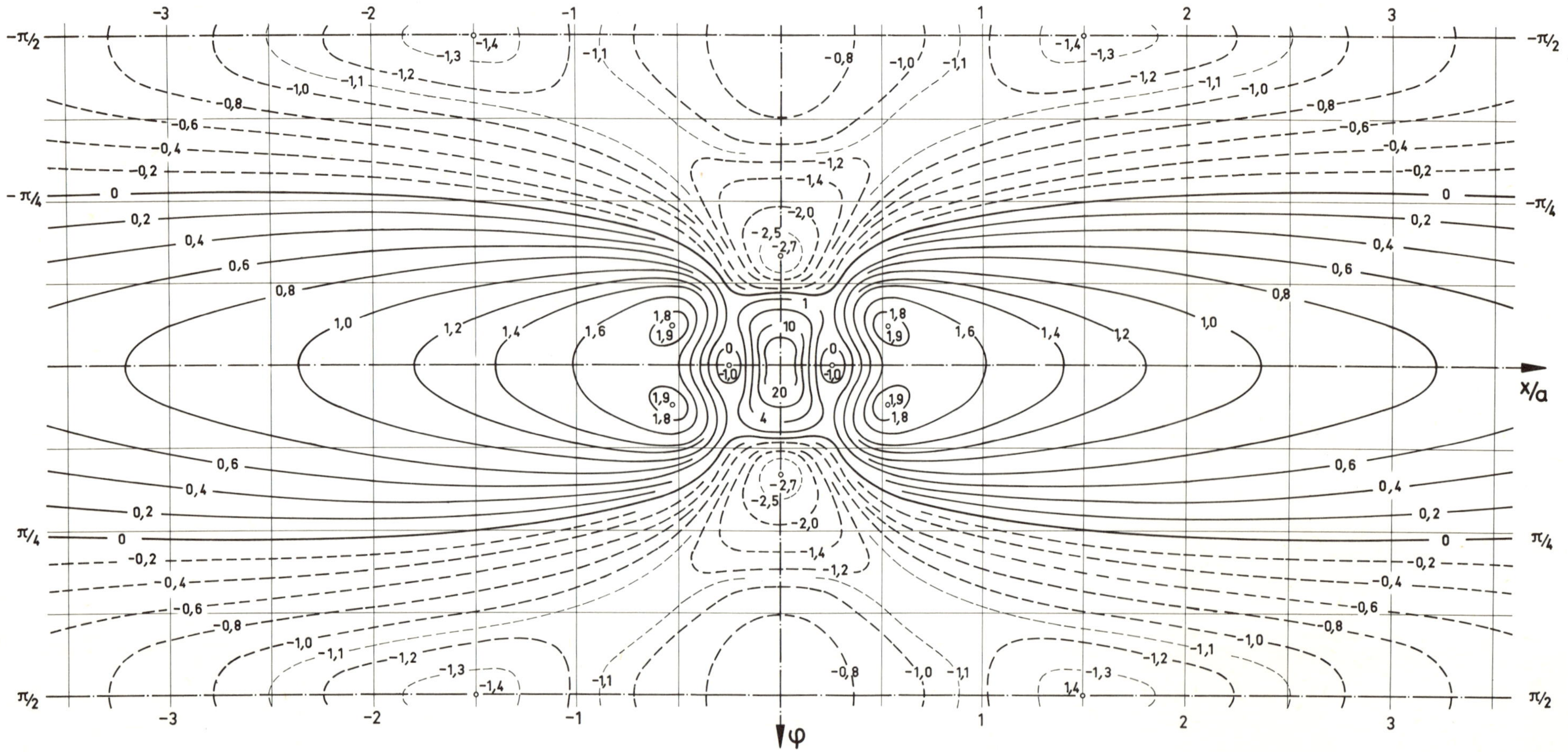

Einflußfläche (Zustandsfläche) des Biegemomentes m_x

Influence surface (condition surface) for the bending moment m_x

$100\,\pi \cdot m_\varphi$	**105**
$k = 1{,}0 \cdot 10^{-4}$	

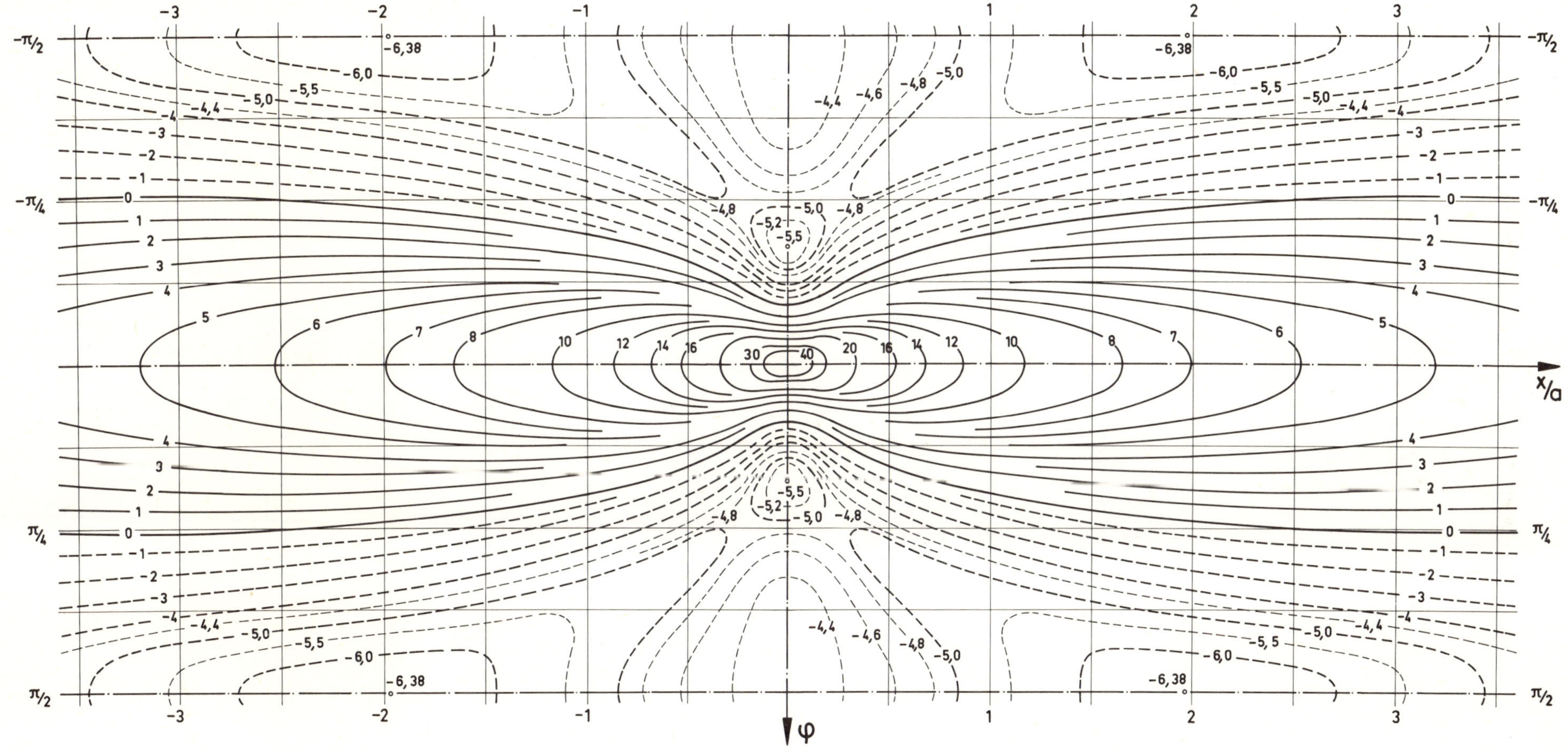

Einflußfläche (Zustandsfläche) des Biegemomentes m_φ

Influence surface (condition surface) for the bending moment m_φ

106 | $100\,\pi \cdot m_{x\varphi}$ / $k = 1{,}0 \cdot 10^{-4}$

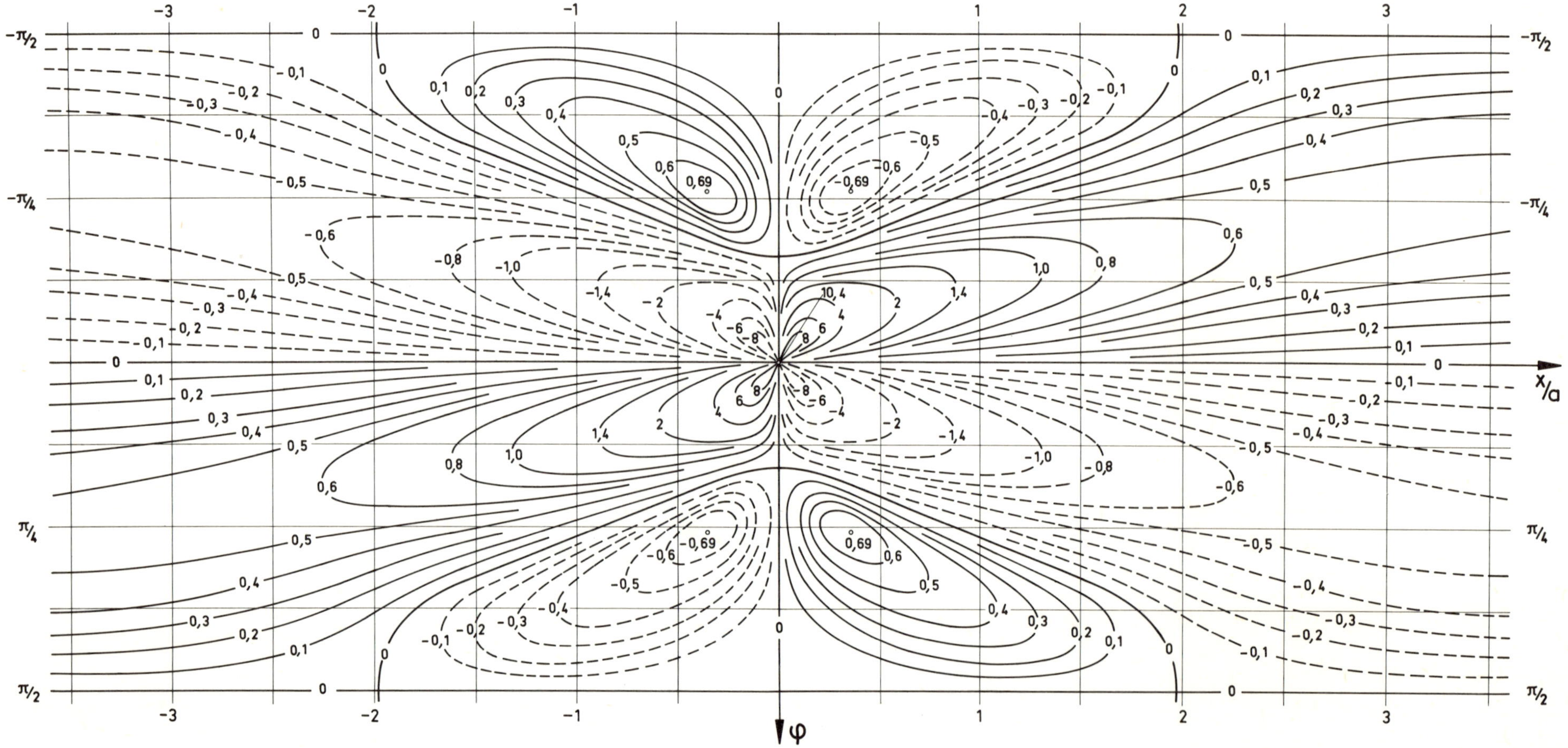

Einflußfläche (Zustandsfläche) des Drillmomentes $m_{x\varphi}$

Influence surface (condition surface) for the twisting moment $m_{x\varphi}$

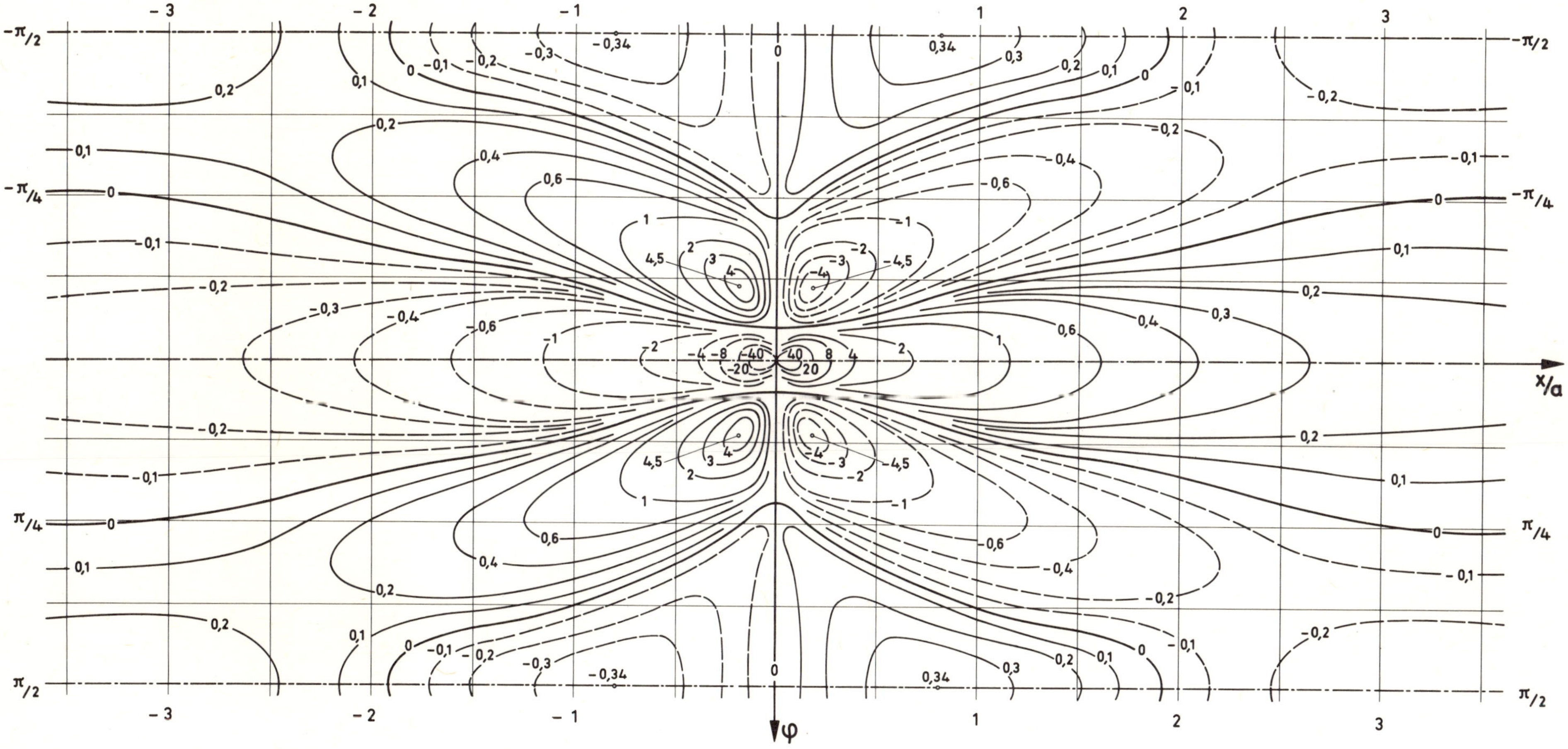

Einflußfläche (negative Zustandsfläche) der Randquerkraft $\bar{q}_x$

Influence surface (negative condition surface) for the transverse boundary force $\bar{q}_x$

108	$10\,\pi\, a \cdot \bar{q}_{\varphi}$
	$k = 1{,}0 \cdot 10^{-4}$

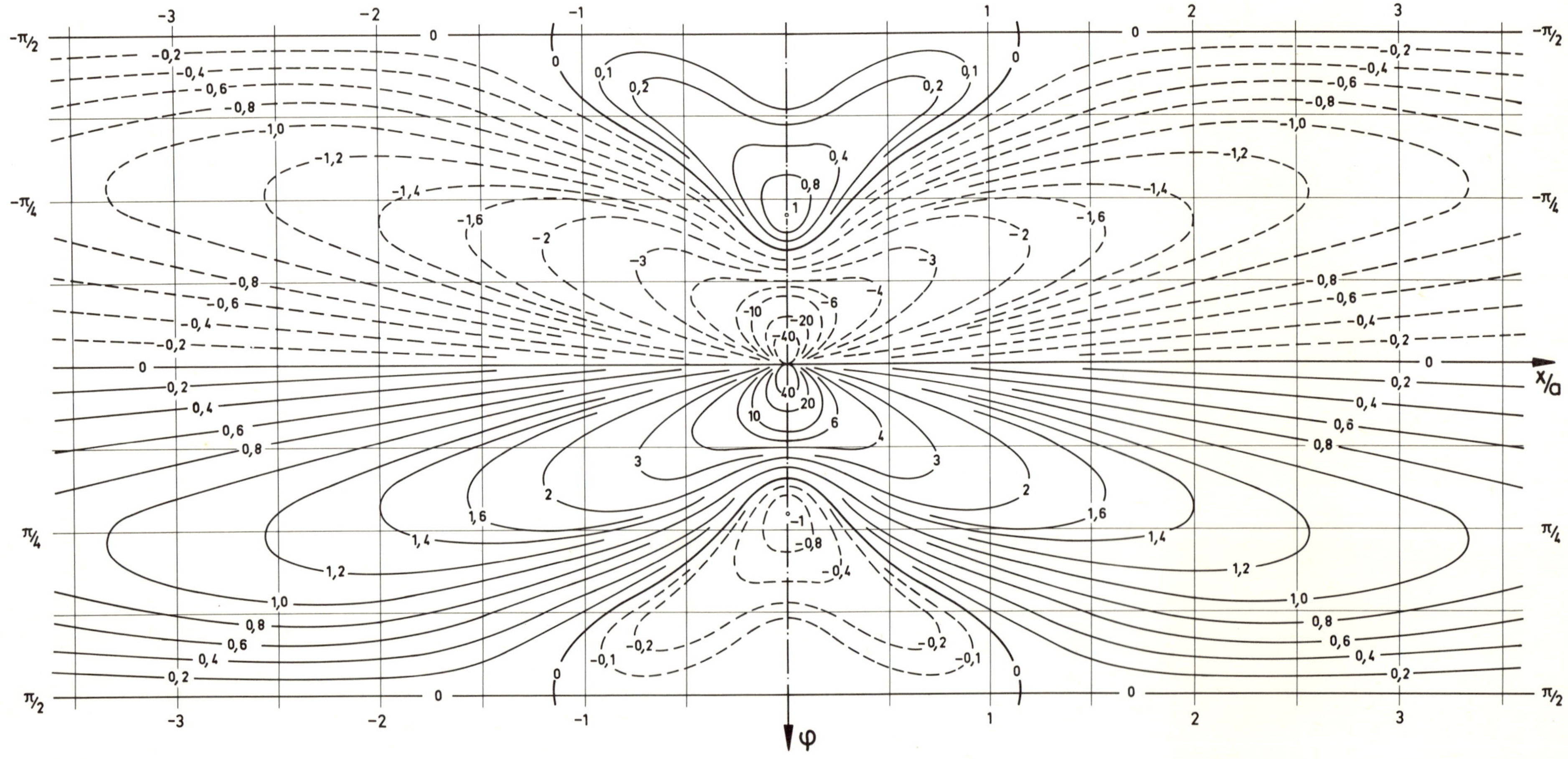

Einflußfläche (negative Zustandsfläche) der Randquerkraft $\bar{q}_{\varphi}$

Influence surface (negative condition surface) for the transverse boundary force $\bar{q}_{\varphi}$